XV International Scientific-Technical Conference "Dynamics of Technical Systems" (DTS-2019)

IOP Conference Series: Materials Science and Engineering
Volume 680

Rostov-on-Don, Russia
11 - 13 September 2019

ISBN: 978-1-7138-0932-6
ISSN: 1757-8981

Printed from e-media with permission by:

Curran Associates, Inc.
57 Morehouse Lane
Red Hook, NY 12571

Some format issues inherent in the e-media version may also appear in this print version.

Institute of Physics
Dirac House, Temple Back
Bristol BS1 6BE UK

Phone: 44 1 17 929 7481
Fax: 44 1 17 920 0979

techtracking@iop.org

Additional copies of this publication are available from:

Curran Associates, Inc.
57 Morehouse Lane
Red Hook, NY 12571 USA
Phone: 845-758-0400
Fax: 845-758-2633
Email: curran@proceedings.com
Web: www.proceedings.com

TABLE OF CONTENTS

SECTION 4. FUNDAMENTAL METHODS OF SYSTEM ANALYSIS, MODELING AND OPTIMIZATION OF DYNAMIC SYSTEMS

IOP Conf. Series: Materials Science and Engineering **680** (2019) 011001 doi:10.1088/1757-899X/680/1/011001

XV International scientific-technical conference "Dynamics of technical systems" (DTS-2019)

Conference was held at September 11-13, 2019, in Rostov-on-Don, Russia, on the base of Don State Technical University.

The Conference was aimed to exchange experience and identifying the state-of-art scientific developments in the study of technical and technological systems, covering the fundamentals of mechanics, dynamics and tribology, and also problems of non-linear dynamics and synergy in the technical systems, the problems of mechanics of heterogeneous and composite materials, as well as the fundamentals of system analysis, mathematical modelling and information processing.

The conference presented scientific reports in the following sections

1. Fundamentals of mechanics, dynamics and tribology of machines
2. Non-linear Dynamics and Applied Synergetics in technical systems
3. Issues of creation and mechanics of heterogeneous and composite materials
4. Fundamental methods of system analysis, modeling and optimization of dynamic systems
5. Dynamics of buildings and structures

The main direction of the Conference was to further integrate the efforts of scientists, engineers and designers to create, implement, and industrial applications of recent advances in science and high technologies at the enterprises of various branches of mechanical engineering, including MIC enterprises and other industries.

IOP Conf. Series: Materials Science and Engineering **680** (2019) 011002 doi:10.1088/1757-899X/680/1/011002

Peer review statement

All papers published in this volume of *IOP Conference Series: Materials Science and Engineering* have been peer reviewed through processes administered by the proceedings Editors. Reviews were conducted by expert referees to the professional and scientific standards expected of a proceedings journal published by IOP Publishing.

IOP Conf. Series: Materials Science and Engineering **680** (2019) 012001 doi:10.1088/1757-899X/680/1/012001

Computer Simulation of Kinetics of Parallel Mechanisms

O V Mkrtychev and A V Kartygin

Novorossiysk Branch of the Belgorod V G Shukhov State Technological University, Myskhakskoe shosse 75, Novorossiysk 353919, Russia

E-mail: oleg214@ya.ru

Abstract. The paper considers the issues of kinematics and dynamics of a parallel mechanism and computer simulation of its movement. To solve the problems of dynamics of parallel mechanisms, the method of Lagrange equations of the second kind was used. In computer modeling, the Matlab and Simulink application were used.

1. Introduction

Now in engineering, the urgent task is to design mechanisms for mechanical treatment of the internal surfaces of cavities of complex shape, for example, in aviation and space technology, in disaster technology, mechanisms are needed to provide access to the internal volumes of destroyed buildings and structures, to carry out repair and restoration work in pipelines, to perform certain surgical operations, etc. [1-8]. There are many projects devoted to the development of such robots and many are working on solving the problems that arise in this process [9-16].

In this article, a structural, kinematic and dynamic analysis of a tripod - a manipulator based on parallel mechanics was conducted. The questions of kinematic and dynamic analysis of a plain parallel mechanism were investigated. The use of parallel mechanisms complicates control tasks and leads to more complex control systems for such machines. Therefore, it becomes important to develop appropriate algorithms for solving problems related to the movement of parallel mechanisms. The authors faced problems of this type in solving a number of theoretical and applied tasks [17-24]. Well-known methods were used for the solution [2, 25-30].

For example, after obtaining a solution in a model using the method of Lagrange equations of the second kind, according to [15, 16], a model was created using the Denavit—Hartenberg method [30]. A comparison of these two solutions made it possible to make certain of applicability and reliability of the obtained numerical solution.

2. Mathematical model

Manipulators can be built on the basis of parallel mechanisms such as tripods. An example of such a tripod, which consists of a fixed base, a movable platform, and 4 legs, each of which consists of 2 rods and an active translational kinematic pair is shown in Fig. 1. To estimate the number of degrees of freedom W of the platform, the well-known Grubler and Somov—Malyshev formulas are used. Tripod shown in Fig. 1, has 3 degrees of freedom (two rotational and one translational).

According to [15], the generalized coordinates $q_{1\text{-}3}$ were introduced, which are the lengths of the rods *Aa*, *Bb*, *Cc*, respectively (Fig. 1).

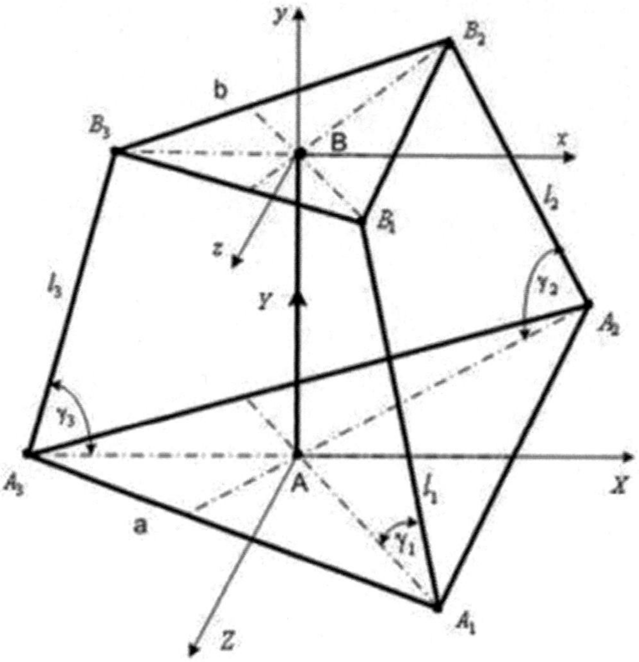

Figure 1. A tripod coordinate system (figure taken from [15]).

It was believed that the fixed platform (base) is horizontal and the points A_i also form a regular triangle with side a and center at point A. The legs A_iB_i have lengths li> 0 and are inclined to the base plane at angles γ_i, $(0 < \gamma_i < \pi/2)$. The distance between the point Bi and the base (plane AXZ) is denoted by h. The coordinate system $AXYZ$ was connected with the base, and the coordinate system $Bxyz$ was connected with the platform (Fig. 1).

The position of the joints A1-3 in the coordinate system $AXYZ$ and the joints $B_{1\text{-}3}$ in the coordinate system $Bxyz$ is determined by the vectors. The position of the platform relative to the base is set by Euler angles $\varphi_{1,2}$ and vector B. The geometric relations between the coordinate systems $AXYZ$, $Bxyz$ can be represented as a 4×4 matrix of homogeneous transformations

$$T = T(h,\ \varphi_1, \varphi_2) = \begin{pmatrix} \cos\varphi_1 & -\sin\varphi_1 & 0 & 0 \\ \sin\varphi_1\cos\varphi_2 & -\cos\varphi_1\cos\varphi_2 & \sin\varphi_2 & h \\ \sin\varphi_1\sin\varphi_2 & -\cos\varphi_1\sin\varphi_2 & \cos\varphi_2 & 0 \\ 0 & 0 & 0 & 1 \end{pmatrix}.$$

Then the generalized coordinate l_i, as a function of distance h and angles $\varphi_{1,2}$, is determined by the expression

$$l_i = l_i(h, \varphi_1, \varphi_2) = \sqrt{\Sigma\left(A_{ij} - [B_{ij}]\right)^2}, \quad i, j = 1,2,3. \tag{1}$$

where the formulas for A_{ij} and B_{ij} are given in [15, 16]. Expressions for the rates and accelerations of the ends of the legs l_i are easy to find if we take the derivatives on h, $\varphi_{1,2}$ of the formulas obtained.

According to the same scheme, the equations of movement of the platform are composed using the Lagrange equations of the second kind:

$$\frac{d}{dt}\left(\frac{dE}{d\dot{\varphi}_i}\right) - \frac{dE}{d\varphi_i} = Q_i, \quad i = 1,2,3, \tag{2}$$

where E is the kinetic energy of the system; $q_1 = h$, $q_2 = \varphi_1$, $q_3 = \varphi_2$ are generalized coordinates; Qi is the generalized force on the i generalized coordinate.

If the mass of the platform is M, then its kinetic energy

$$E = \frac{1}{2}[Mv_B^2 + J(\dot{\varphi}_1)^2 + J(\dot{\varphi}_2)^2],$$

where $v_B = \dot{h}$ is the velocity of the B-platform CG.

The forces acting on the platform, as well as the radius vectors $\boldsymbol{R}_B$, $\boldsymbol{R}_i$, (i = 1, 2, 3), points of their application in the projections on the axis of the coordinate system *AXYZ*:

$$\boldsymbol{P} = (0, -Mg, 0)^T, \boldsymbol{F}_i = (F_i \cos\gamma_{i1}, F_i \cos\gamma_{i2}, F_i \cos\gamma_{i2})^T,$$
$$\boldsymbol{R}_B = (0, h, 0)^T, \boldsymbol{R}_i = ([B_{i1}], [B_{i2}], [B_{i3}])^T,$$

where $\cos\gamma_{ij} = \cos\gamma_{ij}(\varphi_1, \varphi_2) = \frac{[B_{ij}]}{l_i}$

Further, using the elements of the Jacobi matrix of vectors $R_i = R_i(h, \varphi_1, \varphi_2)$, the values of the generalized forces $Q_{1,2,3}$ were obtained. These values were substituted in (2), and the desired ODE system was obtained for solving the problems of dynamic tripod analysis.

$$M\ddot{h} = -Mg + F_1\cos\gamma_{12} + F_2\cos\gamma_{22} + F_3\cos\gamma_{32},$$
$$J\ddot{\varphi}_1 = \frac{b}{2}F_1(-\cos\gamma_{11}\sin\varphi_1 + \cos\gamma_{12}\cos\varphi_1\cos\varphi_2) +$$
$$+\frac{b}{2}F_2(-\cos\gamma_{21}\sin\varphi_1 + \cos\gamma_{22}\cos\varphi_1\cos\varphi_2 + \cos\gamma_{23}\cos\varphi_1\sin\varphi_2) +$$
$$+rF_3(\cos\gamma_{31}\sin\varphi_1 + \cos\gamma_{32}\cos\varphi_1\cos\varphi_2 - \cos\gamma_{23}\cos\varphi_1\sin\varphi_2),$$
$$J\ddot{\varphi}_2 = \frac{b}{2}F_1[\cos\gamma_{12}(-\sin\varphi_1\sin\varphi_2 + \cos\varphi_2) + \cos\gamma_{13}(\sin\varphi_1\cos\varphi_2 - \sin\varphi_2)] +$$
$$+\frac{b}{2}F_2[-\cos\gamma_{22}(\sin\varphi_1\sin\varphi_2 + \cos\varphi_2) + \cos\gamma_{23}(\sin\varphi_1\cos\varphi_2 - \sin\varphi_2)] +$$
$$+rF_3(\cos\gamma_{32}\sin\varphi_1\sin\varphi_2 - \cos\gamma_{33}\sin\varphi_1\cos\varphi_2),$$

where $F_{1\text{-}3}$ are external forces.

To solve the resulting system of equations numerically on a computer, standard mathematical software packages were used. A similar algorithm was applied to tripods with various structural and kinematic schemes.

3. Computer model

For computer simulation, a model was created using block diagrams of the SimMechanics, which is an extension of the MatLab. Fig. 2 shows a fragment of the $A_1B_1C_1$ leg of the tripod model under consideration.

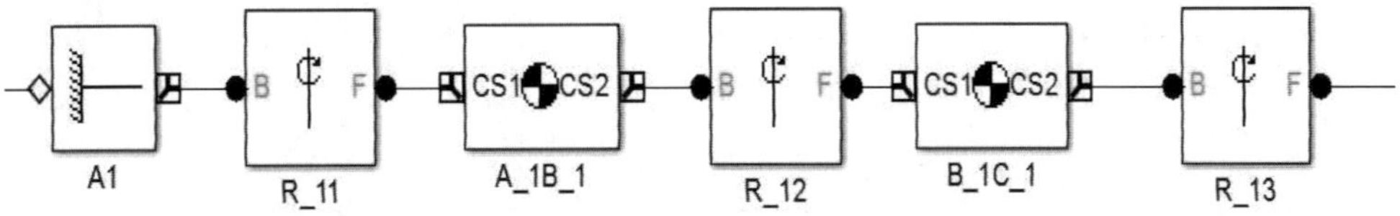

Figure 2. Fragment of the block diagram of the model of a **planar** parallel mechanism with three identical legs, which simulates the leg $A_1B_1C_1$ with the R-pairs connecting platforms and links.

According to [15], the generalized coordinates $q_{1\text{-}3}$ were introduced, which are the lengths of the rods *Aa*, *Bb*, *Cc*, respectively (Fig. 1).

Figures 3 and 4 show the results of comparing the numerical solution and computer simulation for two different tripod block diagrams. Numerical calculations and computer modeling were carried out using the mathematical application program package Matlab. In each figure, the results of the angular acceleration calculations for link A_1B_1, A_2B_2, and A_3B_3 go from top to bottom respectively. The green color shows the results of a numerical solution of the Lagrange equations of the second kind. Red color shows the results of computer simulation.

The tripod block diagram in Fig. 3 shows three identical RRR-legs (each leg consists of two half-legs connected by three rotational pairs). The block diagram of the tripod in Fig. 4 also shows three identical RPR-legs of the (in the middle legs are interconnected by one prismatic pair). The central link *AB* is absent in both schemes.

IOP Conf. Series: Materials Science and Engineering **680** (2019) 012001 doi:10.1088/1757-899X/680/1/012001

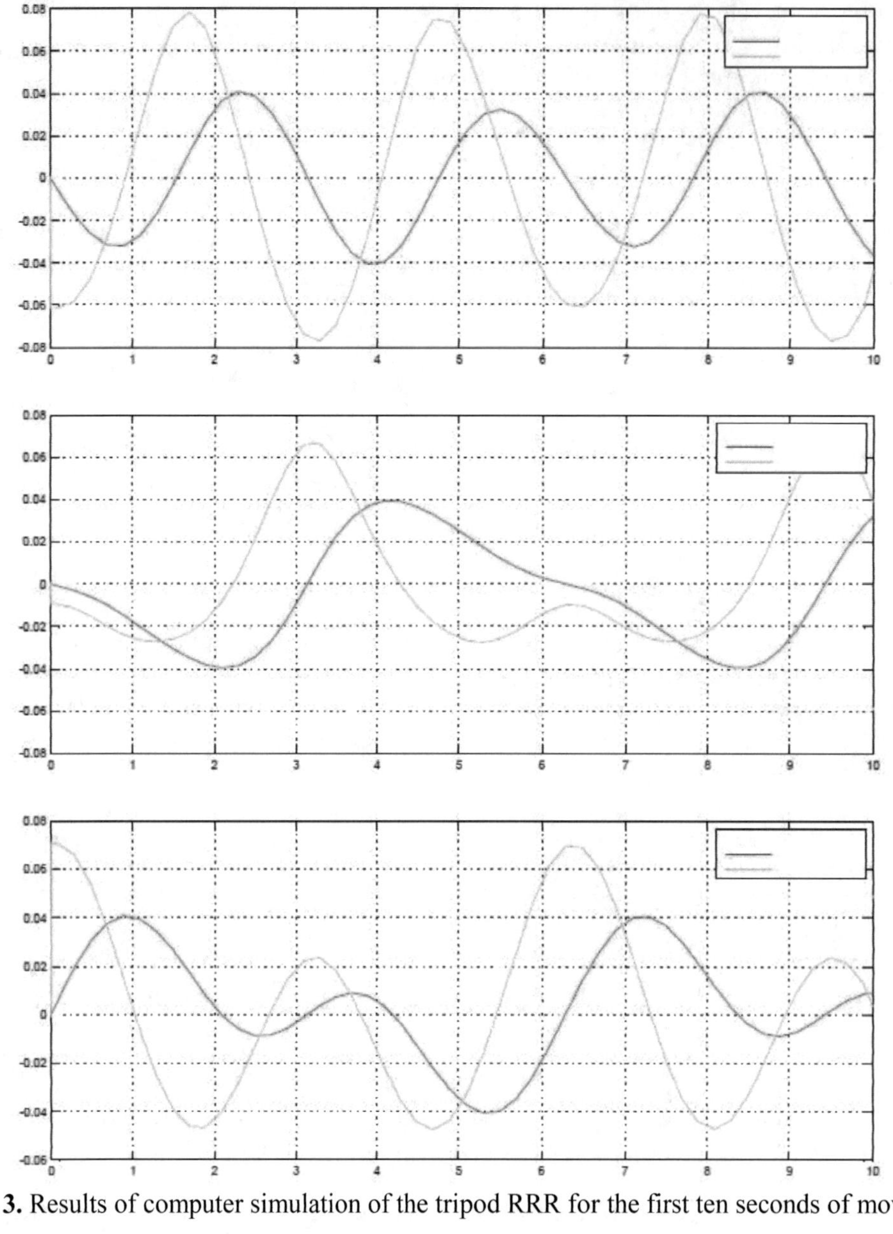

Figure 3. Results of computer simulation of the tripod RRR for the first ten seconds of movement.

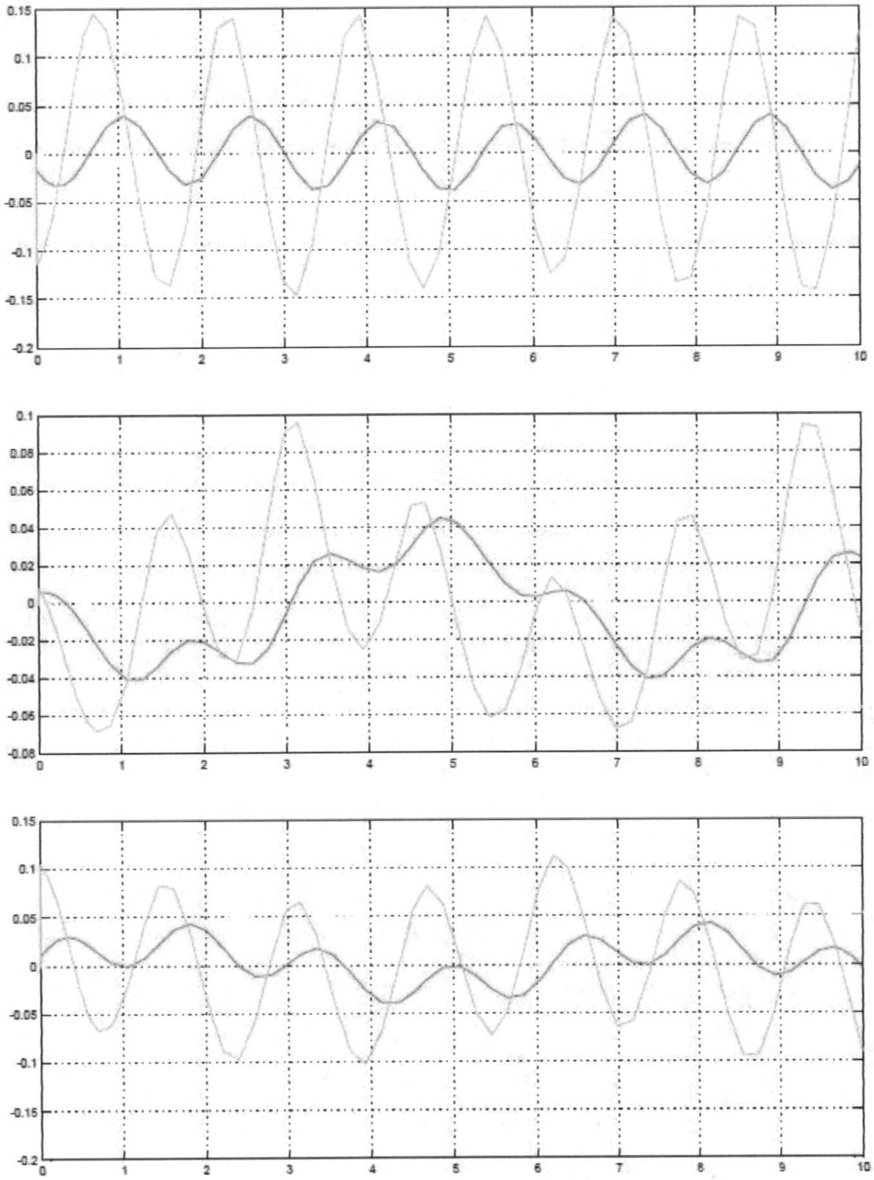

Figure 4. Results of computer simulation of the tripod RPR for the first ten seconds of movement.

4. Conclusions

Using the Grubler and Somov—Malyshev formulas, an analysis of the number of degrees of freedom of these schemes is performed. Although the expressions for **rate** and accelerations of the ends of the legs l_i (B_i points) of the tripod with 2 degrees of freedom, the tripod with 3 degrees of freedom is easy to find by differentiating and twice differentiating the corresponding expressions by generalized coordinates, however, these expressions are too cumbersome and unsuitable for practical use. The study of the velocities and accelerations of the ends of the $l_{1\text{-}3}$ legs is easier to do using mathematical modeling and using modern software.

The following results were obtained for a tripod with 2 degrees of freedom: the inverse kinematics problem was solved; the direct problem of kinematics is solved - equations are obtained that determine the length of the legs as a function of the generalized coordinates; A mathematical model of the dynamics of the mechanism is developed in the form of a system of ordinary differential equations with respect to generalized coordinates. Similar results were obtained for a tripod with 3 degrees of freedom.

Based on the analysis of the movements of the "trunk" type manipulator, a dynamic analysis of the tripod with 2 and 3 degrees of freedom was performed. For a tripod with 2 degrees of freedom, equations are obtained that determine the length of the legs as a function of the generalized coordinates; a model of the dynamics of the mechanism is developed in the form of a system of ordinary differential equations for generalized coordinates. Similar results were obtained for a tripod with 3 degrees of freedom. The results obtained are of theoretical and practical interest, and can also be used to build mathematical models.

References

[1] Merlet J P 2000 *Parallel Robots. Solid mechanics and its applications* (Merlet-Kluwer Academic Publishers).
[2] Glazunov V A, Koliskor A Sh and Kraynev A F 1991 *Spatial Mechanisms of Parallel Structure* (Science, Moscow) (in Russian)
[3] Parallel mechanisms information center (http://www.parallemic.org/)
[4] Volkomorov S V, Kaganov Yu T and Karpenko A P 2010 Information technology, Application **5** 1 (in Russian)
[5] Yamnikova O A, Yamnikov A S and Solyankin D Yu 2011 Science intensive technologies in mechanical engineering **5** 3 (in Russian)
[6] Sementsev A M 2011 Science intensive technologies in mechanical engineering **5** 14 (in Russian)
[7] Govorov I V 2011 Science intensive technologies in mechanical engineering **5** 21 (in Russian)
[8] Swirad S 2018 MATEC Web of Conferences **249** 03002
[9] Kosbolov S, Yeleukulov S Y, Atalykova A, Zhauyt A, Yestemessova G and Yussupova S 2018 MATEC Web of Conferences **22** 01020
[10] Goryanina K I, Lukyanov A D and Katin O I 2018 MATEC Web of Conferences **226** 02015
[11] Furch J and Tran C V 2018 MATEC Web of Conferences **234** 020002
[12] Hongqiang Wan, Peiying Han, Shuai Ge, Fancong Li and Simiao Zhang 2018 MATEC Web of Conferences **237** 01014
[13] Lankin A M, Lankina M Y and Lankin M V 2018 MATEC Web of Conferences **226** 02021
[14] Garcia A, Cuan-Urquizo E, Roman-Flores A and Vazquez E 2018 MATEC Web of Conferences **249** 03004
[15] Kaganov Yu T, Karpenko A P. 2009 Science and education **10** (in Russian)
[16] Kaganov Yu T, Karpenko A P. 2009 Science and education **11** (in Russian)
[17] Mkrtychev O V 2012 Theory of mechanisms and machines **10** 1 (in Russian)
[18] Mkrtychev O V 2013 Theory of mechanisms and machines **11** 1 (in Russian)
[19] Mkrtychev O V 2012 Bulletin of Belgorod V. G. Shukhov State Technology University **2** 211 (in Russian)
[20] Mkrtychev O V 2014 Bulletin of Belgorod V. G. Shukhov State Technology University **1** 217 (in Russian)
[21] Mkrtychev O V 2012 Modern engineering. Science and education **2** 96 (in Russian)
[22] Kartygin A V, Mkrtychev O V and Starchik Yu Yu 2018 Fundamentals of mechanics **3** 69 (in Russian)
[23] Mkrtychev O V 2018 Mechanics and mechanical engineering. Science and practice **1** 14 (in Russian)
[24] Mkrtychev O V, Gerasimov M D and Kartygin A V 2017 Construction and road building machinery **12** 17 (in Russian)
[25] Poduraev Yu V Mechatronics: basics, methods, applications 2007 (Mashinostroenie, Moscow)
[26] Volkomorov S V, Karpenko A P and Leletko A M 2008 Science and education **10** (in Russian)

IOP Conf. Series: Materials Science and Engineering **680** (2019) 012001 doi:10.1088/1757-899X/680/1/012001

[27] Zenkevich S L and Yushchenko A S 2000 Robot Control. Fundamentals of manipulation robots management (Publishing House of Moscow N. E. Bauman State Technical University, Moscow) (in Russian)
[28] Korendyasev A I, Salamander B L, Tyves L I et al. 1989 Robotized manipulation systems (Mashinostroenie, Moscow) (in Russian)
[29] Smirnov V A, Petrova L N and Fedorov V B 2006 Bulletin of SUSU **11** 24 (in Russian)
[30] Denavit J and Hartenberg R S 1955 J. Appl. Mech. **77** 215

Experimental studies of rotatory rolling of long thin-wall pipes

A M Lavrentiev

Lecturer, Kamyshin Tecnological Institute (branch) of the Volgograd State Technical University, str. Lenina 6A, Kamyshin 403874, Russia

E-mail: Lamvstu@gmail.com

Abstract. The article considers the application of rotatory rolling of the external surface of a long thin-wall pipe. The author studied the possibility of correcting the form of a thin-wall pipe outer diameter by means of rotatory rolling with cylindrical rollers having a rectilinear generator and set for a self-tightening angle with a simultaneous decrease of the processed surface roughness. Further, the author provide a photo of the measuring device used to define the circular deviation of a thin-wall pipe surface. A polar diagram of the pipe surface as-delivered and its profile record were made. The article provides a photo showing the adjustment process of a rotatory reeling machine with three deformation rollers for treating long thin-wall pipes in the system of a general purpose lathe. The experiment results are presented in the form of a polar diagram and the treated surface profile record. It is shown that after rotatory rolling the circular deviation decrease to 0,1 mm while the surface roughness decreased from Ra 1,83 um to Ra 0,23 um.

1. Introduction

The industry produces a large number of products (household carts and baby carriages, shopping carts, stacks, furniture pieces, curtain rods, stair railing and balustrades, hand rails, etc.) the structural parts of which are nickel coated long thin-wall pipes with the diameters 16–50 mm and wall thickness 0,5–3mm made in compliance with the 14th accuracy degree and circular deviation within 0.1 mm, the surface roughness Ra0,12 – R0,16um [1].

As workpieces for such pipes seamless cold-deformed pipes are generallyused. As compared with electrically welded pipes seamless cold-deformed ones have smaller circular deviations: in the as-delivered state such pipes have an external surface profile deviation within 0,3% of their diameter. Consequently, to get ready products of these pipes, one can simply use centerless grinding with the purpose of eliminating the defect layer with light gauge and subsequent polishing. However, the cost of seamless cold-deformed pipes is higher than the cost of longitudinal electrically welded pipes, thus, it is more reasonable to use a longitudinal electrically welded pipe as a workpiece.

When manufacturing longitudinal electrically welded pipes, their geometrical form and cross-section are subject to deviations, scale is formed on the surface while at their transportation and storage one can also observe rust formation. All the listed defects must be eliminated in the process of pipe machining. The key operations of the machining industrial process for small diameter pipes (16–50mm) are the following: defect layer elimination, provision of the accuracy of diameter dimensions and external surface form and required elevation parameters for surface roughness up to the value Ra0.16um [2]. However, abrasive machining is a labour-intensive and low-productive machining method resulting abrasive dust generation. Breathing in such dust oftentimes leads to such professional diseases as asthma, flint disease, mechanic bronchitis, etc.

Therefore, the main task at the machining of an external cylindrical surface of thin-wall pipes is the selection of more advanced and high-productive machining methods providing for successive performance of the following operations: defect layer elimination, geometrical form correction and decrease of the treated surface roughness. One of such method can be combined processing of thin-wall pipes with rolling and wire brushing.

Theoretical research conducted in the previous papers [3, 4, 5]on rolling in the planned range necessary for solving the set tasks allowed obtaining theoretical dependencies to define the parameters of the contact zone and stress state in contact between deformed reeling machine rollers and the rolled pipe surface. The analysis of the obtained theoretical dependencies for the definition of efficient structural, power and process-dependent parameters of a rotatory reeling machine providing for a required deformation source in compliance with diametric dimensions of the machined pipe and its wall thickness as well as rolling process engineering support allowed developing technical documents, manufacturing a pre-production prototype of a rotatory reeling machine and conducting experimental studies.

Experimental studies of the impact the deforming rollers design parameters and machining modes have were conducted with the help of a specially developed and manufactured reeling machine installed on a turning lathe which provides for a wide range of planned process-dependent mode parameters. Previously, the design parameters of deformed rollers were studied at an upgraded mount [6] alongside with their penetration depth and self-tightening angle.

The process of improving the outside diameter form accuracy in thin-wall pipes by means of rotatory rolling is influenced by a vast number of factors which is why it is necessary to conduct experimental research complementing and specifying theoretical research.

On the basis of theoretical research conducted [6] and the mathematical dependencies obtained it is established that the penetration depth of a cylindrical deforming roller into the machined surface of a thin-wall pipe depends on the strain force and the angle of its installation in relation to the machined part axis (self-feeding angle). One should define the law of contact stress distribution along the contact area and, in its turn, obtain the contact zone form and dimensions. The intensity and distribution law of occurring contact stresses in the contact area influences the value of the pipe wall deflection and its elastic recovery.

At the rolling of a thin-wall pipe with a cylindrical deforming roller having a rectilinear generator the depth of its penetration into the machine part surface in compliance with the dimensions of the deformed part itself is one of the factors defining the machining results. In this paper it has a particular significance as the author are developing and studying a new design of a reeling machine which has not been applied yet in manufacturing practice [7].

According to the results of theoretical research, average pressure in the contact area (which is an ellipsis) depends on its geometrical parameters changing in compliance with the deformed part dimensions, thin-wall pipe outside diameter, self-feed angle and increases pro rata with the deformed part depth of penetration into the surface. In connection with the aforementioned in this paper the author studied the impact of the surface penetration depth of a deforming part in the form of a cylindrical roller with a rectilinear generator at the work with a thin-wall pipe as well as the self-feed angle impact on the rotatory rolling parameters.

2. Research materials and methods

At planning and preparation of experimental studies the author selected a necessary number of measurements at a set error with the use of a certain measurement method and specific device [8,9].

For conducting the experimental research of the rotatory rolling the pipe diameters were assumed as delivered. In this case the pipe diameter was assumed to be equal to 33,5 mm.

Before conducting rotatory rolling the author has made a polar diagram of the pipe external surface by means of measuring the surface profile polar ray. Changing the polar ray value was performed at the measurement device (Figure 1) consisting of a control device 1, dial test indicator 3 installed inside a magnetic display rack 2. The external surface profile in a cross-section was defined by polar ray measurement at the rotation of the pipe 4 in the centers installed in the front and rear tailstocks 5 and 6 correspondingly. The value measured at this device showed the deviation from a set parameter which is defined by a polar ray vibrations of the measured surface, its shift in relation to centers and deviation from turning motion. This method has significant drawbacks as in fact one measures radial motion variation by which it is impossible to judge on the measured surface form. In addition, it does not take into account the deviation from turning motion. Also, it is likely that the total of the form error and turning motion uncertainty can result in zero radial motion variation due to similar values but different digits. The main factor decreasing the accuracy of this method measurement is the deviation of actual motion of a part from the turning one. In this case it is reduced to minimum.

Figure 1. The measurement device for defining the circular deviation of the pipe external surface.

Further, the author measured the part profile and put the values in the table. However, the representation of data in a graphical form shows the profile deviation more clearly. Polar diagrams are most frequently used to show the circular surface form.

Making up the pipe surface polar diagrams as-delivered (Figure 2a) was performed as follows: in the region limited by the curve (which is actually a profile) the author selected the O center from which the rays were drawn. The values of the polar ray module were marked on these rays in the required scale. As a result a figure in the form of a closed curve was obtained. In a similar way the author made a polar diagram for the pipe surface after rotatory rolling (Figure 2b).

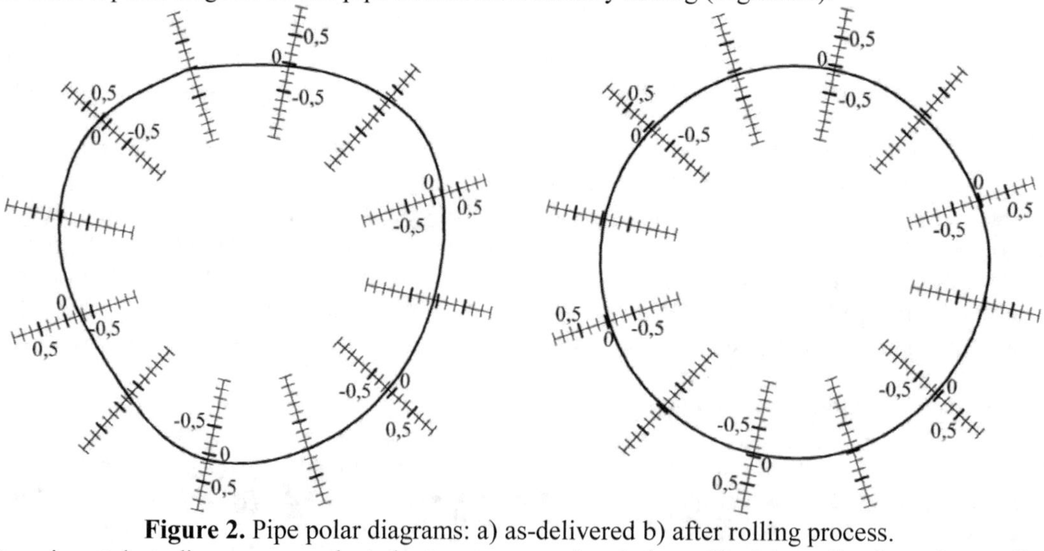

Figure 2. Pipe polar diagrams: a) as-delivered b) after rolling process.

Experimental studies were conducted at a screw-cutting lathe with the application of a reeling machine Figure 3.

IOP Conf. Series: Materials Science and Engineering **680** (2019) 012002 doi:10.1088/1757-899X/680/1/012002

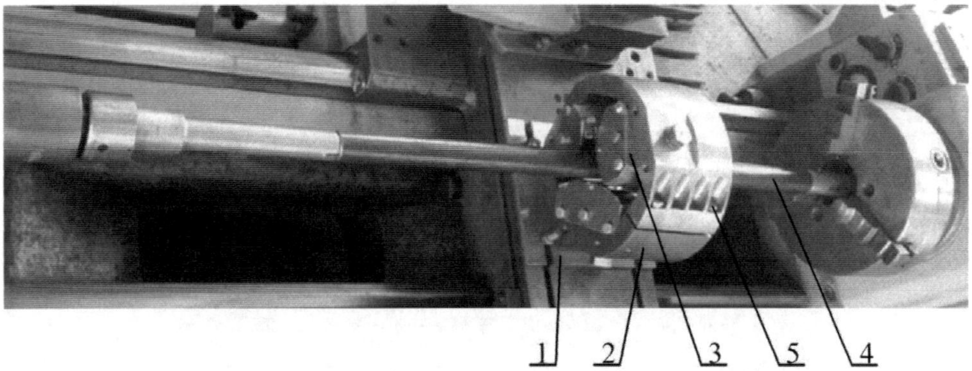

Figure 3. Photo of the manufactured device for thin-wall pipe rolling.

The reeling machine for rotatory rolling of thin-wall pipes is installed on the turning lathe toolhead and consists of the bottom 1 on which one fixes the body 2 with uniformly and circumferentially distributed inserts 3 with the possibility of motion in a radial direction and a purpose of adjusting the penetration depth of rollers into the machined pipe surface 4 as well as the adjustment of the reeling machine for processing pipe diameters (16–35 mm) and the use of deformed rollers with various diameters. Each deformed roller is installed on two road wheels which in their turn are installed on conical axial bearings.

Before machining the pipe is placed between three deforming rollers, and the diameter of the circumference made by the rollers is smaller than the outside diameter by a set depth of roller penetration into the pipe surface.

The adjustment of a necessary penetration depth is conducted with the help of 4 screws 5, two screws allows performing the body release while the other two screws compress the body in a radial direction. For this purpose a longitudinal through groove is made inside the body.

3. Results and Discussion

The polar diagram of the pipe external surface machined at the rotatory reeling machine is given in Figure 2b.

The obtained results of experimental studies show that the circular deviation of the thin-wall pipe external surface decreased to 0,1 mm after machining which complies with the requirements set to the pipes used as structural components of the aforementioned products.

The machined surface roughness decreased from Ra 1,83 um to Ra 0,23 um. Figure 4a shows the crossing line between a non-treated (right) and rolled (left) surfaces. The profile diagrams of the surface before and after machining at the reeling machine are presented in Figure4b and Figure 4c correspondingly.

In compliance with the conducted experiments it is established that the maximum feed angle of deformed rollers should not exceed 30° while the penetration depth should not be more than 0,15 mm.

4. Conclusion

As a result of the conducted theoretical and experimental studies the author proved that the industrial process of rotatory rolling has significant advantages over abrasive machining as there is no necessity to apply labour-consuming polishing, and the set parameters in terms of circular accuracy and roughness can be obtained. Circular deviation decrease occurs simultaneously with the decrease of the machined surface roughness.

The conducted theoretical and experimental studies allowed one to follow them to its logical end, develop recommendations on selecting reasonable structural parameters of rotatory rolling for thin-wall pipes with 16–35 mm diameters at wall thicknesses 1,5–3,5 mm.

a

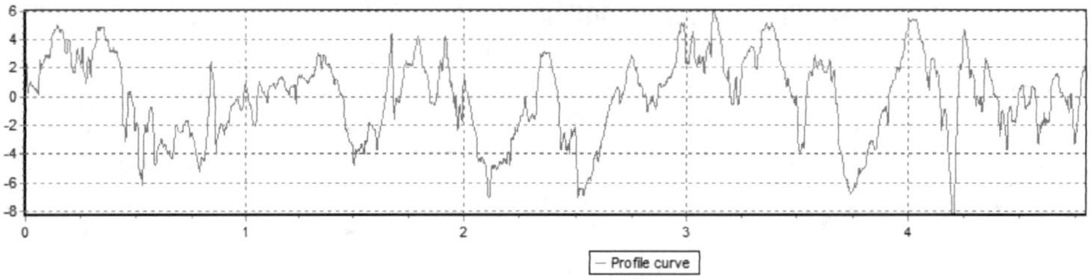

b

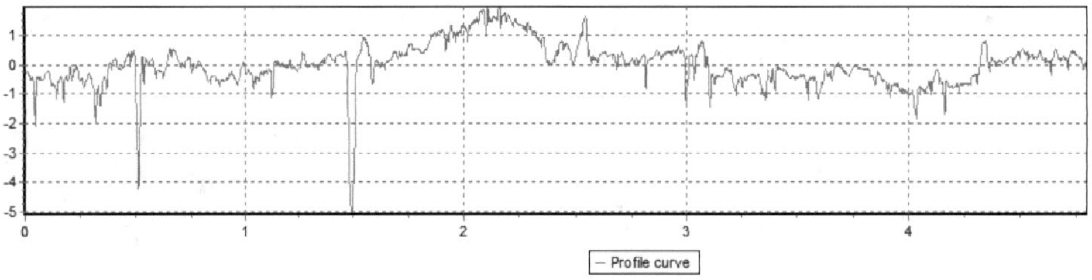

c

Figure 4. Results of experimental research: a) crossing line of a non-machined (right) and rolled (left) surfaces, b) profile diagram of the pipe surface as-delivered, c) profile diagram of the pipe surface after rotatory rolling.

References

[1] Oteniy Ya.N. Processing thin-walled tubes combined cutting and surface plastic deformation / Oteny Y.N., Lavrentiev A.M.,Virt A.E. // Publishes articles on applied and fundamental research. – 2016. – № 12-2. – P. 215-218

[2] Lavrentiev A.M. Analysis of ways of outer diameter of long thin-walled pipes // Modern high technology. – 2018. – № 12-1. – C. 89-94

[3] Lavrentiev A.M. Development of a mathematical model for determining the surface area settings rollers and workpiece witch rotary surface plastic deformation roller with straight and convex images // Fundamental research. – 2016. – № 12-4. – P. 742-747

[4] Oteniy Ya.N. Provision of Form Accuracy at Rotatory Rolling of Long Thin-Wall Pipes/ Ya.N. Oteniy, N.I. Nikiforov, A.M. Lavrentiev, V.A. Alabin //Reinforcement technology and coatings. – 2012. – No. 7. P. 12-16

[5] Oteny Ya.N., Lavrentiev A.M. Development of a mathematical model for determining the surface area settings rollers and work piece witch rotary urface plastic deformation roller with straight and convex images// Fundamental research. – 2016. – № 12-4. – P. 766-770

[6] Lavrentiev A.M. Experimental studies of interrelation of constructive and technological parameters of the deforming rollers and billet at rotational rolling of external cylindrical surfaces // Modern high technology. – 2018. – № 12-2. – P. 289-294

IOP Conf. Series: Materials Science and Engineering **680** (2019) 012002 doi:10.1088/1757-899X/680/1/012002

[7] Sneider Yu.G. Tool for metal fine blanking. – L: Mashinostroyenie, 1971. – 248 p
[8] Sidnyaev N. I. Theory of experiment planning and statistical data analysis. Urait Publishing House, 2018. — 495 p
[9] Tretyak L.N. Basics of Theory and Practice of Experimental Data Processing. M. : Urait Publishing House, 2018. – 237 p

IOP Conf. Series: Materials Science and Engineering **680** (2019) 012003 doi:10.1088/1757-899X/680/1/012003

Experimental research of features of surface treatment using plastic deformation

Ya N Oteny and A M Lavrentiev

Professor, Kamyshin Technological Institute (branch) the Volgograd State Technical University of Kamyschin, 6A, Lenin St., Kamyshin 403874, Russia.

E-mail: Lamvstu@gmail.com

Abstract. This study is devoted to the processing of external cylindrical surfaces by surface plastic deformation. The slip-page of tapered rollers during rolling of cylindrical samples made of steel 35 and 45 has been studied. A diagram is presented for determining the slippage during treatment by surface plastic deformation, as well as a section of a diagram chart with signal recording. It is according to this section that the pulses from current collectors in contact with roller surfaces and workpieces assigned to a specific recording length have been counted. The depth of penetration of the deforming roller into the workpiece and its effect on the roughness of the treated surface have been determined. The results of experimental studies are presented as graphs of dependence of slippage change on self-feed and penetration angles, of roughness on deformation rate as well on penetration and self-feed angle. It has been shown that the deformation force remains practically unchanged at self-feed angles ranging from 0 to 3 degrees. Therefore, a decrease in surface roughness of the treated samples can be explained by the presence of slippage of the contacting surfaces of the rolled workpiece and the deforming roller. It has also been found that if one installs a tapered roller with a smaller base in the feed direction, the roughness is slightly reduced, compared to installing the same deforming element with the same base in the direction opposite to the feed.

1. Introduction

When treating external cylindrical surfaces, one of the most high-performance and cost-effective types of processing is surface plastic deformation (SPD), which is being increasingly used. As a result of SPD, the hardness of the surface layer increases, favorable residual stresses are formed, roughness decreases [1–3], and, in some cases, the accuracy of workpiece dimensions increases.[4–6]

SPD occurs due to the forced rolling of the deforming element in the form of a solid of revolution along the surface to be treated under the influence of deformation force applied to it. [1]

Depending on the necessary quality of the surface layer and the productivity of SPD, the deforming element may have various shapes and sizes. [5,6]

2. Objective

The dependence of the roughness change on the penetration angle and the presence of a minimum on the graph, when a certain penetration angle is reached, were previously known. Yu.G. Schneider [1] and E.G. Konovalov [2] explain this correlation between the change in roughness and the penetration angle by a change in the contact area. As the penetration angle increases, the contact area decreases. This causes an increase in contact stresses and a decrease in roughness. When a certain contact value is reached, rehardening takes place and the roughness increases. [7,8] This explanation is qualitative

IOP Conf. Series: Materials Science and Engineering **680** (2019) 012003 doi:10.1088/1757-899X/680/1/012003

and is not based on any calculated data or theoretical assumptions. Therefore, it is not the only possible or the complete one.

Experimental studies were carried out to use the slip-page phenomenon to explain the influence of the installation angles of deforming elements on the roughness of the treated surfaces. This may add to the explanation and reveal new features of SPD.

3. Materials and methods

When planning and preparing experimental studies, the required number of measurements was chosen for a given error, using a specific measurement method and a specific device. [9,10]

To conduct experimental studies of geometrical parameters of the contact zone, samples (Figure 1) were manufactured on a 1K62 screw-cutting latheusing steel 45 and 35. To increase the reliability of studies, preliminary and final processing of samples was carried out on the ma-chine without reinstallation, which helped to eliminate the runout of samples.Roughness was measured using the TR200 profilometer.

The processing was carried out by an adjustable roller, which allows to set the penetration and self-tightening angles more accurately, as well as to use a deforming element of random configuration, installed between the track rollers, at a given position relative to an axis of a work-piece.

The effect of the penetration value, the relief angle, the self-tightening angle, as well as the size and configuration of the deforming element on roughness has been studied.

The depth of penetration of the deforming element into the treated surface is one of the main features. It includes formulas for determining the deformation force, geometric parameters of the contact pattern, contact stresses, etc. Today, there are no exact design dependencies that would make it possible to determine the depth of penetration of the deforming element into a surface through the deformation force, contact pattern parameters, etc. Therefore, the penetration depth was determined as follows. During the introduction of the deforming element into the treated surface, a 0,005 mm-thick sheet of carbon copy paper was placed. As a result, after load removal, two prints of different sizes have been obtained on the piece of paper and on the surface of the workpiece. The difference in sizes of these prints determined the elastic recovery of the treated surface.

The slippage of the deforming element relative to the surface of the rolled sample was deter-mined as follows (Figure 1a). An insulated current collector 1 was fixed to the holder mounted on the tail stock of the 1K62 screw-cutting lathe. Current collector 2 was installed in the roller head and was in contact with the surface of the deforming element 5, which was not in contact with the rolled workpiece. If it is necessary to bring into contact the entire surface of the deforming element with the surface of the workpiece, for example, when the generatrix of the tapered deforming roller is parallel to the axis of the rolled workpiece, then the roller has a stepped shape, and the current collector 2 is in contact with the lower step of the roller.A thin layer of insulating material is applied to the surfaces of the deforming element and the rolled workpiece in contact with current collectors 1 and 2.During rotation of the workpiece, the insulating layer acts as a chopper.A voltage is applied to the current collectors, which is a signal with a frequency equal to the number of revolutions of the deforming element and the rolled workpiece. This signal is transmitted to a recorder which records readings on a chart tape. Then, the number of impulses received from the current collectors of the deforming element and the rolled workpiece, assigned to certain processing time, has been determined.

The DC voltage supplied to the probes is transmitted as a stepwise signal, containing data on the roller and the workpiece speed, to a high-speed recording device and is recorded on a diagram chart (Figure 1b). After that, the number of impulses coming from roller probes and the workpiece, assigned to specific record length, is counted.

IOP Conf. Series: Materials Science and Engineering **680** (2019) 012003 doi:10.1088/1757-899X/680/1/012003

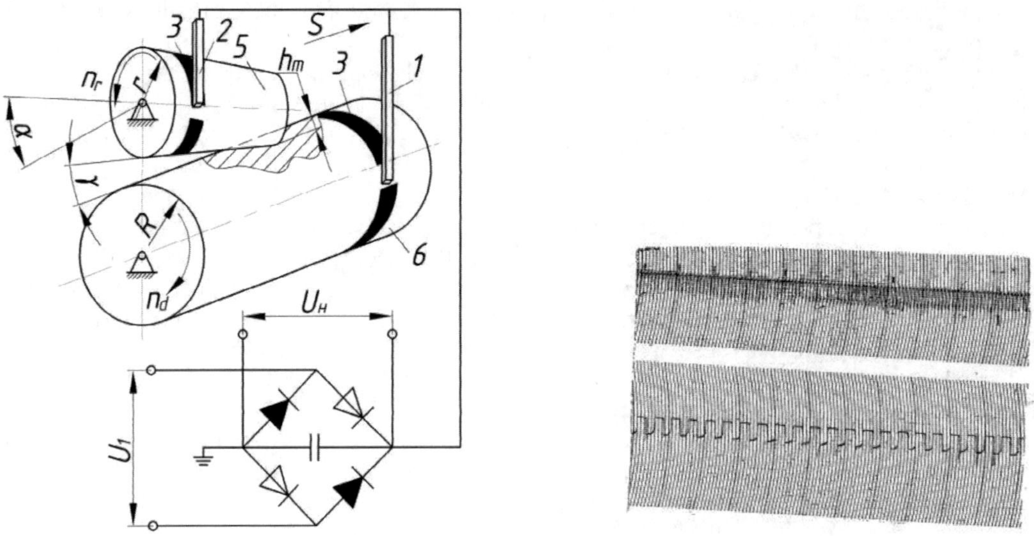

a b

Figure 1. Experimental evaluation of slippage. (1, 2–current collectors, 3, 4–layer of insulating coating, 5–deforming element, 6–rolled workpiece); b) recording of a signal on a diagram chart.

The studies have been carried out to treat the outer surface of a 80 mm cylindrical workpiece by deforming element shaped as a 18 mm tapered roller at various penetration depths and penetration and self-tightening angles.

4. Results and Discussion

Based on the experimental data, dependence graphs of the difference between ratios of the number of deforming elements' revolutions have been obtained experimentally (Figure 2).

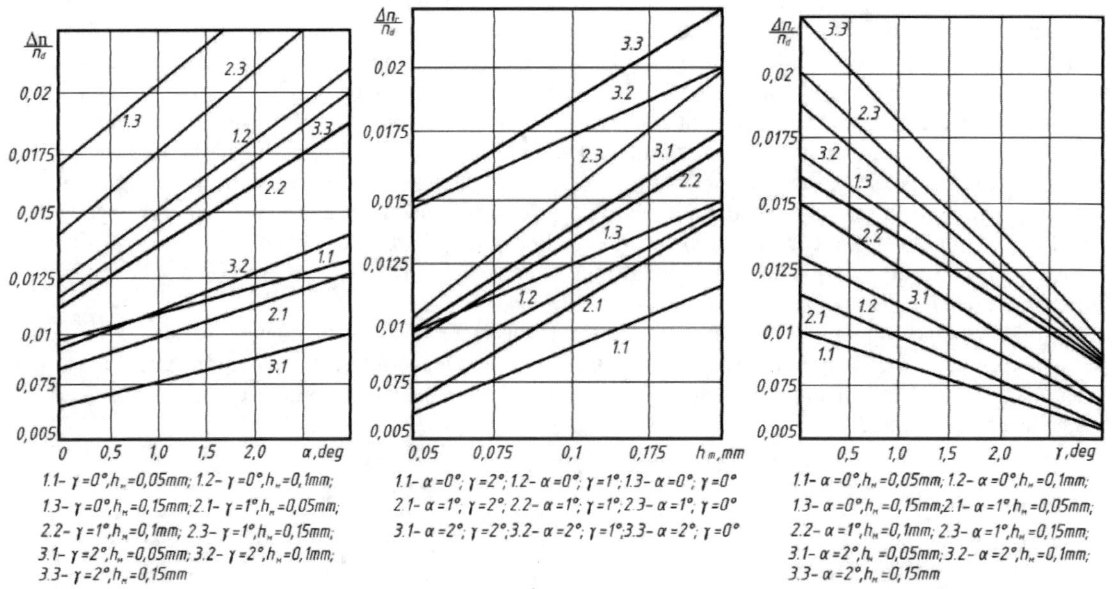

a b c

Figure 2. Dependence of changes in slippage $\Delta n_r/n_d$ on the self-tightening angle α (a), penetration angle γ (b), and penetration depth h_m (c).

As can be seen from Figure 2a, as the self-tightening angle increases, slippage increases, but as the penetration angle increases, slippage decreases (Figure 2b).This is explained by the fact that when the deforming element is installed at the self-tightening angle, circumferential slippage occurs, which is directed opposite to the direction of rotation of the rolled workpiece.A decrease in the slip-page value with an increase in the penetration angle is explained by a decrease in the contact area of the deforming element and the rolled surface. As a result, total slippage also decreases. Slippage also increases as the depth of penetration of the deforming roller into the rolled surface increases (Figure 2c).This points to the fact that the material is being displaced from the contact zone to its less loaded areas in the direction of supply of the deforming element.

Then, roughness of rolled surfaces was measured using the TR220 profilometer. As a result, it was found that the angles of self-tightening and penetration of the deforming element relative to the workpiece axis have a significant effect on the roughness of the treated surface. Figure 3 shows graphs of dependence of changes in the roughness of the rolled surface on the change in penetration (Figure 3a) and installation angles of the deforming element (Figure 3b).It can be seen from the graphs that, as the angle of penetration increases, the roughness of the treated surface first decreases and, having reached the minimum value at an angle of 0,66 degrees, increases.

The decrease in roughness with an increase in the self-tightening angle (Figure 3b) can also be explained by the presence of slippage, since the change of the area of the contact pattern within the range of variable angles of self-tightening, is small. If we consider the graph representing changes of the slippage rate (Figure 3b) and compare it with the graph representing changes in roughness versus the self-tightening angle (Figure 3b), we can see their inversely proportional dependence. However, in [1] it is noted that the rolling speed hardly affects the roughness of the treated surface. This means that in order to determine the influence of slippage between the surfaces of the deforming element and the surface of the workpiece on the obtained roughness, it is necessary to identify discrepancies in smoothing and slipping during rolling.

The difference between slipping and smoothing, in addition to different values of sliding rates, lies in the direction of slipping. During rolling, slippage between the surfaces of the deforming element and the surface of the workpiece is directed towards the feed and perpendicular to the direction of rolling of the deforming element. During smoothing, the sliding of the surface of the indenter is directed around the circumference of the workpiece.

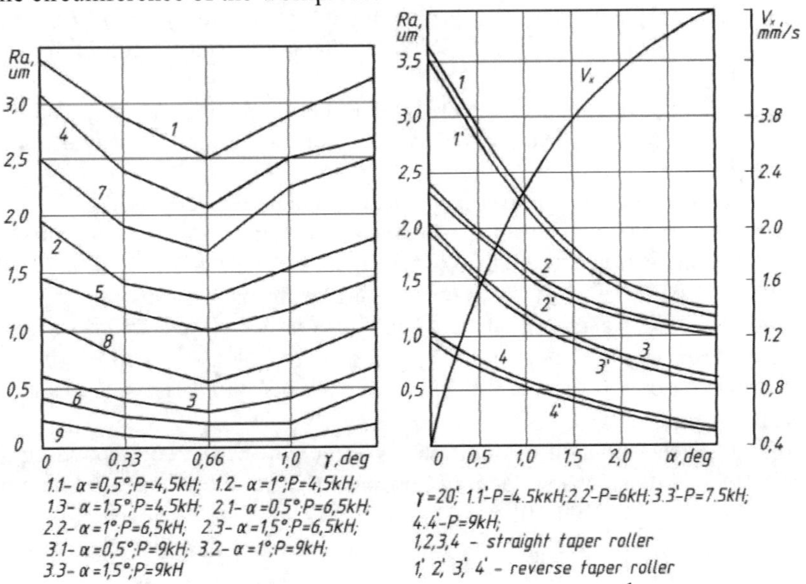

a b

Figure 3. Dependence of changes of the roughness of the treated surface on the penetration angle (a) and the self-tightening angle (b).

One more feature of sliding is a change in its value within the contact zone according to a certain dependence. With that in mind, the differences arising during the formation of the surface layer of the treated workpiece are determined by the nature of slippage of the deforming element relative to the rolled surface. To determine the effect of slippage on the surface roughness of deforming elements of various configurations, experimental studies have been carried out. In these studies, a de-forming element shaped as a tapered roller was installed with larger and smaller bases in the feed direction at the same self-tightening and penetration angles. From the graphs obtained (Figure 3b), it can be seen that when installing a tapered roller with its smaller base in the feed direction, the surface roughness is lower than when installing the same roller with a larger base in the same direction. This phenomenon can be explained by the fact that, when a tapered roller is installed with its smaller base in the feed direction, its radius increases in the direction of that part of the contact where smoothing occurs. This leads to an increase in slippage.

5. Conclusion

Experimental dependences of changes in roughness of the treated surface on the angles of self-tightening and penetration, as well as the location of the roller relative to the workpiece, have been obtained. It was found that, with an increase in the self-tightening angle, the surface roughness of the treated samples decreases. In this case, the contact area and voltage, with a change in the self-tightening angle in the range from 0 to 3 degrees, changes insignificantly. Therefore, the decrease in roughness can be explained by the presence of slippage between the deforming element and the rolled workpiece.

Studying the phenomenon of slippage between the surfaces of the deforming element and the workpiece to explain the influence of penetration and self-tightening angles on roughness can complement the explanation and reveal new features of SPD.

References

[1] Sneider Yu.G. Tool for metal fine blanking. – L: Mashinostroyenie, 1971. – 248 p
[2] Konovalov Ye.G. Rotatory treatment of surfaces with automatic delivery. /Ye.G. Konovalov, P.S. Chistoserdov, A.I. Flomenblit. – Minsk.: "Vysheishaya Shkola", 1976, p.192
[3] Bezyazychny V.F. Industrial processes of machining and physical&chemical processing in mechanical engineering Study guide. – 4th edition, stereotyped. / V.F. Bezyazychny, V.N. Krylov, Yu.K. Charkovsky, Ye.V. Shilkov. - Saint-Petersburg: Lan Publishing House, 2017. – 432p.: ill
[4] Oteniy Ya.N. Provision of Form Accuracy at Rotatory Rolling of Long Thin-Wall Pipes/ Ya.N. Oteniy, N.I. Nikiforov, A.M. Lavrentiev, V.A. Alabin //Reinforcement technology and coatings. – 2012. – No. 7. P. 12-16
[5] Lavrentiev A.M. Development of a mathematical model for determining the surface area settings rollers and workpiece witch rotary surface plastic deformation roller with straight and convex images // Fundamental research. – 2016. – № 12-4. – P. 742-747
[6] Oteny Ya.N., Lavrentiev A.M. Development of a mathematical model for determining the surface area settings rollers and work piece witch rotary urface plastic deformation roller with straight and convex images// Fundamental research. – 2016. – № 12-4. – P. 766-770
[7] G.M.Azarevich, G.Sh. Bernstein, Finishing of external cylindrical surfaces plastic deformation, ONTI, Moscow, 1963
[8] Nikiforov N.I. Determination of deformation force when SPD corresponding start peeling as a result of the perenaklepa surface // Publishes articles on applied and fundamental research. – 2015. – № 12-4. – P. 599-601
[9] Sidnyaev N. I. Theory of experiment planning and statistical data analysis. Urait Publishing House, 2018. — 495 p
[10] Tretyak L.N. Basics of Theory and Practice of Experimental Data Processing. M. : Urait Publishing House, 2018. – 237 p

Control of geometrical parameters of a single layer at additive forming of products by FDM technology

A N Grechukhin, V V Kuts and A V Oleshitsky

Southwest State University, street 50 years of October, 94, Kursk, Russia

Abstract: : Questions are considered about the accuracy of the formation of the surface layer by additive methods. The analysis of works is carried out on a research subject. High values of the error of forming (approximation) are revealed at forming of surfaces of a complex profile by additive methods. The spatial orientation of a single layer is proposed to reduce the error of forming products by additive methods . Mathematical dependences are proposed to determine the value of the parameters of a single directed layer on the results of experimental studies.

1. Introduction

The term additive technology has been entrenched in the scientific community for a long time and it means layer-by-layer cultivation of objects. These technologies are classified as one of the types of new industries and they allow us to launch new products in a short time. The classification of additive shaping methods may vary in the type of material used, they may vary in the state of aggregation of the materials, they may vary in the way the material is sprayed, etc. Additive methods have a common way of shaping. The object is formed in layers and it is formed from top to bottom, and the quality of the surface depends on the geometric parameters of the cross section of a single layer, that is, the error in shaping (approximation) is high when using additive technologies and this is a significant drawback of additive methods [1-8].

Scientific works have been created in large numbers on the study of the mechanisms of the formation of errors in additive shaping methods, but they are devoted to a greater degree to ensure accuracy at the macro level by minimizing the height of a single deposited layer or they are devoted to the selection of the shaping mode [9-13].

Issues to be considered are also about the spatial orientation of the products during shaping, since the physical position affects the formation of the shape of the transverse unit layer, it also affects its geometric characteristics, and it affects the formation of the accuracy of the formed surface [16-19]. The studies were conducted to find the possibility of minimizing the error of additive methods and they were aimed at establishing the components of their error.

2. Analysis of domain research

A large number of scientific papers devoted to the study of mechanisms of error formation in the formation of subtractive and additive methods.

The problems of formation of errors in mechanical processing are studied in [3] and the error of processing was considered as a result of the action of the relevant components:

$$\Delta r_0 = \Delta r_{\text{ст}} + \Delta r_{\text{дин}} + \Delta r_{\text{тем}} + \Delta r_{\text{из.ин}} + \Delta r_{\text{из.ст}}, \quad (1)$$

where the static component of the processing error; the dynamic component of the processing error; and errors caused respectively by thermal deformation and tool wear; error caused by the wear of parts.

The value of the error of forming (approximation) is one of the values of the static component of the error of processing in the formation of parts by additive methods.

Improving the accuracy of additive forming methods is an urgent task, for which it is necessary to determine the geometric parameters of the unit deposited layer, Orient it in space so that the error of forming (approximation) was minimal [13-15].

3. Experimental study

Dimensions and shape parameters must be defined and described mathematically to provide a rational spatial orientation of a single directed layer.

Samples of single layers were sent according to the experimental plan (table 1.1.) – variable parameters: thickness of a single layer h, mm; temperature of extruder heating t, °C; speed of extruder movement v, mm/sec; sections were made, which were investigated on the instrumental MMI-2 microscope. The coordinates of the points on the curves forming the cross section of the unit layer were fixed, the approximation of these points by the ellipse equation was made – this is the shape of the cross section of the unit layer was observed in the samples (Fig. 3), the difference in the coordinates of the characteristic points on the surface of the cross-section profile of the unit layer were determined by the parameters of the width a and height b of the unit deposited layer.

Table 1. The experiment plan.

№	X1		X2		X3	
	Код	*h*, mm/	Код	*t*, °C	Код	*v*, mm/s
1	+	0,22	+	212	+	23
2	-	0,38	+	212	+	23
3	+	0,22	-	228	+	23
4	-	0,38	-	228	+	23
5	+	0,22	+	212	-	47
6	-	0,38	+	212	-	47
7	+	0,22	-	228	-	47
8	-	0,38	-	228	-	47
9	-1,215	0,2	-1,215	210	-1,215	10
10	+1,215	0,4	+1,215	230	+1,215	60
11	0	0,3	0	220	0	35
12	0	0,3	0	220	0	35
13	0	0,3	0	220	0	35
14	0	0,3	0	220	0	35
15	0	0,3	0	220	0	35

The direction of a single layer was carried out on the additive installation of modes. Model Makerbot Replicator 2, single layer material abs plastic.

Figure 1. Samples of a single deposited layer

The combination of the values of geometrical parameters (width and height) was obtained during measurement of the parameters of a single lateral layer.

Table 2. Parameters a– width and b– height, cross-section of a single layer

№	1	2	3	4	5	6	7	8	9	10	11	12	13	14	15
a, mm	0,545	0,8	0,65	0,8	0,565	0,845	0,595	0,755	0,65	0,68	0,605	0,615	0,6	0,615	0,6
b,mm	0,29	0,38	0,2	0,29	0,25	0,29	0,23	0,37	0,15	0,51	0,28	0,32	0,32	0,32	0,32

Processing of the experimental results for the mathematical description of the geometric characteristics of the directed unit layer was made and reflecting the influence of elements of the modes of additive shaping on the width α and height b, mathematical dependences were obtained:

$$a = h(3{,}56h + 8{,}536) - t(0{,}043h - 0{,}02) + v(0{,}0002t - 0{,}04) - 3{,}87 \quad (2)$$

$$b = h(4267 - 5{,}595h) - t(0{,}0006h - 0{,}028) + v(0{,}000312t - 0{,}065) - 31{,}45 \quad (3)$$

where h (mm.) - is the thickness of a single layer; t (°C) - is the heating temperature of the extruder; v (mm/sec.) – moving speed of the extruder.

Graphs of the influence of elements of the regime of additive formation on the value of the parameters of a single layer were built on the results of the analysis of the empirical dependences.

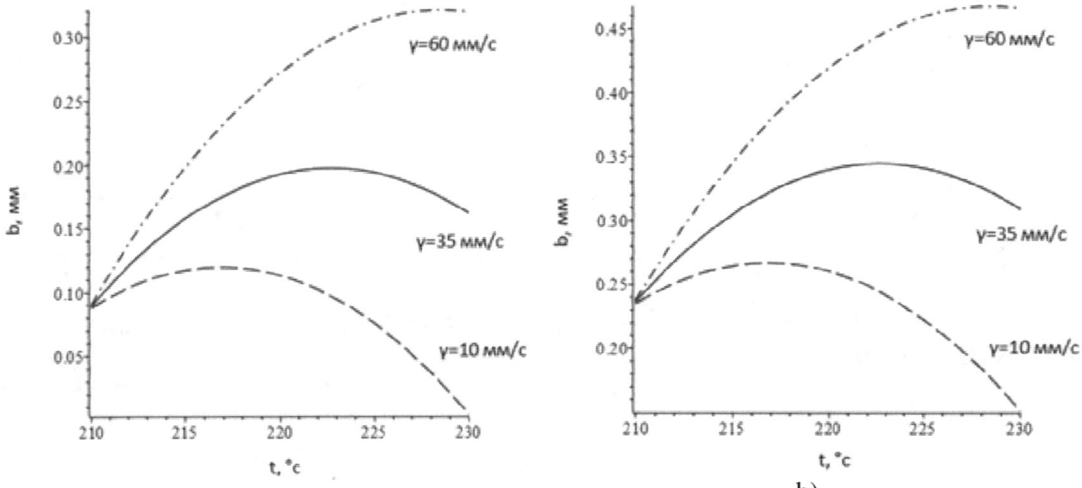

a) b)

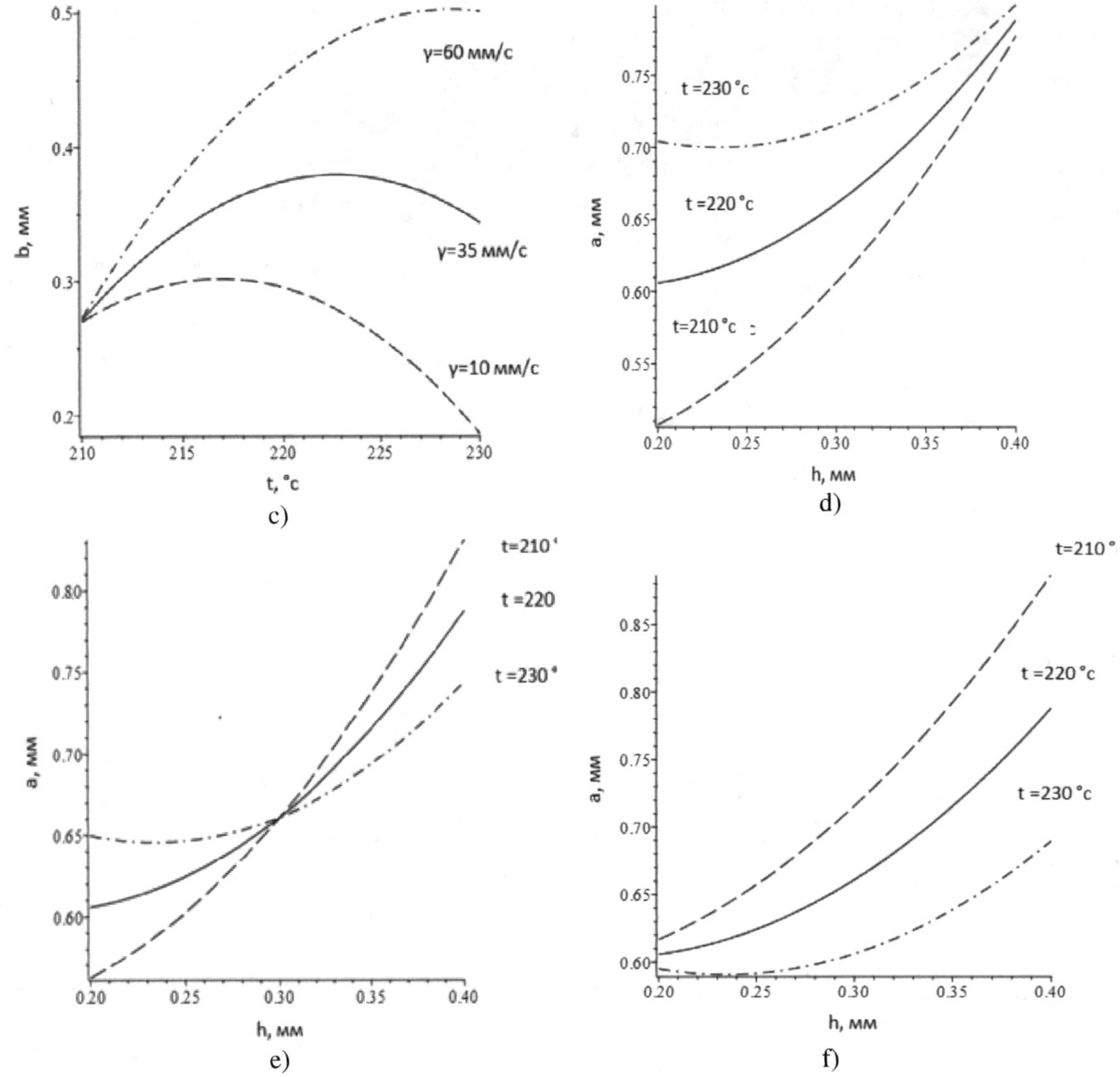

Figure 2 . Graphs of the width of the unit deposited layer a, the height of the unit deposited layer b on the elements of the modes of additive shaping:

a - graph of the height of a single layer b, mm temperature t, °C, v, mm/s, at a constant value of the layer thickness h=0,2 mm;

b - graph of the height of a single layer b, mm temperature t, °C, v, mm/s, at a constant value of the layer thickness h=0,3 mm;

b - graph of the height of a single layer b, mm temperature t, °C, v, mm/s, at a constant value of the layer thickness h=0,4 mm;

g - graph of the width of a single layer a, mm temperature t, °C, h, mm, at a constant value of the layer thickness v=10 m/s;

d - graph of the width of a single layer a, mm temperature t, °C, h, m, at a constant value of the layer thickness v=35 m/s;

e - graph of the width of a single layer a, mm temperature t, °C, h, mm, at a constant value of the layer thickness v=60 m/s.

The results of the analysis revealed the nature of changes in the parameters of width and height of a single deposited layer. The width of a single layer a does not depend on the value of the heating temperature of the extruder t at the thickness of the directed layer h=0,3 mm, which is observed.

4. Conclusion

Mathematical relationship showing the effect of additive elements of modes of formation on the parameters of the cross section of single-directional layers obtained according to the results of the research.

The development of methods and devices to ensure the dynamic spatial orientation of the final link of the forming system of the additive installation is the next stage of research and their application allows to provide the spatial orientation of a single directed layer and the least curvature of the approximated sections of the single layer profile of the forming part.The value of the error of shaping (approximation) will be able to reduce and improve the accuracy of shaping by additive methods.

Acknowledgment

The study was supported by the grant Of the President of the Russian Federation for young scientists MK-6406.2018.8

Reference

[1] Saprykin A A 2006 Increasing the productivity of the process of selective laser sintering in the manufacture of prototypes *Thesis of candidate of technical Sciences* (Yurga: Tomsk Polytechnic University) p 153

[2] Pronikov A, Averyanov O I and Apollo Yu S 1994 *Designing of metal-cutting machines and machine tools* (*Handbook-textbook vol 3*) (Moscow: MSTU them NE Bauman: Mechanical Engineering) p 444

[3] Kuts V V, Razumov M S, Grechukhin A N and Bychkova N A 2016 Improving the quality of additive methods for forming the surfaces of odd-shaped parts with the application of parallel kinematics mechanisms *International Journal of Applied Engineering Research* **11** pp 11832-11835

[4] Dobroskok V L, Abdurayimov L N and Chernyshov S I 2010 Rational orientation of products with their layer-by-layer shaping on the basis of the original triangulation 3d model *Scientific notes of the Crimean Engineering and Pedagogical University* **24** pp 13-21

[5] Singhal S K, Pandey A P, Pandey P M and Nagpal A K 2005 Optimum part deposition orientation in stereolithography *Computer-Aided Design & Applications* **2** pp 319–328

[6] Hong S, Byun Kwan H Lee 2006 Determination of optimal build direction in rapid prototyping with variable slicing *Int. J. Adv. Manuf. Technol* **28** pp 307–313

[7] Hong S, Byun Kwan H Lee 2004 Optimal part orientation of rapid prototyping using a genetic algorithm *Computers & Industrial Engineering* pp 426–431

[8] Hur J, Lee K 1998 The development of a CAD environment to determine the preferred build-up direction for layered manufacturing *Manuf. Technol* **14** pp 247–254

[9] Kim J Y, Lee K and Park J C 1994 Determination of optimal part orientation in stereolithographic rapid prototyping *Technical Report. Department of Mechanical Design and Production Engineering* (Seoul: Seoul National University) pp 356-366

[10] Lan P T, Chou S, Chent Y and Gemmill L D 1997 Determining fabrication orientations for rapid prototyping with stereolithography apparatus *Computer-Aided Design* **29** pp 53– 62

[11] Massod S H, Rattanawong W and Iovenitti P 2003 A generic algorithm for part orientation system for complex parts in rapid prototyping *J. Mater. Process. Technol* **139** pp 110–116

[12] Kuts V V 2012 *Methodology of pre-project studies of specialized metal-cutting systems* Thesis of doctor of technical Sciences (Kursk) p 366

[13] Kuts V V and Ponomarev V V 2017 Construction of a model for the formation of long shafts with a RK profile of a prefabricated disk milling cutter *Fundamental and applied problems of technology* pp 110-115

[14] Grechishnikov V A, Kuts V V, Razumov M S and others 2017 Determination of the error in the shape of a part shaping by a planetary mechanism using the methods of geometric cutting theory *STIN* **4** pp 24-26

[15] Grechishnikov V A, Romanov V B, Pivkin P M and others 2017 Errors in shaping by a planetary mechanism *Russian Engineering Research* **37** pp 824-826

[16] Grechukhin A N, Anikutin I S and Byshkin A S 2018 Management of space orientation of the end effector of generation of geometry system fiveaxis manufacturing machinery for additive generation of geometry *MATEC Web of Conferences* **226** pp 010-021

[17] Grechukhin A.N, Kuts V.V, Razumov M.S 2018 Ways to reduce the error of additive methods of forming *MATEC Web of Conferences* Vol. 226 pp 023-029

[18] Grechukhin A N, Kudelina D V, Razumov M S and others Development of information-analytical system for technological requests monitoring, taking intoaccount regional specifics *International Conference on Actual Issues of Mechanical Engineering* **157** pp 198-202

[19] Grechukhin A N, Kuts V V and Razumov M.S 2019 Calculation of the controlled parameters of the 6-coordinate robot in the process of additive forming of products *Journal of Physics: Conference Series* **1210** pp 1210-1220

IOP Conf. Series: Materials Science and Engineering **680** (2019) 012005 doi:10.1088/1757-899X/680/1/012005

Bearing defects diagnostics using the principal components analysis

E A Abidova, A A Lapkis, A V Chernov

Volgodonsk engineering technical institute branch of NRNU MEPhI, Lenina street 73/94, Volgodonsk 347360, Russian Federation

E-mail: e-abidova@mail.ru

Abstract. In this paper a new method of bearing defects detecting and defect severity assessment based on the results of singular spectrum decomposition (SSD) of the vibration signal is proposed. The presented results of equipment diagnostics in the laboratory conditions have demonstrated the applicability of signal representation in the principal component basis to improve diagnostic sensitivity.

1. Introduction

Bearing defects of the electrically driven mechanisms are commonly detected using vibration analysis as a simple, repeatable and non-destructive method. Common approaches to processing of diagnostic signals such as spectral analysis [1,5], envelope analysis [5,6] are aimed at highlighting defect signs on a background of noise and interferences. On the other side it is shown [4,5] that increase of signal's chaoticness itself can be early and definite sign of the object's defect. The share of nonlinear and stochastic components increases due to the factors like clearances and friction.

Implementing of simple and accurate methods of bearing defects detecting has a practical interest for industry. This research applies a multidimensional analysis of determined and stochastic parameters of a diagnostics object – ball bearing. It is proposed to improve methods of diagnostics with the principal components analysis (PCA).

2. Diagnostics method

The proposed approach has a singular value decomposition (SVD) in its basis. SVD in the calculation method asserting that a matrix [A] of rank L can be decomposed to the three matrices: orthogonal matrix [U], diagonal matrix [S] and transposed orthogonal matrix $[V]^T$ in the following way.

$$[A] = [U_l][S_l][V_l]^T, \tag{1}$$

where $U^TU = I$ and $V^TV = I$, S – diagonal matrix containing singular values (square root of the eigenvalues) that can be presented as $S = \mathrm{diag}(\sigma_1, \sigma_2, \sigma_L)$ where L=min(m,n). The numbers σ_i (i=1,2,..., L) are called singular values of [A]. The singular value defines significance of the corresponding component in the $[V_l]$.

In the diagnostics practice the most of registered signals are rather time series than matrices. So in this paper the operation of hankelization is applied to the signals. The Hankel matrix for the discrete vibration signal xi (i=1,2,...N) can be obtained with the sliding window of width m in the way

$$[A] = \begin{bmatrix} x_1 & x_2 & \cdots & x_n \\ x_2 & x_3 & \cdots & x_{n+1} \\ \vdots & \vdots & \ddots & \vdots \\ x_m & x_{m+1} & \cdots & x_N \end{bmatrix}, \tag{2}$$

where m+n-1 = N.

The Hankel matrix [A] from (2) is used for singular value decomposition according to (1).

The SVD methed gives an opportunity to get the matrix $[\bar{A}]$ m × n of rank l<L reflecting the share of the most significant components. This matrix is obtained as [7]:

$$[\bar{A}] = [U_l][S_l][V_l]^T, \tag{3}$$

where $[\bar{A}]$ is the matrix [A] reconstructed by using commonly the first, most significant components (principal components or PC). The rest of the singular values correspond to the components with smaller share in the initial signal:

$$\sigma_i > \varepsilon,\ i=1,\dots,l\ ;\ \sigma_i \le \varepsilon,\ i=l+1,\dots,L, \tag{4}$$

where ε is significance threshold.

However, with respect to the diagnostic signals «strong components» (i.e. the ones that are multiplied by big singular values) may not represent important diagnostic information. On the contrary, components of the source signal with a defect influence are probable to have smaller values and greater numbers.

Typical vibration signal of a defective bearing can be modeled as

$$x(k)=y(k)+n(k), \tag{5}$$

where y(k) is a determined part and n(k) is «noisy» (stochastic and chaotic) part.

Accordingly the matrix A corresponding to x(k) has two parts: determined data part [A'] containing information about shafts and bearings harmonics etc. (indirect defect information), and noisy data part [N] defined by specific defect influence. The following equation can be obtained from (1) and (2)

$$[A] = [\acute{A}] + N = [U_l \quad U_0] \begin{bmatrix} S_l & 0 \\ 0 & S_0 \end{bmatrix} \begin{bmatrix} V_l^T \\ V_0^T \end{bmatrix}, \tag{6}$$

where [A'] and [N] correspond with y(k) and n(k), S_l contains components with bigger numbers s>e, S_0 – smaller numbers s<e.

The numbers s and e can be obtained by representing [A] on the singular value spectrum (SVS) and choosing the respective rank where singular values tend to zero. So the information about proportion between determined, stochastic and chaotic components in [A] becomes available (Fig. 1). Increase in the share of chaotic components can be considered as an early sign of a defect.

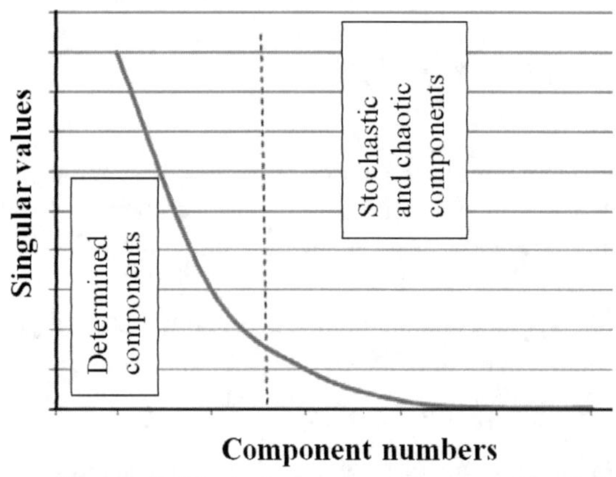

Figure 1. Typical singular value spectrum

The practical problem of the use of SVD in signal processing is the estimation of the thresholds separating the determined information from the variable and noisy one. The SVD method usually includes a reconstruction procedure that is performed by arithmetic averaging along the side diagonals of the matrix [A] or [N]. This procedure is not mandatory in this study.

The principal components selected from the vectors of the matrix U resulting from decomposition (2) define the basis for projecting the diagnostic data. The technical state of an object is connected with the data location in the chosen basis. Opponents of the method, however, argue that the resulting projections allow us to evaluate only the relative state, and not the working ability and especially not a specific defect [10]. Such an objection applies only to the classical algorithm of the method [11], which uses the entire array of data (working and faulty states) to build the basis, and the classification results are relative. Method of PC in the interpretation proposed by Trendafilova [9] assumes the initial construction of the basis and the subsequent projection of the analyzed data upon it. The initial data are the results of diagnosis corresponding to the working state. This data should be presented as a matrix (1).

Another problem of PC method application is the choice of the basis directions and their numbers. In most works [10-12] this problem is solved on the basis of SVS analysis. It is assumed that the components with the highest eigenvalues are sufficiently informative to characterize the data array.

However, it is logical to assume that the singular values of the samples of the working and analyzed states can differ significantly. The assumption is based on the fact that the effect of the fault increases the contribution of greater (stochastic and chaotic) components. In this situation, to ensure selectivity, it is advisable to compare the SVS's and choose the components that correspond to the greatest difference in eigenvalues. Thus, to improve diagnostic data analysis methods, it is proposed to represent diagnostic data in the basis, the components of which reflect the contribution of determined, stochastic and chaotic components. To build the space it is proposed to use the dataset corresponding to the working state of the object.

3. Experimental testing of the diagnostics method

An experiment has been conducted to test the proposed method based on the representation of diagnostic signal in the space of determined, stochastic and chaotic components. During the experiment an abrasive wear of ball bearings was performed using test facility shown at the fig. 2. The level of vibroacceleration has been measured after 10, 20 and then every 20 minutes of wear till 140 minutes from test start.

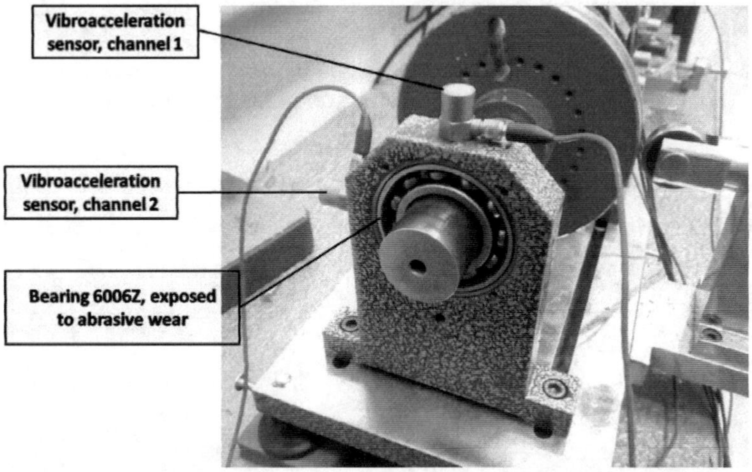

Figure 2. Test facility. Bearing support, view along the shaft axe

Initially, the signals of bearing vibroacceleration were processed with the conventional approaches, e.g. the root mean square (RMS) values of the defect-free bearing and bearing after abrasive wear were calculated. An abrupt increase followed by decrease and subsequent smooth growth has been

IOP Conf. Series: Materials Science and Engineering **680** (2019) 012005 doi:10.1088/1757-899X/680/1/012005

observed on all measuring channels (so the clear match between technical state of tested element and vibroacceleration level was not observed).

Subsequently, the vibration signals were processed using singular decomposition; the singular value spectra (SVS) were obtained. As a result, the form of the spectrum corresponds to the technical state of tested element. SVS of the defect-free bearing shows prevalence of determined components: the first ten components mainly describe the signal and the share of greater components is insignificant.

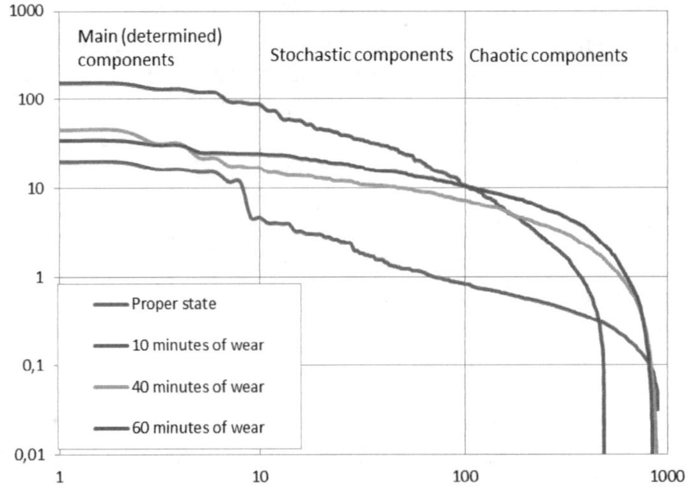

Figure 3. Singular value spectra of the diagnostic signals

The first ten minutes of wear are characterized by growth of the determined and stochastic components (which can be seen on the graph as an increase of the share of components number 1-100). Subsequent wear (next series) is represented in the SVS as a relative decrease of the components 1-100 and increase of the greater components interpreted in this research as chaotic.

Thus, the preliminary processing of the diagnostic data shows that using first PC is not sufficient for correct assessment of equipment degradation status despite this approach being recommended by the literature on PCA application.

The next step in the experiment data processing was projecting the signals onto the principal component basis. Initially, the data has been projected onto the first three components to demonstrate several features of degradation process (Fig. 4a).

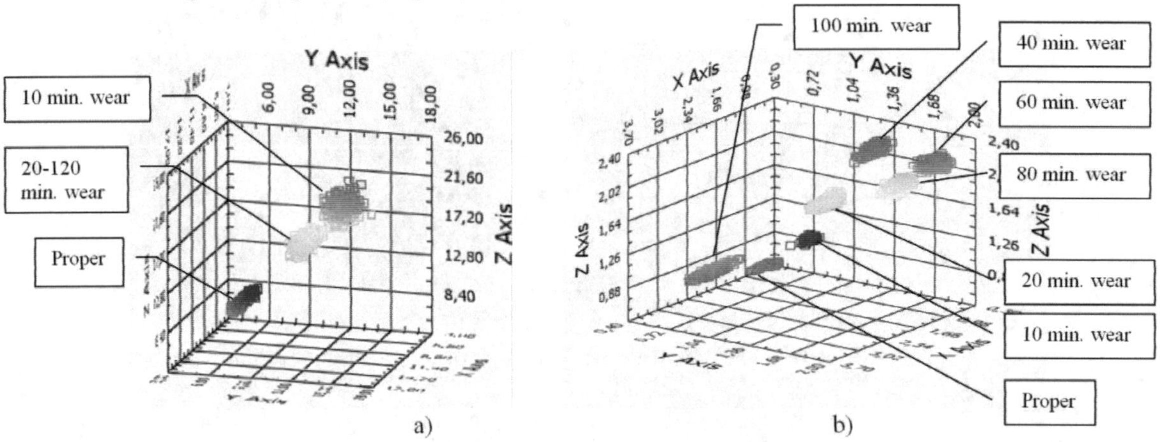

Figure 4. Projecting the signals onto the PCs
a) PC №1-3; b) PC №1,55,110

The first three components at the Fig. 4a coincide with the axes X, Y, Z. The plotted projections show that the initial fault appears as an abrupt change of the projection in the PC basis. The subsequent alignment of moving parts is reflected as a return of the characteristics to their initial location. Such diagnostics results do not allow to distinguish fault stages, and degradation can be erroneously considered as an improvement of the bearing's technical state.

In order to improve diagnostics accuracy, we propose to select as a basis for projection three PC's which reflect determined, stochastic and chaotic components in the signals. Based on the signal's SVS we have chosen the components №1, 55, 110 (coincide with the axes X, Y, Z at the Fig. 4b).

Seven clusters plotted on the Fig. 4b represent stages of defect development: initially, an increase of X and Y (determined and stochastic) components with the subsequent growth of Z (chaotic) component has been observed. This the stages of degradation process can be differentiated. The described experiment demonstrates an ability to improve diagnostics accuracy by building reference basis considering the shares of determined, stochastic and chaotic components.

4. Summary

As development of a bearing defect is driven with determined, stochastic and chaotic factors, the corresponding signal's (informative) features have to be identified to improve diagnostics sensitivity and selectivity. The conventional approaches are unable to identify the chaotic dynamics in the signals. So, identification of informative elements based on the PCA is a simple and effective strategy.

It is proposed:

- to determine a proportion between determined, stochastic and chaotic components using SVS.
- to represent the signal in the reference basis of principal components corresponding with the biggest difference between singular values obtained by signal decomposition (it is more effective than using only the first components with the greatest values).

The results of the experiment on bearing abrasive wear using the proposed method have demonstrated that defect signs were (more) successfully distinguished as compared with the most common approach (evaluation of RMS vibroacceleration) and even with conventional PCA algorithm. The result obtained indicates the possibility to implement the proposed method for the diagnostics of electromechanical equipment in industrial environment.

References

[1] Randall R.B. 1987 *Frequency analysis* (Copenhagen: Bruel & Kjaer; 3rd edition)
[2] Chernov A.V., Pugachiova O. Yu., Abidova E.A. 2011 Obrabotka diagnosticheskoj informacii pri ocenke tekhnicheskogo sostoyaniya elektroprivodnoj armatury AES *Inženernyj vestnik Dona (Rus)* 3(2011)499 URL: ivdon.ru/ru/ magazine/archive/n3y2011/499
[3] Sejnov S.V., Goshko A.I., Adamenov A.K. et al 2012 *Tekhnicheskoe diagnostirovanie armatury AES (Rus)* (Moscow: Mashinostroenie)
[4] Abidova E.A., Hegaj L.S., Chernov A.V., Draka O.E., Pugachiova O. Yu. 2016 Diagnostirovanie elektroprivodnoj armatury s ispol'zovaniem entropijnyh pokazatelej Global Nuclear Safety vol. 4 (21), pp. 69-77.
[5] Y.G. Xu, L.L. Li, Z.J. He, Approximate Entropy and its applications in mechanical fault diagnosis *Information and Control* 31 (6) (2002): pp. 547–551.
[6] Nemirko A.P., Manilo L.A., Kalinichenko A.N., Volkova S.S. 2010 Entropijnye metody ocenki urovnya anestezii po EEG–signalu *Informatsionno-upravliaiushchie sistemy (Rus)* vol. 3 pp. 69-73
[7] Pomerantsev A. Metod glavnyh komponent (PCA) *Rossijskoe hemometricheskoe obshchestvo (site)* URL: http://rcs.chemometrics.ru/Tutorials/pca.htm
[8] Reza Golafshan, Kenan Yuce Sanliturk. SVD and Hankel matrix based de-noising approach for ball bearing fault detection and its assessment using artificial faults. Mechanical Systems and Signal Processing 70-71 (2016). pp. 36–50
[9] Al Bugharbee H., Trendafilova I. 2017 A methodology for fault detection in rolling element bearings

IOP Conf. Series: Materials Science and Engineering **680** (2019) 012005 doi:10.1088/1757-899X/680/1/012005

using singular spectrum analysis *International Conference on Engineering Vibration 2017 (ICoEV 2017)* URL: https://strathprints.strath.ac.uk/62663

[10] Leont'eva N.L. 2013 Mnogomernaya gusenica, vybor dliny i chisla komponent gusenicy *Machine Learning and Data Analysis (Rus)*, 2013. vol. 1, №5. pp. 5-15.

Study of adaptive hydraulic drive of rotary drilling machine

M A Lemeshko, S G Stradanchenko, A G Iliev, I A Zanina

Don State Technical University (branch) in Shakhty, Shevchenko street 147, Shakhty 346500, Russia Federation

E-mail: Lem -mikhail@ya.ru

Abstract. The article deals with drilling machines in which the ratio of the load between the power to rotate and the power to feed is carried out automatically due to the structure of the adaptive machine, with the provision of self-regulation for drilling modes close to optimal. Changing drilling conditions automatically results in a change in the ratio of speed to feed force, providing maximum performance while protecting against sudden overloads, for example, when encountering solid inclusions in the drill rock, or when shading a well. The ratio of frequency of rotation and feed rate determines the specific depth of cut (specific feed) and determines the performance of the working process of the drilling machine. A typical example of such a machine is a drilling machine rotary type with hydraulic rotation and feed.
The cutting theory determines the rational drilling conditions, known rational specific feed, rational speed, minimum and maximum feed force, depending on the strength of the material being drilled and the degree of blunting of the cutting tool.
The block diagram of such a drilling machine is considered and the analysis of the influence of parameters and characteristics of individual hydraulic elements on the drive setting on the rational mode of operation is given. The research stand and its structure are given, some results of experimental researches are given. It is established that the chokes of the drive settings can adjust the degree of adaptation, change its sensitivity to external influences and change the range of operating conditions for the rational functioning of adaptive drilling.

1. Introduction

Known drilling machines in which the drilling process is carried out by automatically maintaining the ratio of two connected movements. These are the two main movements of the Executive body-the drill rod: forward movement in the direction of drilling and rotational movement.

Classic drilling machines designed for mining or construction work; represent an extensive class of machines, with a variety of power and performance. These machines are designed for specific conditions of their application. But for all these machines there is a rational relationship between the two working movements.

The mining and construction drilling machines are adjacent to Metalworking machines, in which the drilling process is provided by at least two interconnected movements. Moreover, in most cases, their ratio determines the performance of Metalworking, especially the specific energy consumption of cutting and productivity, as well as the temperature conditions of cutting and the purity of the surface. The resource of the cutting tool depends on how close the considered ratio is to the rational value. The current level of development of the cutting theory regulates the optimal conditions of Metalworking for different materials and conditions of the machine.

IOP Conf. Series: Materials Science and Engineering **680** (2019) 012006 doi:10.1088/1757-899X/680/1/012006

2. Topicality

Obviously, it is important to automatically maintain a rational ratio of the two working movements of the Executive body of the drilling machine in the conditions of variability of strength, viscosity and other characteristics of the drilling material. In this regard, the concept of drilling process control without the involvement of automation and computer technology is known [1]. This method is known, it was developed by a number of researchers and tested in industry [2,3]. Let's take a closer look at some of the known studies.

3. Practical relevance

The peculiarity of the control of the working process of machines with an Executive body performing two working movements is the account of random and systematic external influences to which the control system must respond to an adequate (adaptive) change in the operating mode, one or both control channels. In this case, the control criterion can be both a rational ratio of working movements and their parameters, and the transition of the drive to a special, for example, emergency (protective) mode.

For this purpose, it is advisable to use the methods of hydraulic automation, and in particular to use an automated hydraulic drive, which uses hydraulic differentials [4]. Thus a feature of such schemes is the implementation of feedback via silo-instantaneous kinematic links [5]. Experience in the development, research and application of this type of drives has shown the ability to provide high performance and stability of these machines [6,7].

An example is a device that is designed for roughing parts of a complex profile by milling (Fig.1). This device automatically provides stabilization of the thickness of the removed layer during machining of curved surfaces [8].

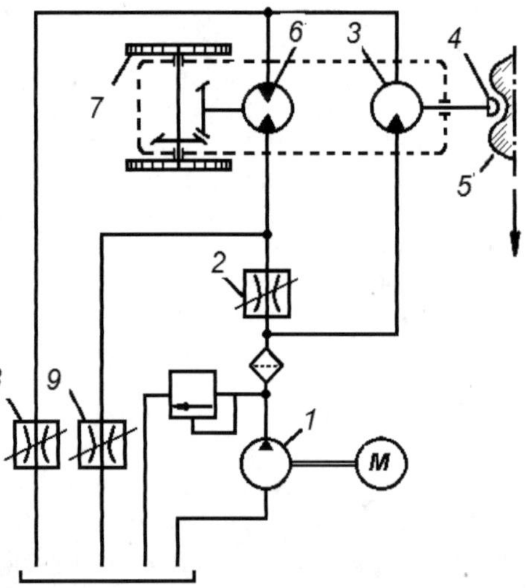

Figure 1. Schematic diagram of the device to stabilize the thickness of the removed layer when machining curved surfaces. (SU A. S. No. 483224): 1-motor, 2-throttle feedback, 3 -actuator, 4 - mill, 5 - surface processing, 6 - reversible feed drive, 7-feeder; 8,9 - throttles setup

Next (Figure 2) the block diagram of this device is shown.

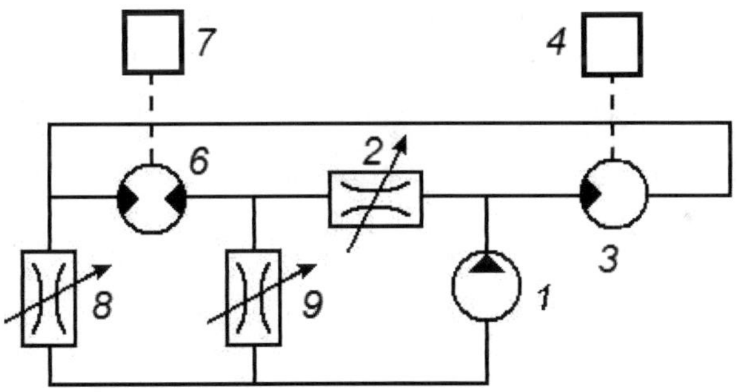

Figure 2. Block diagram of the device to stabilize the thickness of the removed layer in the machining of curved surfaces

For figure 3 shows the hydraulic scheme of the adaptive drilling machine [9], the drive structure of which contains two hydraulic differentials. The first differential is formed by the hydraulic hose from the exit of the hydraulic pump, a hydraulic hose connected to the choke DR1 and the hydraulic hose going to the hydraulic motor HM. The second differential is a hydraulic cylinder.

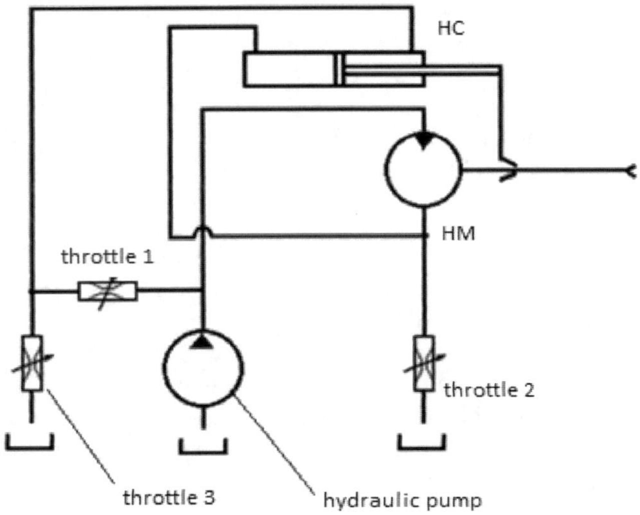

Figure 3. Hydro scheme adaptive drill machine "UBG-1A»

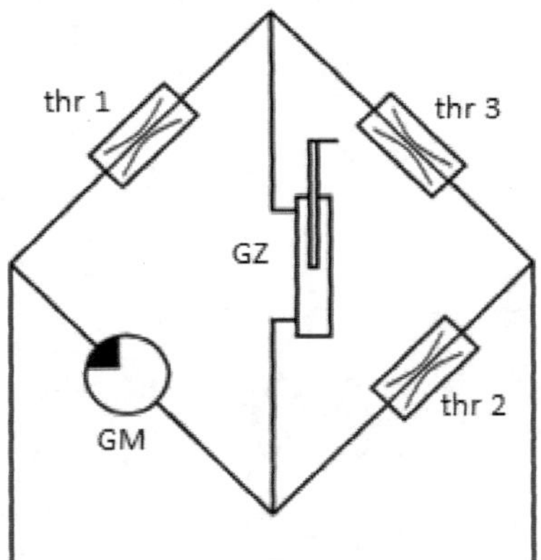

Figure 4. Structural scheme of adaptive hydraulic drill machine "UBG-1A»

As can be seen from the comparison of the above schemes, the drive of the adaptive drill, machine and drilling machine can be represented by the bridge inclusion of a hydraulic motor or hydraulic cylinder in the diagonal of the bridge.

Obviously, depending on the ratio of the shoulders of the bridge, the direction of flow in the diagonal can be direct or reverse. The principle of operation of the adaptive drilling machine "UBG-1A" is given in the article [10].

During the research was used as a standard drill machine rotary type "UBG-01-000" Lyudinovo aggregate plant and the adaptive drill machine "UBG-1A", the drive which was re-made from gidrobiontov model of the drilling machine. The drive scheme of a typical drilling machine is shown in figure 5.

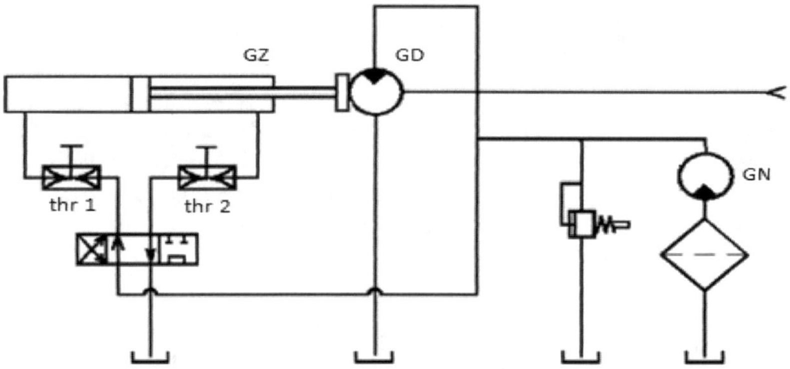

Figure 5. Hydraulic scheme of drilling machine " UBG.01-000» where HP- hydraulic pump; YM - hydraulic motor; throttle1, throttle2 - throttles settings; HC-hydraulic cylinder

As can be seen in the above diagram, the speed of the drill rod of a typical drill machine is not regulated, and the feed control is performed by the reverse spool and the throttles throttle1, throttle2. The experiments used the real characteristics of the hydraulic cylinder, hydraulic motor and switching equipment of a typical machine "UBG.01-000".

The feature of drill machines that have adaptive drive, is their ability to automatically maintain a rational ratio of the speeds of these movements without additional automatio.

To solve the problems of establishing the degree of adaptation to changing drilling conditions and determining the rational conditions of regulation, a research stand was developed, the scheme of which is shown in figure 6. The General view of the stand is shown in Fig. 7.

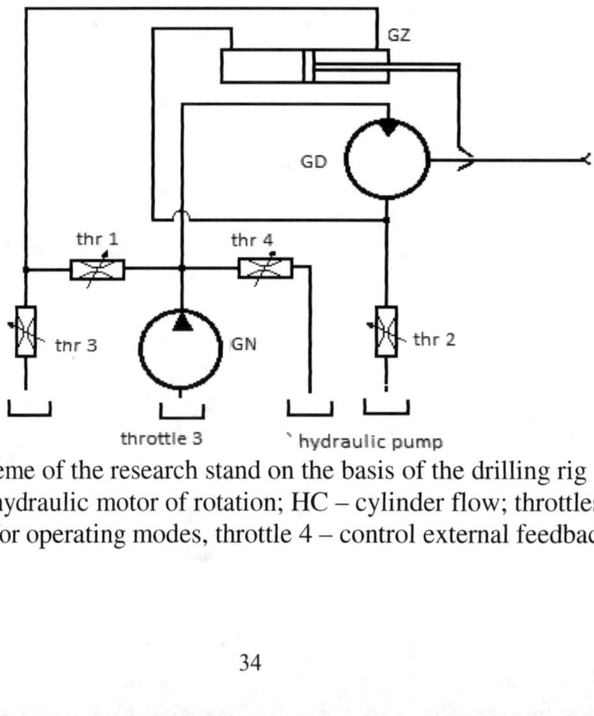

Figure 6. Hydraulic scheme of the research stand on the basis of the drilling rig "UBG.01": GP – hydraulic pump; HM – hydraulic motor of rotation; HC – cylinder flow; throttles1,2,3 – throttles settings for operating modes, throttle 4 – control external feedback

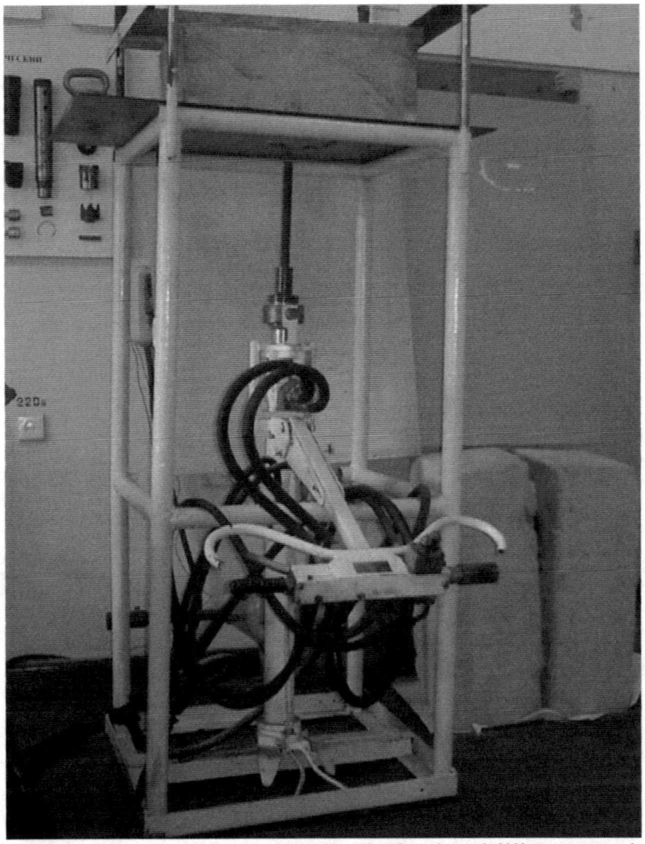

Figure 7. General view of the stand for the study of adaptive drilling control systems

The developed stand made it possible to Supplement and clarify the known research in the field of adaptive rotary drilling machines, to outline ways of their improvement. Table 1 provides information on the research drilling rig used. The table shows the nominal operating parameters and modes. The stand is a control system for the hydraulic cylinder feed and hydraulic motor rotation of the drill rod at different throttle settings, which allows you to adjust the ratio of flow between the rotation and feed drives at different settings, which corresponds to different drilling modes. By adjusting the throttles, it is possible to achieve redistribution of flows in hydraulic circuits, which will bring the measurement conditions closer to the modeling conditions of this system [11].

Table 1. Technical characteristics of the stand «UBG-1».

Installation parameters	Parameter value
Torque on spindle, NM at a differential oil pressure, MPA	154 / 14-20
Spindle speed, vol./min / at oil consumption, l / min	345-805 / 30-70
Feed stroke, mm	1360
Maximum feed force, N	5000
Dimensions (H*H*L), mm	1280*700*890
Weight, kg	80
Working fluid	min. oil viscosity 200-330 SST
Bore diameter, mm	28-45

The studies obtained data to determine the adequacy of the previously developed model of adaptive drilling machines, studied the factors affecting the accuracy of adaptation and the degree of approximation of the automatic control process to the rational conditions of drilling. The following studies were carried out:

1. The influence of the fortress is drilling rocks on the testimony of the drilling process
2. The influence of the settings of the chokes on the performance of the actuator and the indicators of the drilling process.

The influence of the degree of contact strength on the performance and performance of the drilling process on the specific feed - S , torque - M_{rot} shown in figure 8.

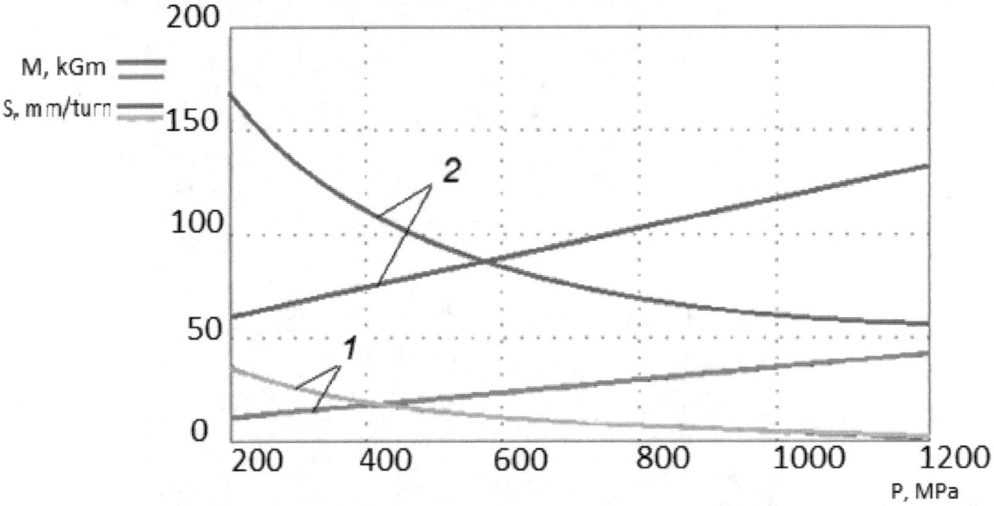

Figure 8. Dependence of torque (M) and specific feed (S) on rock strength (Rk), at feed force: 400H (1) and 800H(2)

The influence of the degree of wear of the cutting edge on the performance and performance of the drilling process at different rock strength are shown in figure 9.

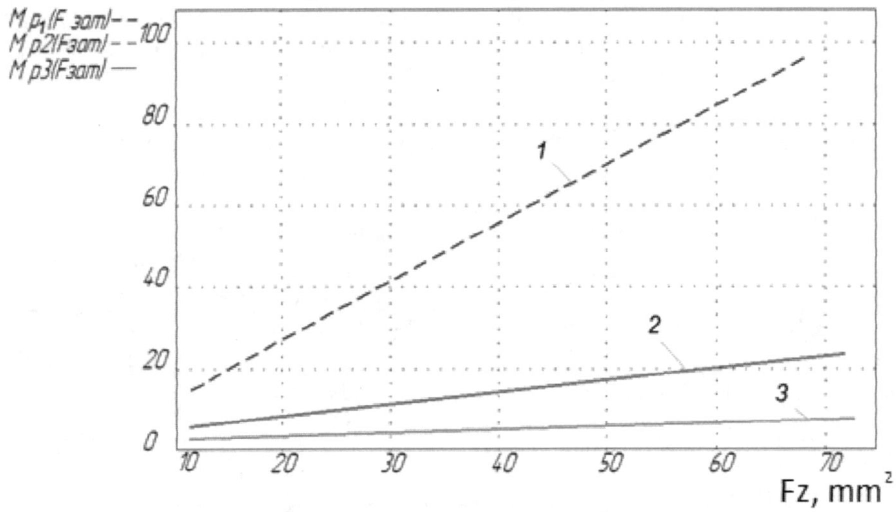

Figure 9. The Dependence of torque from the area of bluntness of the cutter at the castle rocks (1) Rk_1 =1188MPa, (2) Rk_2 =250 MPa, (3)Pk_3 = 44MPa

The possibility of setting the throttles to regulate the threshold value of the feed force and automatic transition to the reverse movement of the feed cylinder rod in the case of jamming of the drill rod was experimentally investigated. The influence on the control characteristics of one of the most important in the scheme of the throttle "DR1" – which is the "shoulder" of the first hydraulic differential. The value of the flow section of the throttle determines the degree of sensitivity of the adaptation scheme.

For Figure 10, 11 examples of dependence of flows on the moment of resistance to rotation at different values of resistance of the throttle "DR1"(degree of opening) are given.

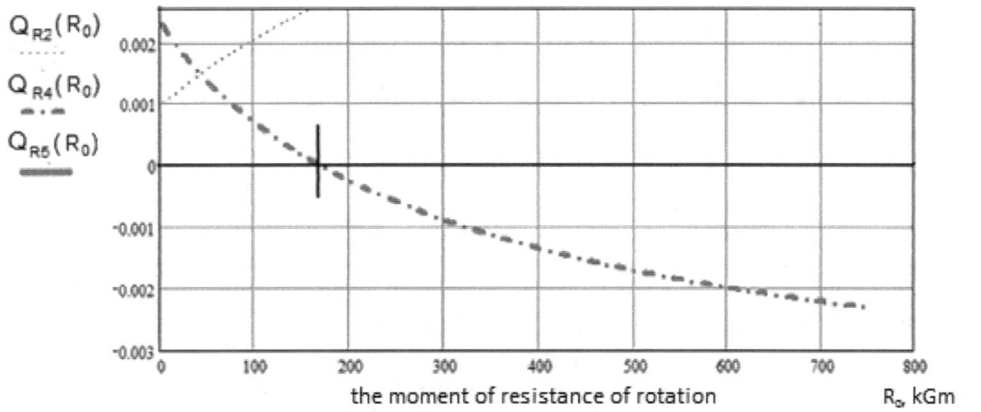

Figure 10. Dependence of flows in the hydraulic lines from the moment of resistance to rotation, when the degree of closing throttle Dr1: 4%

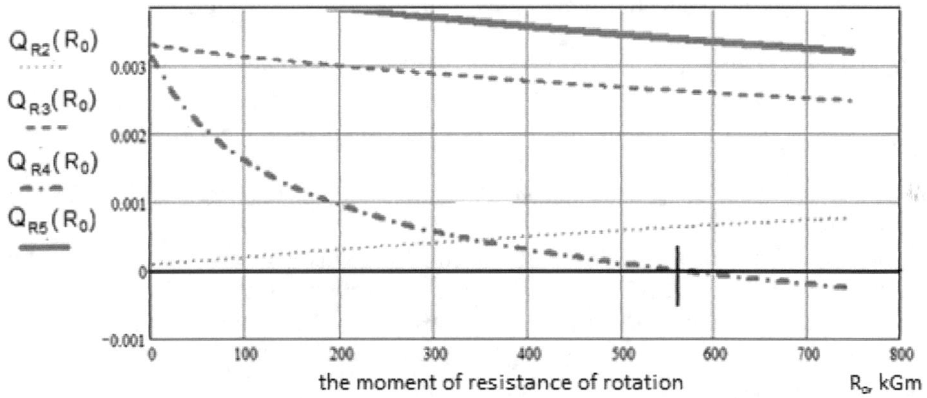

Figure 11. Dependence of flows in the hydraulic lines from the moment of resistance to rotation, when the degree of closing throttle Dr1: 80%

As can be seen in the graphs at a relatively low hydraulic resistance of the throttle DR1 (figure 10), the reverse automatically occurs when the torque resistance to rotation increases at about 180 Nm. And with a high degree of throttle closure (figure 11) the reverse is not performed up to 570 Nm. This mode for the operating range is performed without reverse. Adjustment of the throttle DR1 provides adjustment of the degree of adaptation.

4. Conclusions

1. Feature of the above inclusion of elements of the hydraulic circuit is the solution of automatic control circuit. By increasing the torque on the hydraulic motor rotation, the feed force will automatically decrease, up to the reverse movement. What is necessary, for example, when jamming the rod during drilling.
2. Increasing the strength of the drill rock, would reduce the specific feed and reduce the torque, but in this control scheme, reducing the torque will increase the feed force and stabilize the drilling speed.

IOP Conf. Series: Materials Science and Engineering **680** (2019) 012006 doi:10.1088/1757-899X/680/1/012006

References

[1] Paul. A. Potter (SUBC Engineering Ltd.) Advent of Innovative Adaptive Drilling Methods DOI https://doi.org/10.4043/26379-MS. Document ID OTC-26379-MS. Publisher Offshore Technology Conference Source and electromechanics. Novocherkassk, 2007, no.10, pp. 54-57.

[2] Whitepaper - The adaptive machine | B&R Industrial Automation. URL : https://www.br-automation.com/ru/technologies/.../whitepaper-the-adaptive-machine/ New machine concepts are needed in order to be able to economically produce batches of one. This whitepaper shows the benefits provided by an adaptive

[3] Drovnikov, A.N. Theory and practice of adaptive mechanisms, Part 1, part 2. / News. polytech. in-t - Novocherkassk., Dep. in the VINITI 5.12.83., 3 6570 -83D and 6571 -83D, 445 p. (1983)

[4] Vodyanik G.M., Drovnikov A. N., Streltsov I. Adaptive drive of mining machines on the basis of the Dynamo, hydromachine and mechanical differentials //Mining machinery

[5] Mikhail Lemeshko, Mikhail Molev, and Igor Golovin Hydraulic technological machines with adaptive drive structure/ (ICMTMTE 2018) Sevastopol, Russia, September 10-14, 2018 MATEC Web of Conferences, Volume 224, 02087 (2018) DOI: https://doi.org/10.1051/matecconf/2018224020875

[6] Drovnikov A.N., Lemeshko M.A. Structural diagram of the dynamic model "adaptive drilling rig - slaughter" / Mining information analytical bulletin (scientific and technical journal).. № 8. pp. 147-149 (2003) Offshore Technology Conference Asia, 22-25 March, Kuala Lumpur, Malaysia, 2016

[7] Kuznetsov A. S. Investigation of the two-differential drive of the drilling machine with automatic feed BMW -1 in production conditions. Cand. Diss. Novocherkassk Polytechnic Institute, 1976. 169 p.

[8] A.c. SU № 483224 IPC B23Q5 / 06, B24b5 / 16. Published: 09/05/1975. A device for stabilizing the thickness of the layer to be removed during machining of curvilinear surfaces / Shakhtinsky Technological Institute of Consumer Services // A.N.Drovnikov, G.M. Vodyanik, V.A. Pershin (1975)

[9] Patent RUS No. 2473767. Published on 07/29/2011. Adaptive rotary drilling machine / M.A. Lemeshko, A.V. Trifonov (2011).

[10] Lemeshko M.A.,Trifonov A. V. Research drive adaptive machine "UBG-1A"/ Mining information and analytical Bulletin (scientific and technical journal). 2013. No. 6. P. 202-207.

[11] Lemeshko. M. A, Trifonov A. V. Mathematical Model of Limitations of Adaptive Control of the Rotary Drilling Machines./ Mining Information and Analytical Bulletin (Scientific and Technical Magazine). 2012. No. 2. P. 207–210.

IOP Conf. Series: Materials Science and Engineering **680** (2019) 012007 doi:10.1088/1757-899X/680/1/012007

Research and algorithmization of the process for completing clothes by materials with electric propertie

I Cherunova[1,2], G Kutilina[1], E Milyutin[1]

[1] Department of Design and technology Don State Technical University, Shevchenko street 147, Shakhty 346500, Russia Federation
[2] University of Rostock, Albert-Einstein-Straße, 2, Rostock, Germany

Abstract. In the article research results are presented, which aim to provide of automation algorithm of the process for completing clothes by materials with electric properties. The problem of electrostatic safety is included into the field of particular relevance. Analysis of the list of quality indexes of the special protective clothing allowed to form for automation of design works a group of protective properties of clothing to be used for positioning of the rational suit designated for the employees in the northern regions. Developed Structure of protective properties of clothing for automation of design works; Required properties of special anti-electrostatic clothing for the protection from reduced temperatures; additional quality parameters of materials and physical and mechanical properties. Designed and applied a support algorithm of the process for completing clothing by materials with electric properties. Installed normalizing parameters of materials for their automated forming into packages of details for protective clothing with up-to-date electric properties. The research was made in Don State Technical University within the framework of State Assignment of the Ministry of education and science of Russia under the project 11.9194.2017/BCh.

1. Introduction

Static electricity acquires high pace of spreading both in domestic field and in production. This is related to expanding application of electronic equipment, synthetic materials and substitution of natural materials by synthetic ones due to their inexpensiveness. This in combination with low ambient temperatures, low humidity and particularly when explosive gas mixtures are in the air can lead to occurrence of emergency situation [1].

At the present moment the main methods of protection from static electricity are the following [2]:

1) application of collective protection equipment, including earthing devices, neutralizers, damping devices, anti-electrostatic substances, shielding devices;

2) application of personal protective equipment, such as special anti-electrostatic clothing, special anti-electrostatic shoes, safety anti-electrostatic devices (rings and bracelets), anti-electrostatic equipment for the protection of hands;

3) ionization of premises.

The problem of electrostatic safety is included into the field of particular relevance. Therefore, it is important not only to select the materials correctly, but also to suggest technology for improving anti-electrostatic properties of materials for clothing that will enhance level of human safety at production.

2. Another Topicality, scientific relevancy

Analysis of the list of quality indexes of the special protective clothing allowed to form for automation of design works a group of protective properties of clothing to be used for positioning of the rational suit designated for the employees in the northern regions [3] (Figure 1).

Special anti-electrostatic clothing for protection against reduced temperatures must assure protection from electrostatic charges and fields, low temperatures as well as from general and specific production pollutions attritions [4,5,6].

Analysis of regulatory-technical and methodical literature showed that in the established system for providing human with personal protective equipment considering the rules of their mandatory certification there is no single regulatory document for the fabrication of anti-electrostatic suit for the protection from reduced temperatures [7,8,9]. In this case only averaged recommendations to the materials of similar suits that require detalization and additional research of the issues related to optimizing packages of materials and design parameters of similar protective clothing [10,11].

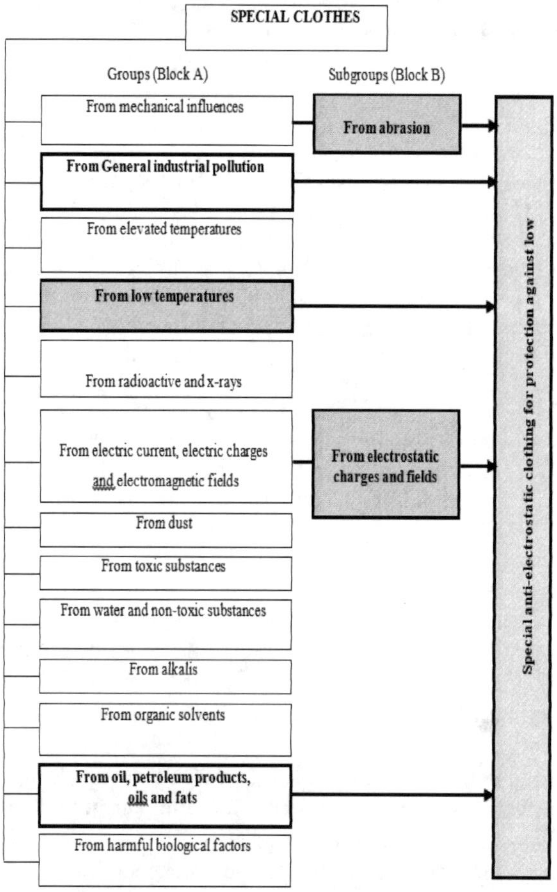

Figure 1. Structure of protective properties of clothing for automation of design works.

Special clothing must comply with complicated set of requirements of protective, hygienic, operational and aesthetic nature. Technical requirements to articles shape the main values of the article properties, and then requirements to the materials for this article. The most rigid requirements are imposed to special protective clothing designated for usage in unfavorable conditions. Clothing must protect the human from hazardous factors which have continuous impact in the course of production or which may arise in case of contingency. Clothing must not hinder movements and must allow to perform certain functional operations [12].

IOP Conf. Series: Materials Science and Engineering 680 (2019) 012007 doi:10.1088/1757-899X/680/1/012007

Design options for sets of materials for warm clothing with electrical properties is represented in Figure 2.

Requirements to special clothing for the protection from reduced temperatures are regulated by BS EN 342:2004 [5].

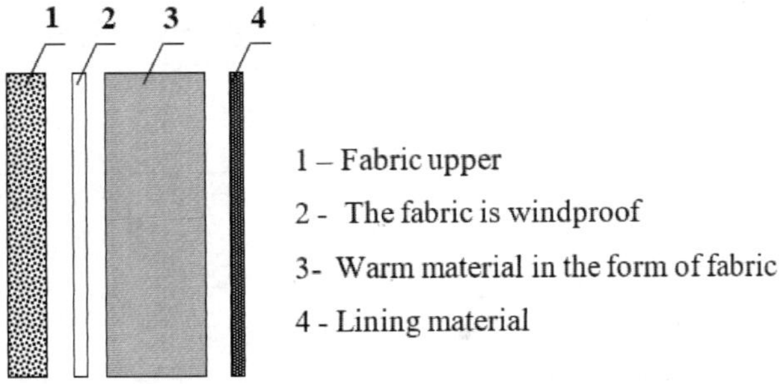

a) Warm material in the form of fabric

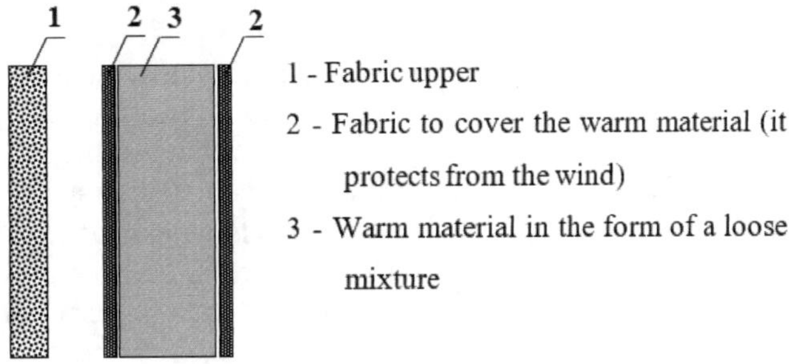

б) Warm material (bulk fiber blend)

Figure 2. Construction of a set of materials for warm clothes with electrical properties

The overall thermal resistance of the set of heat-protective clothing must comply with the level of energy consumption, meteorological conditions (temperature, air moisture, wind speed) and the time of uninterrupted staying in the cold [13-15].

The range of thermal resistance of clothing as a whole which is used under conditions of the Far North must be within the limits 0,73 – 1,02 m2·K/W [5,6]. GOST 12.4.124-83 [16] stipulates general technical requirements to the protective equipment of workers against hazardous and harmful impact of static electricity. GOST 12.1.002-84 [17] introduces limitations and the extreme value of electric field intensity in the vicinity of operational area.

The processes of automated search and making of decisions represent important blocks of works [18]. The quality of received decisions depends on the completeness of the defined information and regulatory base as well as on algorithms used during analysis of data and decision-making. Therefore, a certain interrelation appears which becomes the ground for statement of task about the necessity to develop computer methods for reveal priority options in the up-to-date range textile materials of electrostatic protection.

3. Theoretical part

In order to automate the process of defining materials for the set of clothing with special anti-electrostatic properties the primary question is to consider complex protection of human by special anti-electrostatic heat-protective clothing . A group of the main properties of materials represented in Table 1. that form a package can be defined for the forecast conditions of operation.

Table 1. Required properties of special anti-electrostatic clothing for the protection from reduced temperatures.

Required properties of special anti-electrostatic clothing for the protection from reduced temperatures	
Protective properties	Characteristics of the suit (Block C)
From attritions	1.Materials with increased density 2.Reinforcing pads in the points of increased deterioration
From general production pollutions	3.Design of article with maximum coverage of the surface of human body by a layer of clothing 4.Properties of materials with water-repelling effect
From reduced temperatures	5.The package of suit materials must assure thermal resistance in accordance with the requirements of GOST [6] as well as considering actual conditions of operation 6.Materials used in the clothing package must assure protection against wind load to preserve its stable thermal resistance
From electrostatic charges and fields	7. Materials with surface specific resistance of not more than 107 Ohm must be applied for the fabrication of anti-electrostatic special clothing 8. Electric resistance between current-conducting element of anti-electrostatic special clothing and earth must be from 106 to 108 Ohm 9. Whereas the time for charge drainage from items of anti-electrostatic clothing that reduces voltage 10 times must not exceed 2 s
From oil, oil products and fats	10. Grease-, oil-repelling treatments for the materials of the top 11. Oil-resistant pads for the areas of direct clothing contact with oily fractions

Apart from properties listed in Table 1, special anti-electrostatic clothing for the protection from reduced temperatures must be comfortable for a person to perform production activity, i.e. to have minimal weight, not to hinder movements, to have reasonable cost and to possess good operation properties within the entire service life. Therefore to create comfortable conditions of work and to enhance human working efficiency technical requirements are introduced related to values of hygienic, operational and technological properties as well as economic values of the finished article. Physical and mechanical properties that determine the above-mentioned additional requirements are represented in Table 2. They define the next stage of arranging algorithm [19] for defining a set of materials for clothing with the

4. Practical relevance

During selection of textile materials for working clothing considering their diversity one should be guide by comprehensive assessment of all properties and their designation.

A support algorithm of the process for completing clothing by materials with electric properties is represented in Figure 3.

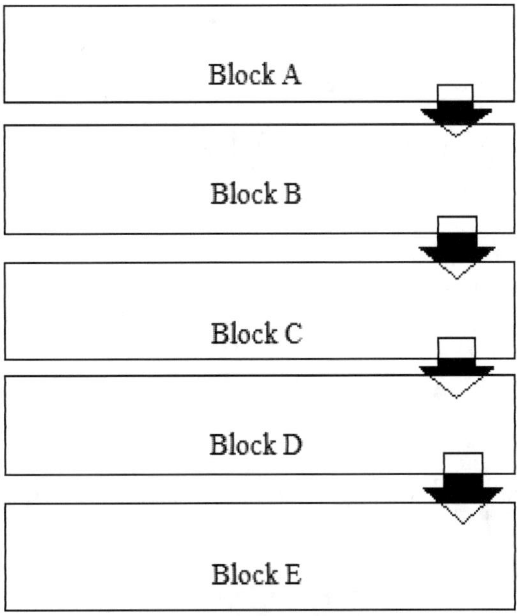

Figure 3. Algorithm of the process for completing clothing by materials with electric properties.

Considering the issue of increasing protection degree for special cold-protective clothing of workers at oil and gas industry a complex analysis was done of regulatory documents introducing limitations for the main criteria of assessing quality and safety of the special protective clothing. The results of the proposed algorithm for CAD are presented in Table 3.

Table 2. Normalizing parameters of materials for their automated forming into packages of details for protective clothing with up-to-date electric properties

Layer of clothing package	Parameters of materials		
	Characteristics	Value	Value of index
Lower (natural fibers)	Weight	Density	200 – 400 g/m3
	Moisture exchange	Hygroscopicity	Not less than 6 %
	Antistaticity	Specific surface electric resistance	not exceeding 1014 Ohm
	Anti-toxicity	Content of free formaldehyde	not exceeding 75 μ/g
	Heat exchange	Thermal conductivity	0,038 – 0,049 W/(m·0C)
Average (thermal insulating)	Weight	Density	300 – 800 g/m3
	Thermal insulation	Thermal resistance	0,50 - 1,02 m2K/W
	Heat exchange	Thermal Conductivity	0,04 – 0,06 W/(m·0C)
	Moisture exchange	Hygroscopicity	Not less than 5 %
	Antistaticity	Specific surface electric resistance	not exceeding 109 - 1012 Ohm
	Air exchange	Air permeability	not less than
Upper (protective)	Weight	Density	145 – 460 g/m3

IOP Conf. Series: Materials Science and Engineering **680** (2019) 012007 doi:10.1088/1757-899X/680/1/012007

Layer of clothing package	Parameters of materials		
	Characteristics	Value	Value of index
	Heat exchange	Heat conductivity	0,04 – 0,06 W/(m·0C)
	Moisture exchange	Hygroscopicity	Not less than 5 %
	Antistaticity	Specific surface electric resistance	not exceeding 107 Ohm
	Resistance to oil impact	Is not soaked during	3 - 6, h
	Fire resistance, no residual smouldering	Time of staying in the flame	Not less than 15 s, not less than 20 - 30 s
	Antitoxicity	Content of formaldehyde	not exceeding

Based on the data of table 3, a suit with thermal and electrostatic protection was developed (Figure 4).

Figure 4. Suit with thermal and electrostatic protection.

The assembly of data received and the developed algorithm of the process for completing clothing by materials with electric properties is an important component of system base to solve the tasks of designing special anti-electrostatic clothing for the protection from reduced temperatures.

Application of such algorithm in general CAD-system at an enterprise of consumer industry will allow to significantly reduce the influence of personal variations of experience and time expenditure for search-analytical and experts procedures by design engineers of up-to-date special clothing.

References

[1] I.Cherunova, S.Tashpulatov, and A. Merkulova, "Development of automation algorithm for step of designing technology of static electricity protection clothing,"International Russian Automation Conference. «RusAutoCon-2018». 2018, DOI: 10.1109/RUSAUTOCON.2018.85018218501821.

[2] A.Merkulova, I.Cherunova, and I.Danilenko, "Investigation of electrostatic fields by modeling methods, "MATEC Web of Conferences, 2017, Vol.132, Article number 04022, DOI: 10.1051/matecconf/201713204022.

[3] I.V.Cherunova, A.V.Merkulova, and Shenikova E.A., "Electrostatic safety. Ways to improve protective clothing. Monograph, "Publishing, "LAP Lambert Academic Publishing, 2011, 128p. ISBN 978-3-8454-1174-3.

[4] G.N. Mathur, H.Raj, N.Kasturiya, "Protective clothing for extreme cold region, " Indian J. Fibre Textile. Res. 1997, Vol. 88, pp.292-96.

[5] BS EN 342:2004 "Protective clothing - Ensembles and garments for protection against cold, "London: BSI Group Headquarters-2008.

[6] GOST R 12.4.236-2011 "System of occupational safety standards. Special clothing for protection against low temperatures. Technical requirements, "Moscow: Publishing STANDARTINFORM, 2011.

[7] M.Jachowicz, Electrostatic Properties of Selected Personal Protective Equipment Regarding Explosion Hazard / Journal of Sustainable Mining, 2013, Vol.12 (1), pp. 27-33.

[8] K.A.Shoush, M.K.Mohamed, Zaini, H., and Ali, W.Y., "Measurement of static electricity generated from contact and separation of clothes and car seat covers, "Int. Journal of Scientific & Engineering Research, 2013, Vol. (10), pp. 33-38.

[9] E.R.Post, and K.Waal, "Electrostatic Power Harvesting in Textiles, "Proc. Electrostatic Power Harvesting in Textiles, Annual Meeting on Electrostatics, ESA, 2010, URL:http://web.media.mit.edu/~rehmi/pdf/ElectrostaticPowerHarvesting.pdf

[10] I.Cherunova , S.Samarbakhsh, and N.Kornev, "CFD Simulation of thermo-aerodynamic interaction in a system human-cloth-environment under very low temperature and wind conditions , "ECCOMAS Congress 2016 - Proceedings of the 7th European Congress on Computational Methods in Applied Sciences and Engineering, 2016, 4, pp. 7703-7710. doi: 10.7712/100016.2366.10854.

[11] L.Zhang, H.Zhang, and Y.Li, "The New Development of CAD in Clothing, "Computer and Information Science, 2008, Vol.1, No.8, pp.104-106.

[12] D.Gupta, "Design and engineering of functional clothing, "Indian Journal of Fibre and Textile Research, 2011. Vol.36. pp.327-335.

[13] K.C.Parsons, G.Havenith, I.Holmér, H.Nilsson and Malchaire J., "The effect of wind and human movement on the heat and vapour transfer properties of clothing, "Annals of Occupational Hygiene, 1999, 43(5). pp.347-352.

[14] B.W.Mangum, and J.E. Hill Thermal, "Analysis--human Comfort-indoor Environments, "Proceedings of a Symposium Held at the National Bureau of Standards, Gaithersburg, Maryland, 1977, Vol.13. 191 p.

[15] I.Cherunova, N.Kornev, and I.Brink, "Mathematical model of the ice protection of a human body at high temperatures of surrounding medium, "Forschung im Ingenieurwesen, Springer, 2012, Vol.76, Issue 3-4. pp 97-103.

[16] GOST 12.4.124-83, "Occupational safety standards System. Means of protection against static electricity. General technical requirements, "M.: Publishing house of standards, 2003.

[17] GOST 12.1.002-84, "System safety standards. Electric fields of industrial frequency. Permissible levels of field strength and requirements for control at workplaces, "Moscow: Publishing STANDARTINFORM, 2009.

[18] I.Dabolina, and A.Vilumsone, "The Role of the Latest Clothing CAD/CAM System Applications in the vol educational Process, "Material Science. Textile and Clothing Technology, 2012,

IOP Conf. Series: Materials Science and Engineering **680** (2019) 012007 doi:10.1088/1757-899X/680/1/012007

№7, pp.63-67.
[19] M. Aihua, L.Yi, L.Xiaonanb, W.Ruomei, and W.Shuxiaoa, "A CAD system for multi-style thermal functional design of clothing / Computer-Aided Design, 2008. No. 40. pp.916–930.
[20] K.Parsons, "Human thermal environments: The effects of hot, moderate and cold environments on human health, comfort and performance, , "2nd ed. London : Taylor & Francis, 2003.

IOP Conf. Series: Materials Science and Engineering **680** (2019) 012008 doi:10.1088/1757-899X/680/1/012008

Information-measuring system to improve the quality of motion control robot manipulators

K I Goryanina[1], O I Katin[1], M A Vernezi[1], D Yu Donskoy[1], L A Tregubenko[1] and N X Chiem[2]

[1] Laboratory of local automation and embedded systems Don State Technical University, Gagarin Square 1, Rostov-on-Don 344000, Russia Federation
[2] Le Quy Don Technical University, Hanoi, Vietnam

E-mail: gorianina.k@yandex.ru

Abstract. The use of robotic systems consisting of mechanical manipulators and control systems is now widely used not only in industrial production, but also in medicine, sorting and even in everyday life and services. In recent years, the market of service robotics has been actively developing. Successful implementation of automation systems depends on solving complex scientific and technical problems primarily in the following areas: machine vision, sensor networks, navigation systems. Of particular importance is the task of increasing the accuracy of manipulation. The article proposes a variant of the formation of information-measuring system of the robot manipulator based on MEMS sensors. And also recommendations on the choice of sensors of the sensor system to control the manipulation system of the robot are given.

1. Introduction

Currently, industrial manipulation robots (MR) are used in various fields of industry to perform a wide range of technological tasks.

They are a complex electromechanical object, which has a number of features. First, MR distinguished by a complex kinematic structure that contains many independent or interconnected links. Secondly, the change in the position of the latter in space affects the physical forces acting on the manipulator. Thirdly, there is a need for synchronous control of a large number of engines.

The interaction of modern robots with the real environment, in addition to hard-coded actions, should be built with the elements of adaptation, i.e. taking into account information about the state and position of working bodies and technological equipment, the state of the environment and objects in the working area. This information, the volume and nature of which are determined by the purpose of the robot, the features of the environment of its functioning and the process, provides information and measurement system, structurally included in the control device of the robot.

It is the organs of "senses" of the robot and largely determines its functionality, operational efficiency and reliability, the complexity of the tasks, as well as the safety of the staff robot, Which has the ability to "feel", it is easier to perform complex actions, to adapt when performing a wide range of tasks. This increases the degree of versatility of the robot, which ultimately, despite the increase in the intrinsic value of the robot, leads to a decrease in the cost of production and maintenance during operation.

IOP Conf. Series: Materials Science and Engineering **680** (2019) 012008 doi:10.1088/1757-899X/680/1/012008

The main requirement for industrial manipulators is the accuracy of positioning and moving along a given trajectory. In this regard, the design of the robot control system should take into account the factors that affect the quality of motion: 1) features of the kinematic scheme, 2) changes in dynamic characteristics.

Thus, the task of improving positioning accuracy is associated with the need for accurate and/or mutually synchronized manipulations by several manipulators. Potentially, this problem can be solved by increasing the sensor equipment of the manipulators and on this basis – the optimization of controlled movements of the robot manipulators.

One of the fundamental problems, the solution of which depends on the success in the creation of adaptive robots, is the use of modern types of sensors, which allow to obtain a large amount of local information about the problem environment in a short time, and the integration of this information into the control system of the robot manipulator.

Thus, the global goal of the work is to improve the quality of motion control of robot manipulators by: development of MIMO model, automatic control system and placement on the manipulators of the array of sensors of the local sensor system.

In the article for a better understanding of the movement of the manipulator will be considered direct and inverse problems of kinematics, as well as the developed system array of sensors of local information.

2. Mathematical description of the kinematic scheme

The characteristics of control systems for complex mechatronic objects, such as manipulation robots, are significantly influenced by their kinematic and dynamic parameters. As a result, it is necessary to give their detailed mathematical description, convenient for use in the implementation of control system.

In the solution of the specified task it is possible to allocate two interconnected directions. The first is to create an accurate kinematic model of the manipulator, allowing to uniquely determine its spatial configuration, which in turn will make it possible to describe the laws of movement of the working body. The second direction is to describe the dynamic characteristics and relationships that exist in the manipulator, which will describe its behavior when moving along a given path.

To plan the trajectory of the manipulator and determine its position in space, it is necessary to solve two main classes of problems in the form of direct and inverse kinematics problems [1].

The solution of the direct problem is used to convert information about the position of the manipulator from its own coordinate system to the working (absolute), which is required to determine the coordinates of the working body of the manipulator.

The solution of the inverse problem is designed to calculate the required spatial configuration of the manipulator by the position of the working body, and is the main problem in planning the trajectory of its movement.

The solution of these problems requires a description of the overall characteristics of the manipulator in a form convenient for their analysis and writing coordinate transformation equations. Of the existing approaches, the main ones are their expression in the form of a system of linear or matrix equations [2, 3].

The method of homogeneous transformations is the most effective for performing numerical transformations. As applied to the description of manipulation robots, the Denavit-Hartenberg parameters (DH-parameters) has become widespread. It allows you to record the kinematics of the manipulator by a set of matrices of spatial transformations of dimension 4x4, expressing the relative position of the CS of individual units. In this case, the kinematic scheme of the manipulator is described by an equation of the form:

$$T = \prod_{i=1}^{n} {}^{i-1}A_i \qquad (1)$$

where:

T - the matrix position of the working body,

A_i - I-junction transformation matrices,

n - number of joints in operation.

2.1 Equation of the direct problem of kinematics

The solution of the direct kinematics problem requires the transformation of the set of coordinates describing the position of the manipulator in its own coordinate system into the coordinates of the working coordinate system.

$$(q_1, q_2, ..., q_n) \to (x_1, x_2, ..., x_n),$$

where:

$q_1, q_2, ..., q_n$ - coordinates in their own coordinate system (attached);

$x_1, x_2, ..., x_n$ - coordinates in absolute coordinate system (absolute).

When using a six-axis manipulator operating in a rotational coordinate system, and applying the previously selected working coordinate system, using the expression (1), we obtain:

$$T = {}^0A_1{}^1A_2{}^2A_3{}^3A_4{}^4A_5{}^5A_6{}^6 = \begin{bmatrix} n & s & a & p \\ 0 & 0 & 0 & 1 \end{bmatrix}, \quad (2)$$

where:

n, s, a - vectors 3x1 orientation of the working body;

p - vector 3x1 position of the working body in the base coordinate system.

The individual transformation matrices will take the form:

$$ {}^0A_1 = \begin{bmatrix} \cos\theta_1 & 0 & -\sin\theta_1 & 0 \\ \sin\theta_1 & 0 & \cos\theta_1 & 0 \\ 0 & -1 & 0 & 0 \\ 0 & 0 & 0 & 1 \end{bmatrix}, {}^1A_2 = \begin{bmatrix} \cos\theta_2 & -\sin\theta_2 & 0 & \alpha_2\cos\theta_2 \\ \sin\theta_2 & \cos\theta_2 & 0 & \alpha_2\sin\theta_2 \\ 0 & 0 & 1 & d_2 \\ 0 & 0 & 0 & 1 \end{bmatrix}, $$

$$ {}^2A_3 = \begin{bmatrix} \cos\theta_3 & 0 & \sin\theta_3 & -\alpha_3\cos\theta_3 \\ \sin\theta_3 & 0 & -\cos\theta_3 & \alpha_3\sin\theta_3 \\ 0 & 1 & 0 & 0 \\ 0 & 0 & 0 & 1 \end{bmatrix}, {}^3A_4 = \begin{bmatrix} \cos\theta_4 & 0 & -\sin\theta_4 & 0 \\ \sin\theta_4 & 0 & \cos\theta_4 & 0 \\ 0 & -1 & 0 & d_4 \\ 0 & 0 & 0 & 1 \end{bmatrix}, \quad (3)$$

$$ {}^4A_5 = \begin{bmatrix} \cos\theta_5 & 0 & \sin\theta_5 & 0 \\ \sin\theta_5 & 0 & -\cos\theta_5 & 0 \\ 0 & 1 & 0 & 0 \\ 0 & 0 & 0 & 1 \end{bmatrix}, {}^5A_6 = \begin{bmatrix} \cos\theta_6 & -\sin\theta_6 & 0 & 0 \\ \sin\theta_6 & \cos\theta_6 & 0 & 0 \\ 0 & 0 & 1 & d_6 \\ 0 & 0 & 0 & 1 \end{bmatrix}, $$

where:

$\theta_i, d_i, a_i, \alpha_i$ - DX-parameters (figure 1):

θ_i - the attached angle by which you want to rotate an axis x_{i-1} around an axis z_{i-1} so that it is aligned with the axis x_i;

d_i - the distance between the intersection of the axis z_{i-1} with the axis x_i and the origin of the $(i-1)^{th}$ and coordinate system;

a_i - linear offset (shortest distance between axes z_{i-1} and z_i);

α_i - angular offset – the angle by which you want to rotate an axis z_{i+1} around an axis x_i so that it is aligned with the axis z_i.

The parameter values d_i, a_i, α_i are determined by the orientation of the absolute coordinate system and the selection of the reference point.

The values θ_i in this case correspond to the attached coordinates of the manipulator and describe the position of the manipulator in its own coordinate system.

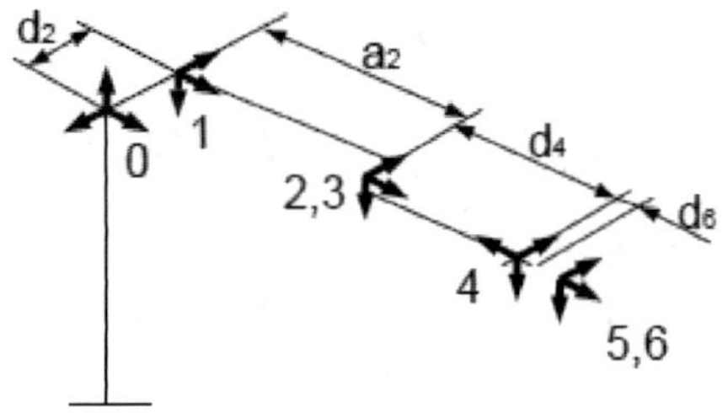

Figure 1. Manipulator DH parameters ratio

2.2 Equations of inverse kinematics problem

In contrast to the direct kinematics problem, the inverse one has much more approaches to its solution, of which the main ones are the methods of inverse transformations and biquaternions [4, 5].

Let's consider the main features of each method.

The inverse transformation method consists in determining the values of the individual attached coordinates based on the existing transformation matrices (3). As a result, ready-made analytical expressions can be obtained.

Solution of kinematics by means of biquaternions is carried out by kinematic configuration using dual quaternion and compute the values of joined coordinates by the unit quaternion transformations. This method makes it possible to describe the kinematic configuration of the manipulator in a more compact form compared to the matrices (3), but requires the implementation of more complex computational algorithms compared to the method of inverse transformations.

Based on the need to implement an algorithm for solving the inverse kinematics problem in the control system of the manipulation robot, which has limited computational capabilities, it is necessary to choose an approach with the possibility of optimizing the calculations.

2.3 Manipulator dynamic structure

When moving the manipulator there are changes in its spatial configuration, which in turn leads to a change in the forces acting on its individual elements. To account for these changes is required to describe the dynamic characteristics of the MR and their relationship with the kinematic structure of the manipulator.

The task of describing the dynamic characteristics of the MR can be divided into the following stages:

1) determination of the moments acting on separate links;

2) description of the set of links between the individual links;

3) presentation of dynamic characteristics in a convenient form for further analysis.

In the process of motion, the forces acting on the manipulation robot create two main types of moments: 1) gravitational moments, 2) centrifugal and Coriolis moments.

Gravitational moments are determined by the influence of gravity on the links of the manipulator. They act independently of the speed and are an expression of the potential energy of the manipulator.

Centrifugal and Coriolis moments occur when the manipulator moves in space. Since there are kinematic connections between individual links, the forces acting on them are determined not only by their own speeds, but also by the speeds of other links.

Generalized moments of dynamic interaction arise as a result of changes in the position of the links in their accelerated motion relative to each other.

The General equation of dynamics of the manipulator [6-8] can be written as:

$$\tau(t) = D(q(t))\ddot{q}(t) + h(q(t), \dot{q}(t)) + c(q(t)), \tag{4}$$

$\tau(t)$ - vector $n \times 1$ of generalized moments in manipulator joints;

$q(t)$ - vector $n \times 1$ of adjoint variables of the manipulator;

$\dot{q}(t)$ - vector $n \times 1$ of generalized velocities;

$\ddot{q}(t)$ - vector $n \times 1$ of generalized accelerations;

D - matrix $n \times n$ of inertia (kinetic energy) of manipulator;

h - vector $n \times 1$ of Coriolis and centrifugal moments;

c - vector $n \times 1$ of gravitational moments;

n - number of joints in MR.

Gravitational moments are determined by the effect of gravity on the individual links of the manipulator. They depend on the mass, position and orientation of the manipulator in space. Each element of the vector of gravitational moments c_i can be expressed as:

$$c_i = \sum_{j=1}^{i} F_{gJ} r_{ij} , \tag{5}$$

where:

F_{gJ} - gravity acting on a link j;

r_{ij} - distance vector between the centers of gravity links i and j.

Thus, to determine the gravitational component of the moment, it is necessary to describe the kinematic structure of the manipulator and determine the positions of the centers of gravity of individual units.

3. Information-measuring system

The manipulator can be considered as an analogue of the human hand. Its design largely determines the capabilities of the robot. These capabilities are much lower than the capabilities of the human hand, which has 27 degrees of mobility or, if you do not take into account the movement of the fingers, 12 degrees of mobility. The number of degrees of mobility of the manipulator is limited and usually does not exceed 7.

In existing manipulators, as a rule, only the sense of orientation (or rather – the relative relative position of the manipulator elements relative to each other) and, sometimes – a sense of force/weight due to the control of the torque forces (usually – the magnitude of the current drives the manipulator).

All this naturally (and significantly) limits the sensitive capabilities of existing robotic manipulators and makes their use in everyday life and special areas limited.

The main function of the information-measuring system are feedback devices designed for active control in the process of the parameters of the state of the robot and process equipment, as well as the parameters of the environment and objects in the working area.

In General feedback devices include systems:

1) control of parameters of the robot state (positions and speeds of movement of working bodies and elements of mechanisms, efforts in elements, emergency blocking, diagnostics and forecasting of a resource of work);

2) perception and analysis of information about the external environment (tactile, visual, location, etc.);

IOP Conf. Series: Materials Science and Engineering **680** (2019) 012008 doi:10.1088/1757-899X/680/1/012008

3) ensuring safety (registration of the spatial position of the robot and its individual parts, the location of personnel and equipment in the work area). Primary information these systems receive from feedback sensors, or sensitive elements, which are the most important components of feedback devices.

Internal state sensors are used to generate signals in the feedback circuits by the position and speed of the manipulator links, by force and torque.

External state sensors are designed to measure parameters in the far and near zones and for tactile measurements. These sensors are divided into contact and non-contact. The contact sensors measure when an object is in contact with a touch, slip, or torsion process. The principle of operation of contactless sensors is based on the determination of changes in the acoustic or electromagnetic fields of interaction with the object.

To ensure the absolute orientation of the manipulator fingers in the gravitational field of the Earth and measure the angle of rotation of the manipulator relative to a fixed coordinate system associated with its base, MEMS accelerometers and gyroscopes are installed on each finger of the manipulator.

Also, to determine the mutual orientation of the fingers of the manipulator, it is advisable to use magnetic sensors consisting of a sensing element and an electromagnet. The use of such sensors is important when visual inspection is difficult or impossible.

To determine the surface temperature, low-inertia film sensors are installed, which are installed on the inner surface of the phalanx of the fingers or palm.

When working with small and fragile objects, the proximity and touch sensor is of great importance. Sensors due to their simplicity can be placed in several places on the inner surface of the hand of the manipulator and used in the capture and fixation of various objects.

Research is conducted on an anthropomorphic manipulator manufactured by Android technology. The module with the manipulator movement system is equipped with a metal anthropomorphic grip with five structural links, a vision system and specialized software. It is designed to practice manipulative actions with various objects. Open architecture allows you to make changes to existing SOFTWARE or create new ones for specific tasks, testing your own developments on a robotic system. The anthropomorphic manipulator module is shown in figure 2.

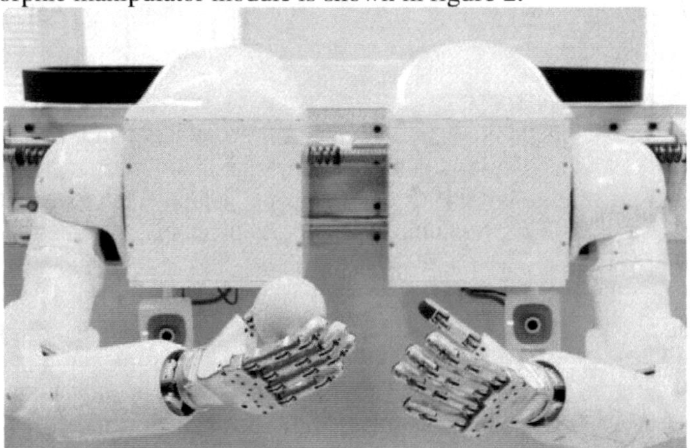

Figure 2. Manipulator system module

Most of the above sensors exist in the form of MEMS-integrated circuits with output to digital bus type I2C and / or SPI, providing easy docking with modern interfaces of computing devices. Certain types of sensors (primarily tactile) will require the development of their own control circuits. The greatest difficulty lies in the processing and coordination of information received from this heterogeneous array of sensors, its analysis, and the development of control actions.

IOP Conf. Series: Materials Science and Engineering **680** (2019) 012008 doi:10.1088/1757-899X/680/1/012008

4. Conclusion

The proposed principles of hardware architecture allow us to develop a control system for manipulative robots that meets the requirements of modern industry. By integrating heterogeneous array of sensors automatic control system of a manipulator must be moved without colliding with the surrounding objects; the retention of objects with a single force; the determination of the temperature of the object clamped in the manipulator or object that it approaches the manipulator; determining the color of the object, which is approaching the manipulator.

The proposed new concept will improve the quality and accuracy of movement of manipulators when performing a wide range of technological tasks.

References

[1] Lebedev P.A. 1967 *Kinematics of spatial mechanisms p 280*

[2] R. S. Hartenberg and J. Denavit, *A kinematic notation for lower pair mechanisms based on matrices* Journal of Applied Mechanics, vol. 77, pp 215–221

[3] D'Souza A., Vijayakumar S., Schaal S. 2001 *Learning inverse kinematics*. Proceedings. 2001 IEEE/RSJ International Conference on / IEEE. Vol. 1. pp 298–303

[4] Tolani D., Goswami A., Badler N. I. 2000 *Real-time inverse kinematics techniques for anthropomorphic limbs*. Graphical models. Vol. 62, no. 5. pp 353–388

[5] I. L. Kantor, A. S. Solodovnikov. 1973 *Hypercomplex numbers*. p 144.

[6] Tertychny-Dauri V. 2012 *Dynamics of robotic systems*. p 128

[7] Avagimov A. G. 1989 *About special terms and conditions single-loop spatial mechanisms with several degrees of freedom* pp. 11-18

[8] Kolovsky M. Z., Sloushch A. V. *Fundamentals of industrial robot dynamics* p 240

IOP Conf. Series: Materials Science and Engineering **680** (2019) 012009 doi:10.1088/1757-899X/680/1/012009

Study of the influence of the basic spectra of discrete light sources on the seeds of greenhouse cultures

D Yu Donskoy[1], M A Vernezi[1], O I Katin[1], K I Goryanina[1], S G Studennikova[1] and M Petkovic[2]

[1] Laboratory of local automation and embedded systems Don State Technical University, Gagarin Square 1, Rostov-on-Don 344000, Russia Federation
[2] University of Kragujevac, Čačak, Serbia

E-mail: dand22@bk.ru

Abstract. This article refers to studies conducted on the seeds of vegetable and green crops with low useful mass, which require light stimulation at all periods of development. Research was conducted using RGBW LEDs. There is a large number of scientific studies proving the effectiveness of the technology of photostimulation of seeds and seedlings of various crops. However, there is no consensus on what the positive effect of the variable light field on a biological object is based. This is due to the complexity of the process of photosynthesis, where the productivity of photosynthesis is considered as an integral response to the influence of external conditions. The study of the influence of various light modes on plants is necessary to search for performance criteria. The found efficiency criteria will provide an opportunity to systematize the many well-known results of the impact of variable light conditions on plants in order to identify plants as a biological object by optical properties, as well as to use to standardize technological lighting in greenhouse production.

1. Introduction

Existing methods of light stimulation of seeds and seedlings are almost the same. Usually, ultraviolet luminaires or "full-spectrum" LED arrays with a predominance of red and blue components of the spectrum are used. This is due to the fact that at the moment there is no experimental base on which it is possible to form an idea of the characterization of the influence of various light spectra on the seeds of vegetable and green cultures.

Existing studies often lead to the conclusion that the entire spectral range affects plants and the use of individual spectra does not provide the desired effect during germination and, conversely, can lead to plant degradation. Thus, the most suitable solution is to model modified sunlight using discrete light sources, but for this it is necessary to prove the correctness of this theory.

To find criteria for the effectiveness of the impact of different light spectra, studies are required with the same climatic indicators.

To carry out these tests, a setup was developed that forms the required ratio of spectral components using discrete light sources, while maintaining constant microclimate conditions for the purity of experiments (figure 1). [1-2]

Figure 1. System for experimental studies on photostimulation of vegetable and green cultures

2. Method of increasing seed germination by photostimulation with discrete light sources

We considered the mathematical solution to this issue in the article "Simulation, identification and dynamic control of the luminaire of the synthesized spectrum" and gave an example of approximating RGBW LEDs to the spectrum of sunlight on the soil surface [3].

Nowadays, more modern LEDs with a wide spectral range are used (sometimes they are called full-spectrum). The use of these LEDs for approximation to modified sunlight allows us to reduce the total square error by 4.41% (figure 8). [4-5].

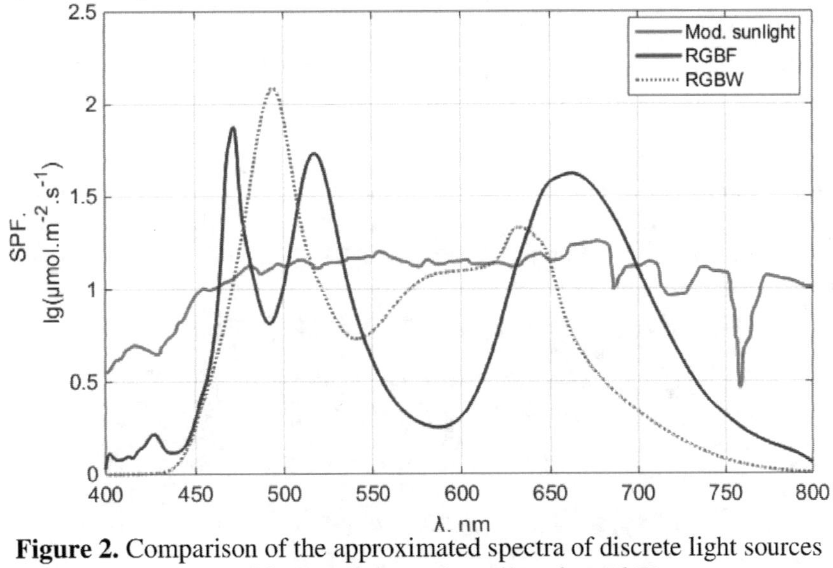

Figure 2. Comparison of the approximated spectra of discrete light sources to modified sunlight on the soil surface [6-7].

We also determined approximation coefficients for modified sunlight at a depth of 3, 6, 9 mm and compiled graphs of the obtained spectra (figure 9a, 9b, 9c, respectively).

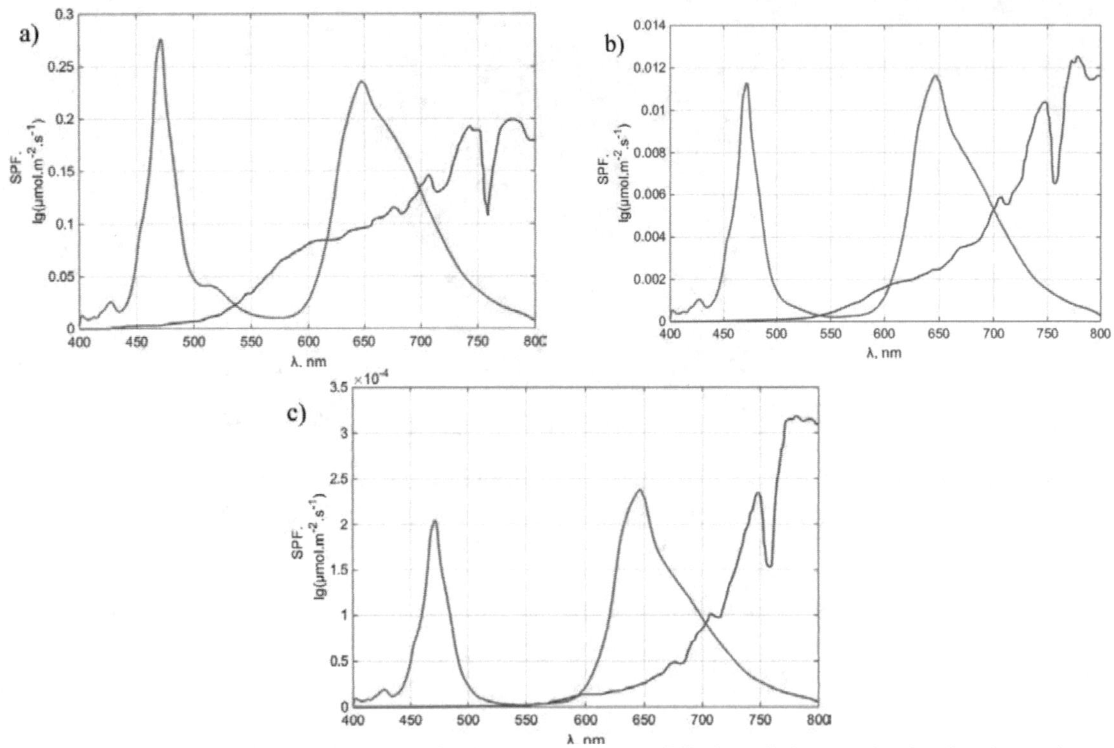

Figure 3. Graphs of approximation of RGBF LEDs to modified sunlight at a depth of a) 3mm, b) 6mm, c) 9mm.

3. The study of RGBW spectra on radish seeds

Experimental studies were carried out at a constant temperature of 27 ° C, substrate humidity of 100%, and light intensity of 1500 lux. The concentration of CO2 throughout the test period was in the range of 600-1100 ppm, there was also no direct air flow to the seeds, which prevented weathering. In the role of the substrate was foam rubber ST 18/20 (10 mm) (figure 2).

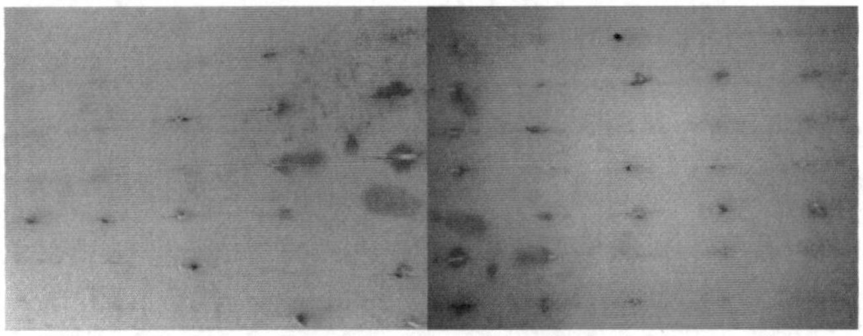

Figure 4. ST 18/20 (10 mm) with seeds during testing.

For the high speed of the first experiments, the object of the study was radish seeds, because the average time of their development to the seedling stage is about 3-4 days. All seeds from one batch. The spectral characteristics of discrete light sources can be seen in Figure 3.

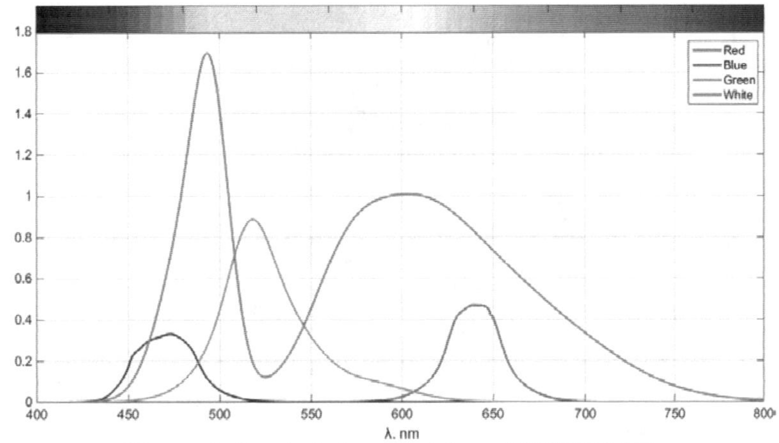

Figure 5. Spectral characteristics of RGBW LEDs.

When studying the effects of the blue spectrum of LED lighting, we obtained the following results. The average root length is 7.41 mm, the total sprout length is on average 14.8 mm, the average maximum diameter is 1.3 mm, and the color is rgb (166,162.63). The germination rate is 85%. An example of a seedling as a result of testing in Figure 4.

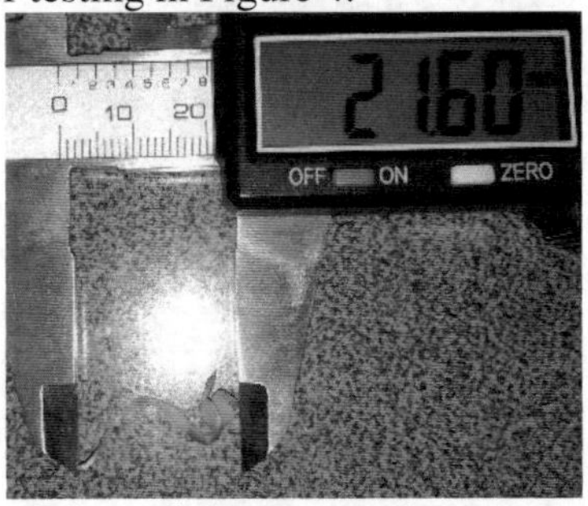

Figure 6. One of the radish sprouts when exposed to blue light.

When studying the effects of the red spectrum of LED lighting, we obtained the following results. The average root length is 8.1 mm, the total sprout length is on average 16.7 mm, the average maximum diameter is 0.78 mm, and the color is rgb (176,177.58). The germination rate is 56%. An example of a seedling as a result of the tests in Figure 5.

Figure 7. One of the radish sprouts when exposed to red light.

When studying the effects of the green spectrum of LED lighting, we obtained the following results. The average root length is 6.74 mm, the total sprout length is on average 10.92 mm, the

IOP Conf. Series: Materials Science and Engineering **680** (2019) 012009 doi:10.1088/1757-899X/680/1/012009

average maximum diameter is 1 mm, and the color is rgb (112,131,13). The germination rate is 73%. An example of a seedling as a result of the tests in Figure 6.

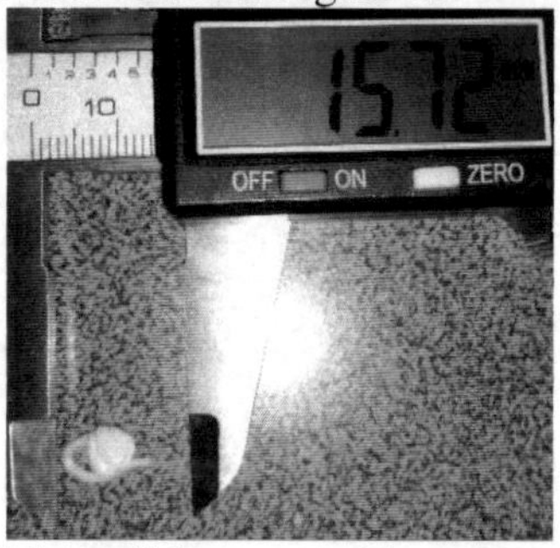

Figure 8. One of the radish sprouts when exposed to green light.

When studying the effects of the white spectrum of LED lighting, we obtained the following results. The average root length is 7.37 mm, the total length of the sprout is on average 15.1 mm, the average maximum diameter is 1.1 mm, the color is rgb (112,131,13). The germination rate is 84%. An example of a seedling as a result of the tests in Figure 7.

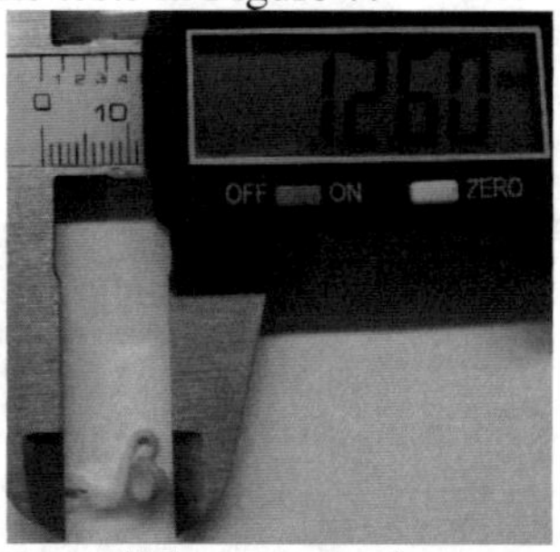

Figure 9. One of the radish sprouts when exposed to green light.

Thus, these studies showed how radish seeds behave under different exposure spectra. The most dense root and the best germination of seeds developing under the influence of blue LEDs, the greatest length is seen when preparing seeds with red LEDs, and average values are seen when exposed to white LEDs.

4. Results

Figure 9 shows that there is an exaggeration of 400 to 550 wavelengths, which can be corrected by using a light filter, but this was not available. The practical application of these methods of light stimulation of seeds of vegetable and green crops, while maintaining favorable microclimate conditions, increases the percentage of seed germination relative to conventional methods of germination up to ~15%.(figure 9) The overall quality of seedlings as a result of stimulation also increases, and the color index of seedlings with rgb is normalized (236,224,106). to color gradients rgb (130,157,40) and rgb (146,90,75).

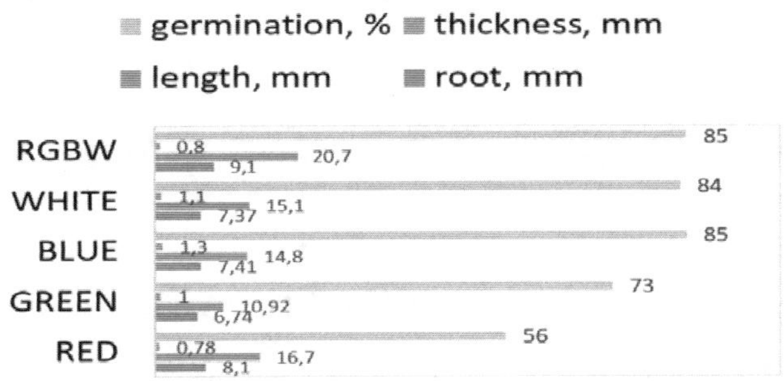

Figure 10. One of the radish sprouts when exposed to green light.

Acknowledgments
I thank Don State Technical University for its help in conducting research.

References

[1] Donskoy D.Yu., Lukyanov A.D., Vernzi M.A., Katin O.I «Development of automated systems for intensifying the technological process of seed preparation using agrophotonic methods».- Modern informatization problems in the technological and telecommunication systems analysis and synthesis (MIP-2019'AS): Proceedings of the XXIV-th International Open Science Conference (Yelm, WA, USA,January 2019)/ Editor in Chief Dr. Sci., Prof. O.Ja. Kravets. - Yelm, WA, USA:Science Book Publishing House, 2019. – 5 p. (299-303)

[2] Klyuchka E.P. Structural-technological scheme of variable irradiation installation for photostimulation of greenhouse plant seeds / E.P. Klyuchka, A.A. Degtyarev, D.S. Tyschenko, B.Yu. Sorokin. - IX International Conference on European Science and Technology Germany, Munich official website http://sciencic.com 9th International Scientific and Practical Conference "European Science and Technology" May 28-29, 2015 Munich, Germany www.sciencic.com – p. (28 ... 31).

[3] Donskoy D.Yu., Lukyanov A.D., Vernzi M.A. «Simulation, identification and dynamic control of the luminaire of the synthesized spectrum».- MATEC Web of Conferences [Electronic resource]. - 2018. - Vol. 226. - Article number 02030. - (XIV International Scientific-Technical Conference "Dynamic of Technical Systems" (DTS-2018); Rostov-on-Don, Russian Federation, September 12-14, 2018).- URL: https://doi.org/10.1051/matecconf/201822602030.

[4] Massa, Gioia D., Hyeon-Hye Kim, Raymond M. Wheeler, and Cary A. Mitchell. "Plant Productivity in Response to LED Lighting", HortScience horts 43, 7: 1951-1956, accessed Jul 10, 2019, https://doi.org/10.21273/HORTSCI.43.7.1951

[5] Jakubiak, Mateusz & Gdowska, Katarzyna. (2013). INNOVATIVE ENVIRONMENTAL TECHNOLOGY APPLICATIONS OF LASER LIGHT STIMULATION. Energetics and automation. 3. 14-21.

[6] J. Li, L.Y. Yin, M.A. Jongsma, C.Y. Wang, Effects of light, hydropriming and abiotic stress on seed germination, and shoot and root growth of pyrethrum (Tanacetum cinerariifolium), Industrial Crops and Products, Volume 34, Issue 3, 2011, Pages 1543-1549, ISSN 0926 -6690, https://doi.org/10.1016/j.indcrop.2011.05.012

[7] A.V. Patsukov, A. P. Mishanov, S. A. Rakutko, A. E. Markov, V. N. Sudachenko "Influence of the optical radiation spectrum on the quality of tomato seedlings" UDC: 628.941.8: 581.14

IOP Conf. Series: Materials Science and Engineering **680** (2019) 012010 doi:10.1088/1757-899X/680/1/012010

Analysis of the possibility of levitation of ferromagnetic bodies in a static magnetic field

M A Vernezi[1], K I Goryanina[1], O I Katin[1], D Yu Donskoy[1], A D Lukyanov[1] and A N Tooraj[2]

[1] Laboratory of local automation and embedded systems Don State Technical University, Gagarin Square 1, Rostov-on-Don 344000, Russia Federation
[2] Instrumentation Lab, School of electrical and computer engineering, University of Tehran, Tehran, Iran

E-mail: slavikvernezi@mail.ru

Abstract. The hypothesis of the existence of three-dimensional regions near the sources of the magnetic field, in which it is possible to obtain a stable position of the magnetic object in a stationary field. The hypothesis is based on the existence of two types of points: points with zero field level and points with local extremum of this field. The obtained distributions of magnetic forces act stabilizing on the object under study, in particular, create the effect of levitation of ferromagnetic bodies in the gravitational field of the Earth. Examples of configurations of the field sources that generate many such points. Systems of current windings providing creation of one-and two-dimensional configurations of areas near sources within which magnetic levitation is possible are analyzed.

1. Introduction

The idea of magnetic levitation – the balancing of body weight by magnetic forces-is currently being vigorously debated in connection with advances in high-temperature superconductivity. The experimental data obtained, however, indicate that the practical implementation of this idea is limited by the distances between the levitating body and the field sources and the critical values of the magnetic field induction for the superconducting systems used.

The known methods of magnetic levitation are based either on the diamagnetic properties of superconducting materials or on the dynamic effect of field pulses on the ferromagnetic body[1]. The creation of stationary magnetic levitation by such methods is difficult for two reasons. The first reason is related to the purely technical difficulties of creating powerful fields, maintaining the superconducting state of materials, ensuring compliance with the level of the field and the mass of levitated objects. The theoretical injunction (Earnshaw's theorem [2]) to create steady state magnetic object through the field of stationary sources.

The studies given by the authors indicate the possibility of obtaining a spatial zone outside the field sources, within which it is possible to create a stable state of a magnetic object, in particular, a ferromagnetic body in the gravitational field of the Earth [3]. The idea of such magnetic levitation is associated with the presence in the distribution of the magnetic field of sources of special points, the field level in which is zero. In work [4] these points are called points with zero level of modulus of tension, in [1]- "magneticl wells", in [5] – magnetic" bulge", in works of the author [6-11] – points of magnetic anomalies.

This article discusses the results of research on obtaining spatial zones outside the field source in order to create magnetic levitation in a static magnetic field.

2. One-dimensional levitation of a ferromagnetic body in the field of a permanent magnet

The paper [4] presents the results of experimental studies of the magnetic field (MP) of a ring permanent magnet magnetized axially. A feature of the distribution of its MP is the presence of two points with a local minimum magnetic field along the axis of the magnet. The presence of these points is explained by the properties of the permanent magnet material. For Figure 1 shows the layout of the permanent magnet and the induction curves B_e, given in [4], as well as the curves of the axial component B_z and induction module B, calculated by the authors of this paper.

This figure shows that in addition to the points with zero magnetic field P_{a1}, P_{a2} points of local extremums are located in the direction of the same axis P_{m1}, P_{m2}. For fig. 1, b shows the location of these points on the induction curve.

In this paper we used the author's methods for calculating the MP distributions of a permanent magnet, which is due to the possibility of their implementation in the mathematical systems Pascal and Maple [14, 15]. Differences of calculation formulas are connected only with coordinate system during calculations.

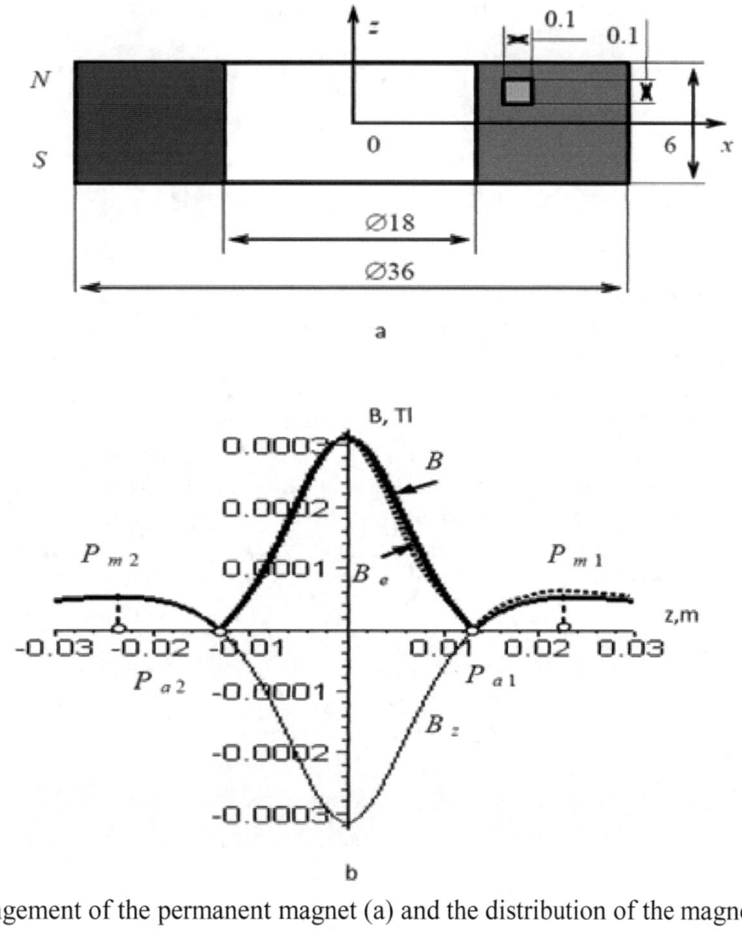

Figure 1. The arrangement of the permanent magnet (a) and the distribution of the magnetic field induction B_e along axis (b)

IOP Conf. Series: Materials Science and Engineering **680** (2019) 012010 doi:10.1088/1757-899X/680/1/012010

The authors ' calculations of the components of the magnetic force acting in the axial plane of a permanent magnet on a ferromagnetic body reveal two areas between pairs of neighboring points-the zero field and the local extremum, where magnetic levitation of a ferromagnetic body is possible. The distribution of force (ponderomotive) and on one of these sites is shown in the curve in Figure. 2.

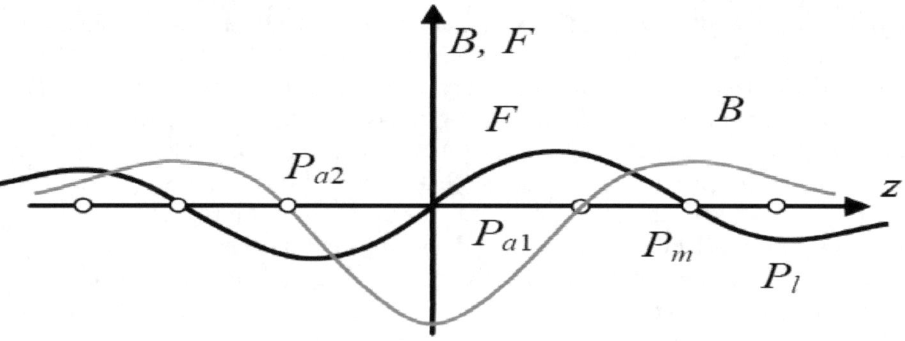

Figure 2. Diagram power distribution F

The resulting distribution of force can be explained as follows. At these points, the MP energy is zero: at the points of zero field due to zero induction, at the point of local extremum due to zero derivative of energy.

Mathematical description of these positions is given by the well-known formula [16]:

$$F = -\frac{dW}{dz}, \tag{1}$$

where

F – ponderomotive force;

W– magnetic energy;

z – axis of the permanent magnet.

As seen in Fig. 2, plot P_{a1}, P_m, – P_l the radial force distribution function is characterized by a negative slope, and the point P_m is the point of stable equilibrium (along the z axis). The same area with a negative slope is present for the left branch of the function with a point P_{a2}.

Power stationary MP permanent magnet on the ferromagnetic test body was tested experimentally. The experimental setup is the following (Figure 3). In a horizontally located permanent magnet 1, a test tube 2 was placed, inside of which was a ferromagnetic ball 3. When moving the tube with the ball in the vertical direction and reaching the areas between the special points, the ball hung in positions 3 and 4. The walls of the tube kept the ball from lateral movements. Thus, experimental confirmation of one-dimensional (along the axis of the annular permanent magnet) levitation of a ferromagnetic body was obtained.

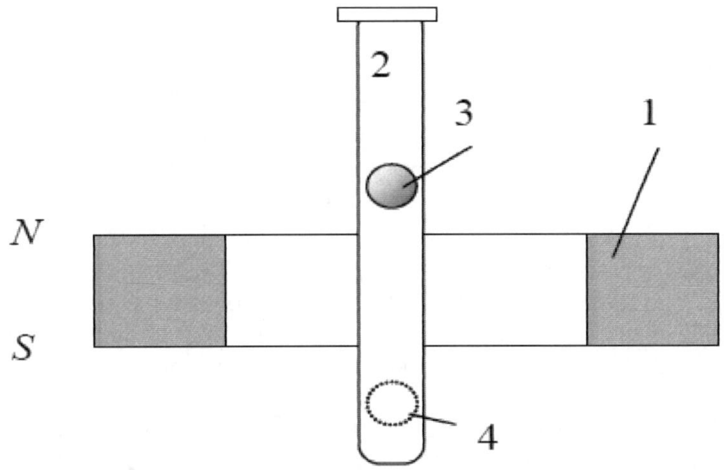

Figure 3. Scheme of experiment on one-dimensional levitation of a ferromagnetic body in the magnetic field of a ring magnet

3. Local minimum point theory

Aside from the material properties of the permanent magnet, its geometry and the nature of the magnetization, by the first question, the problem can be formulated as follows: for the minimum number of MP sources it is necessary to determine the position of points of zero field and a local extremum of the field.

It is obvious that one point source of the magnetic field-dipole does not have such points in the distribution of its own MP in outer space.

In [6] considered a system of two shifted dipoles, which generates these points. The scheme of mutual arrangement of dipoles and their orientation is shown in Figure. 4.

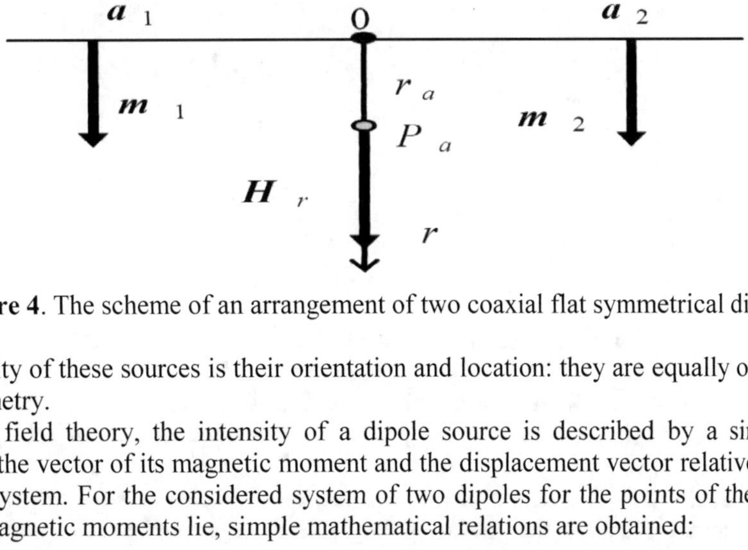

Figure 4. The scheme of an arrangement of two coaxial flat symmetrical dipoles

The peculiarity of these sources is their orientation and location: they are equally oriented and have a plane of symmetry.

In magnetic field theory, the intensity of a dipole source is described by a simple expression associated with the vector of its magnetic moment and the displacement vector relative to the origin of the coordinate system. For the considered system of two dipoles for the points of the plane in which the vectors of magnetic moments lie, simple mathematical relations are obtained:

$$H_r = \frac{m}{2\pi(r^2 + a^2)^{5/2}}(2r^2 - a^2)e_r, \tag{2}$$

$$H_r = \frac{m}{2\pi(r^2 + a^2)^{5/2}}\left|2r^2 - a^2\right|, \tag{3}$$

where H_r, H – is the radial component and the module of the vector field strength in any observation point P_a; m – module of the vector of the magnetic moment of the dipole source; r – is the radius of the observation point on-axis sources; and – the distance (offset) of the source from the axis r; e_r ORT-vector axis sources.

The coordinate of the zero field point is determined from the condition

$$H = |H_r| = 0. \tag{4}$$

$$r_{m1} = a\sqrt{6}/2 \approx 1{,}2247\,a;\; r_{m2} = -a\sqrt{6}/2 \approx -1{,}2247\,a;\; H = 0{,}202m/2\pi a^3. \tag{5}$$

More subtle studies show a fundamental difference in curves B_e, B_z, B and their behavior near the zero-field point. The zero-field point is a saddle point, not a minimum (extreme) point. In comparison with the model [6], the proposed model of MP permanent magnet allows to calculate the position of the zero field point (7) and takes into account the sign of induction or tension in (8).

Since the zero field point is equal to the saddle point for a pair of dipoles, at points near it the forces acting on the ferromagnetic body do not provide a state of stable equilibrium for lateral deviations from the axis.

The property of a pair of displaced dipoles to generate these singular points is preserved in a permanent magnet due to symmetry transformations [17], which is illustrated in Figure 5.

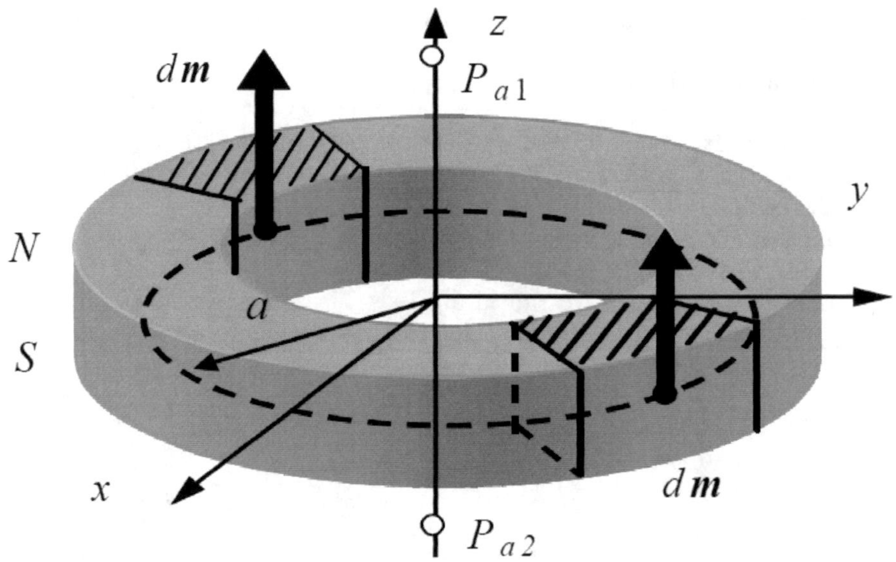

Figure 5. Scheme of experiment on one-dimensional levitation of a ferromagnetic body in the magnetic field of a ring magnet

Elementary magnetized volumes of a permanent magnet located symmetrically to the axis (z) are represented by a pair of vectors with a magnetic moment dm. Integral methods are widely used to calculate the external magnetic field generated by a uniformly magnetized permanent magnet in the axial direction [6, 7]. In this case, the volume of the permanent magnet is divided into elementary sources in the form of selected volumes, within which the magnetization vector is considered to be constant, and the resulting field is determined by the overlap of the fields of elementary sources. The intensity of the external magnetic field at any point can be represented by the following expression [18, 19]:

$$H = \frac{B}{\mu_0} = \frac{1}{4\pi}\sum_{j=1}^{N}\sum_{v=1}^{l_j}(n_{vj} \times M_j) \times \int_{\Delta S_{vj}} \frac{r}{r^3 dS}, \tag{6}$$

where B – the induction vector, μ_0 – magnetic constant, N – number of allocated volumes ferromagnetic element; j, l_j, M_j – number, number of faces and magnetization vector selected volume;

$v-j$-th face number selected volume; n_{vj}– external normal vector, drawn to the plane of the v-th face of j-th selected volume; ΔS_{vj} – the surface area of the v-th face of the j-th selected volume; r is the vector, drawn from observation point to midpoint (geometric center) of each selected volume's.

In the case of the considered ring permanent magnet, uniformly magnetized along the axis ($M_j = M$) and placed in the center One hundred seventeen coordinate systems from the ratio (9) are obtained the following expressions are axial and radial component of the field strength in arbitrary observation point $P(z, \varphi)$:

$$H=\frac{M}{2\pi}\left\{\lambda_1\int_{-1/2}^{1/2}\int_{-\pi/2}^{\pi/2}\frac{(\xi_1-\lambda_1\eta)\sin\phi}{\left[1+(\xi-\lambda_1\eta)^2+p_1^2-2\rho_1\sin\phi\right]^{3/2}}d\phi d\eta-\lambda_2\int_{-1/2}^{1/2}\int_{-\pi/2}^{\pi/2}\frac{(\xi_2-\lambda_2\eta)\sin\phi}{\left[1+(\xi-\lambda_2\eta)^2+p_2^2-2\rho_2\sin\phi\right]^{3/2}}d\phi d\eta\right\}; \quad (7)$$

For figure 6, a-b shows the distribution of magnetic induction $B = \mu_0 H$ in the central section permanent magnets having the same outer radius and different size internal radius and height.

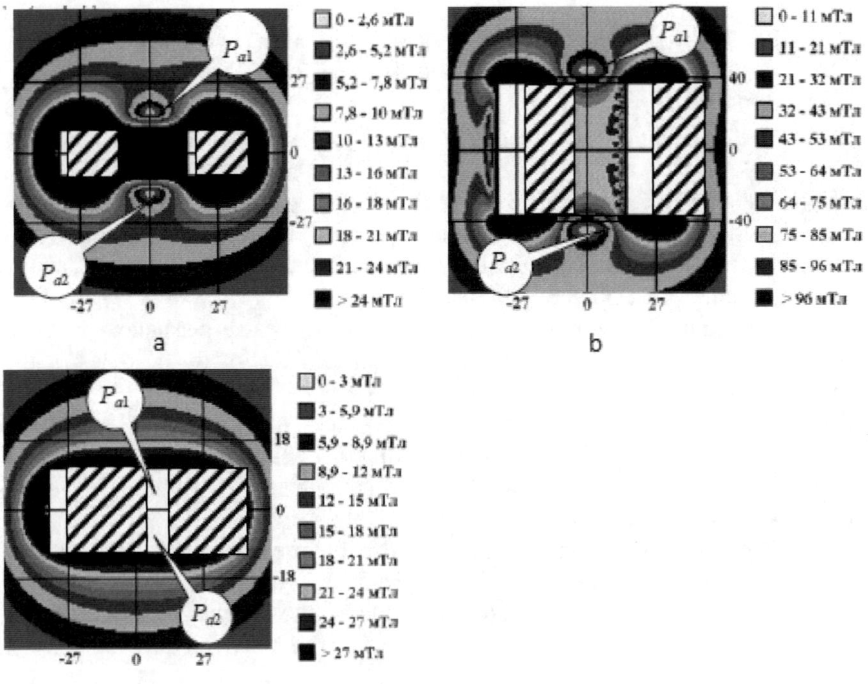

Figure 6. Field distribution in the Central plane permanent magnet

Calculations are carried out in accordance with ratios (10). The results are shown in Fig. 6. When calculating the volume of a permanent magnet it was divided into elementary volumes with a diameter of 0.1 the field level is used to construct distributions were averaged within the flat square cells the size of the side is 2 mm, and the number of gradations of the field for all the permanent magnets of the selected constant, 10. Central rectangular area the permanent magnet is not painted over. As seen for fig. 6, a-b, zero field points are y ring permanent magnet regardless of its geometries. The latter determines only the position zero field points: points can be placed on its axes are both outside the scope of the permanent magnet and in the inner free space.

IOP Conf. Series: Materials Science and Engineering **680** (2019) 012010 doi:10.1088/1757-899X/680/1/012010

4. Conclusion

The field distributions of the ring axial magnetized permanent magnet are simulated and compared with known experimental and computational data. It is established that such magnets have special points in the distribution of the magnetic field outside the volume and analytical relations for the points of the zero field and the local extremum of the field along the axis of the magnet are obtained. It is proved that the presence of these points creates conditions for one-dimensional magnetic levitation of a ferromagnetic object.

Based on the simulation of the field distribution of a pair of coaxial dipoles arranged symmetrically to the plane, it is shown that such a system is a carrier of the properties of one-dimensional magnetic levitation. Sets of such pairs of dipoles constructed using symmetry properties inherit this property. As supporting examples schema modeling using these sources of a permanent magnet. The use of current circuits expands the possibilities of modeling source systems to obtain sets of zero-field points and local extremums forming magnetic levitation regions. The configuration of current circuits with a flat magnetic levitation region is obtained.

References

[1] V. S. Mikhalevich, V. M. Rashkovan, D. Y. Khusainov, O. G. Deborin. "Magnetic potential well" - stabilization effect of superconducting dynamical systems. Kyiv: Naukova Dumka, 1991, 336 p.

[2] S. Earnshaw. On the nature of the molecular forces // *Trans. Cambridge Phil. Sociaty.* 1842, v. 7, p. 97-113.

[3] V. S. Lupikov, N. N. Chernyshov, E. Mvogo. Possible magnetic levitation of magnetic bodies nanoscale gradient magnetic field current windings // international scientific.-tekhn. Conf. "Physico-chemical bases of formation and modification of micro-and nanostructures". Kharkiv: nftc mon and NAS of Ukraine, 2007, pp. 109-110.

[4] E. A. Vasilyev. Spatial distribution of magnetic induction near a permanent magnet and the" effect " of attraction of a high-temperature superconductor to the poles of the magnet. 1989, vol. 49, V. 2, pp. 111-113.

[5] The world in magnetic ring // Elektropanorama. 2000, July-August, p. 52.

[6] B. V. Klimenko, M. A. Lyubchik, V. S. Lupikov, etc. Features of magnetic field intensity distribution near the surface of electrical equipment / / Bulletin of Kharkiv state Polytechnic University. 1999, No. 75, pp. 109-115.

[7] V. S. Lupikov. Compensation of the magnetic field in the presence of magnetic anomalies near the surface switchgear // Technon elektrodynamika. 2000, No. 6, pp. 18-24.

[8] B. V. Klimenko, V. S. Lupikov. Features of reducing the external magnetic field near the surface of the switchgear / / Bulletin of Kharkiv state Polytechnic University. 2000, No. 84, pp. 117-122.

[9] V. S. Lupikov. Theoretical justification of the United dipole model of the external magnetic field of electrical equipment / / Visnik National technical University "KPI". 2001, No. 17, pp. 95-102.

[10] B. V. Klimenko, V. S. Lupikov, E. I. Baida, V. L. Emel'yanov. Investigation of magnetic anomalies near the ring permanent magnet / / Visnik National technical University "KPI". 2001, No. 16, pp. 74-79.

[11] B. V. Klimenko, V. S. Lupikov, V. L. Emelyanov, etc. Algorithm for determining the position of magnetic anomaly points near the surface of the switchgear / / Visnik National technical University "KPI". 2001, VIP. 16, pp. 87-92.

[12] M. McCaig. Permanent magnets for repulsion systems // Electrical review. 1961, No. 9, p. 77-89.

[13] Sagava, M., N. Fujumura. New material of permanent magnets on a base of Nd and Fe // J. Appl. Phys. 1984, №6, p 2083-2087.

IOP Conf. Series: Materials Science and Engineering **680** (2019) 012010 doi:10.1088/1757-899X/680/1/012010

[14] N. N. Chernyshov. Approximate calculation of the volume of the permanent magnets on the required rigidity of the system components to control vibration devices Technon elektrodynamika. 1998, No. 4, pp. 78-79.

[15] E. I. Canoe. Calculation of magnetically controlled switching device on a permanent magnet / / Elektrotekhnika I elektromekhanika. 2002, No. 2, pp. 12-14.

[16] L. A. Bessonov. Theoretical foundations of electrical engineering. M.: "High school", 1973, 752 p.

[17] V. S. Lupikov, O. A. Gilyarovsky, S. V. Vyrovets, D. L. Martynenko. The symmetry of the magnetic sources and the set of points of zero field // Visnyk Sovereign pateince University: Naukow of Pratsi KDPU, 2004, V. 42, pp. 24-28.

[18] P. A. Kurbatov, S. A. Arinchin. Numerical calculation of electromagnetic fields. M.: Energoatomizdat, 1984, 168 p.

[19] N. N. Chernyshov, L. D. Bogdanov. Mathematical models of elements of systems with permanent magnets for the calculation of their power characteristics and optimization of the magnetic system: Preprint. National Academy of Sciences of Ukraine, Institute of electrodynamics. No. 819. Kiev, 1998, p. 34

IOP Conf. Series: Materials Science and Engineering **680** (2019) 012011 doi:10.1088/1757-899X/680/1/012011

The use of statistical analysis methods in assessing long-term prognoses in patients with acute coronary syndrome and contrast-induced nephropathy

N Kobzeva[1], V Terentev[1], I Zolotuhina[2]

[1] Rostov State Medical University, lane 29 Nakhichevanskyi, Rostov-on-Don 344000, Russia
[2] Don State Technical University, Gagarin's square 1, Rostov-on-Don 344000, Russia

E-mail: kobzeva.nataliya1@gmail.com

Abstract. Thus, using a rank correlation analysis, a preliminary analysis of the relationship between the increase / decrease in blood creatinine, blood urea, GFR (Cockcroft-Gault) and GFR (CKD-EPI) after coronary angiography with vascular events was performed. A Kaplan-Meier survival analysis demonstrated the statistically significant effect of contrast-induced nephropathy on the risk of death from myocardial infarction in patients with long-term acute coronary syndrome.

1. Introduction

Survival analysis is a set of statistical methods that allow you to evaluate any events that occur after a certain initial event (for example, the start of treatment) without the occurrence of a certain event (for example, it can be death, a relapse of a disease or a condition). This probability is called the probability of survival. At the end of the study and censoring variable, indicating the completeness of the data [1].

The STATISTICA package implements the procedure for constructing survival curves using the Kaplan – Meyer method [2], which is as follows. It is estimated that there are n patients. Let dti be the number of deaths at time ti, and nti the number of people observed at time ti, where i is the number of the time at which the event occurred. The survival curve is given by the following function:

$$\hat{S}(t) = \prod_{t_i \in [0,t]} \left(1 - \frac{d_{t_i}}{n_{t_i}}\right). \tag{1}$$

2. Methods

The study included 323 patients with acute coronary syndrome who were treated in the cardiology department of the regional vascular center, of which 217 (67.1%) were men, 106 (32.9%) were women. The average age of the examined patients was 59.6 ± 9.2 years.

All patients underwent standard biochemical studies, including the assessment of lipid metabolism: total cholesterol; high and low density lipoprotein cholesterol; atherogenic index; triglycerides, creatinine and urea levels.

The complex of instrumental studies included recording ECG at rest, Holter ECG monitoring, ultrasound of the heart, ultrasound examination of the kidney, coronary angiography to assess the degree of damage to the coronary vessels, selective angiography of the renal vessels.

Statistical analysis of the results was carried out using a set of applied statistical programs Microsoft Office Excel 2010 (Microsoft Corp., USA) and STATISTICA 10.0 (StatSoft Inc., USA). In order to assess the type of data distribution, Kolmogorov-Smirnov analysis was used; for values of $p > 0.05$, the distribution was considered not different from the normal one.

Descriptive statistics were carried out to determine the following features: the data were presented as M ± SD (M is the arithmetic mean, SD is the standard deviation) with a normal distribution and in the form Me [Q1; Q3] (Me is the median, Q1 and Q3 are the first and third quartiles) with abnormal distribution. In the normal distribution of the sample, when comparing two independent samples, the Student criterion was used, and when different from the normal one, the Mann-Whitney test and $\chi 2$ or the Leuven test with the determination of F. We also used logistic regression analysis with the calculation of relative risks (OR) and determination $\chi 2$, the relationship was considered statistically significant at a value of $p < 0.05$.

Kaplan-Majer survival analysis was performed with the Gehan's Wilcoxon test, the test was considered statistically significant at a p value of <0.05.

3. Results

At the first stage, a preliminary analysis of the relationship between the increase / decrease in blood creatinine, blood urea, GFR (Cockcroft-Gault) and GFR (CKD-EPI) after coronary angiography with vascular events was performed. For this purpose, a rank correlation analysis was used. It was shown that an increase in creatinine did not have a statistically significant relationship with vascular events (Table 1).

Table 1. Results of rank correlation analysis of Δ creatinine and vascular events

Sign	Spearman r	T (N-2)	p
Nonfatal myocardial infarction	-0,09	-0,89	0,37
Non-fatal brain stroke	-0,11	-1,12	0,26
Fatal myocardial infarction	-0,06	-0,62	0,53
Fatal brain stroke	0,09	0,99	0,32
Fatal and nonfatal myocardial infarction	-0,10	-1,09	0,28
Fatal and non-fatal cerebral stroke	-0,05	-0,52	0,60
Fatal vascular events	-0,01	-0,10	0,92
Nonfatal vascular events	-0,11	-1,21	0,23
Fatal and non-fatal vascular events	-0,10	-1,08	0,29

However, as blood urea increased after coronary angiography, there was an increase in the likelihood of developing fatal vascular events and non-fatal ones (table 2).

Table 2. Results of rank correlation analysis of Δ urea and vascular events

Sign	Spearman r	T (N-2)	p
Nonfatal myocardial infarction	0,13	2,0	0,12
Non-fatal brain stroke	-0,12	-1,26	0,21
Fatal myocardial infarction	0,22	2,36	0,02
Fatal brain stroke	0,02	0,17	0,87
Fatal and nonfatal myocardial infarction	0,01	0,11	0,91
Fatal and non-fatal cerebral stroke	-0,09	-1,02	0,31
Fatal vascular events	0,21	2,21	0,029
Nonfatal vascular events	0,23	2,49	0,014
Fatal and non-fatal vascular events	-0,05	-0,5	0,61

he decrease in GFR after coronary angiography was associated with the subsequent development of non-fatal myocardial infarction, as well as non-fatal vascular events (table 3). In general, a decrease in GFR also influenced the development of all vascular events.

Table 3. The results of rank correlation analysis Δ GFR and vascular events

Sign	Spearman r	T (N-2)	p
Nonfatal myocardial infarction	-0,25	-2,71	0,008
Non-fatal brain stroke	0,11	1,17	0,24
Fatal myocardial infarction	0,07	0,71	0,48
Fatal brain stroke	-0,09	-0,97	0,33
Fatal and nonfatal myocardial infarction	0,24	2,61	0,01
Fatal and non-fatal cerebral stroke	0,05	0,53	0,59
Fatal vascular events	0,02	0,19	0,85
Nonfatal vascular events	-0,26	-2,79	0,006
Fatal and non-fatal vascular events	-0,22	-2,38	0,019

The decrease in GFR determined by the CKD-EPI method after CAG was also associated with the development of nonfatal myocardial infarction, nonfatal vascular events, all vascular events, as well as fatal and nonfatal myocardial infarction (table 4).

Table 4. Results of rank correlation analysis Δ GFR (CKD-EPI) and vascular events

Sign	Spearman r	T (N-2)	p
Nonfatal myocardial infarction	-0,28	3,1	0,003
Non-fatal brain stroke	0,10	1,1	0,28
Fatal myocardial infarction	0,10	1,1	0,26
Fatal brain stroke	-0,09	-1,1	0,29
Fatal and nonfatal myocardial infarction	-0,29	3,25	0,002
Fatal and non-fatal cerebral stroke	0,04	0,46	0,65
Fatal vascular events	0,05	0,57	0,57
Nonfatal vascular events	-0,27	3,05	0,003
Fatal and non-fatal vascular events	-0,26	2,91	0,004

n addition, it was shown that nephrosclerosis, defined as a difference in kidney length of more than 15 mm, is a factor associated with the development of non-fatal myocardial infarction, fatal and non-fatal myocardial infarction, as well as non-fatal vascular events and all vascular events (table 5).

Table 5.-Results of a rank correlation analysis of nephrosclerosis (difference> 15mm) and vascular events

Sign	Spearman r	T (N-2)	p
Nonfatal myocardial infarction	0,21	2,19	0,03
Non-fatal brain stroke	0,002	0,02	0,98
Fatal myocardial infarction	0,02	0,20	0,84
Fatal brain stroke	0,14	1,49	0,14
Fatal and nonfatal myocardial infarction	0,19	2,04	0,044
Fatal and non-fatal cerebral stroke	0,07	0,75	0,45
Fatal vascular events	0,09	0,94	0,35
Nonfatal vascular events	0,19	1,99	0,049
Fatal and non-fatal vascular events	0,22	2,34	0,02

Nephrosclerosis, determined by a decrease in kidney size less than 90 mm, had an effect only on fatal myocardial infarction (table 6).

Table 6. Results of rank correlation analysis of nephrosclerosis (<90mm) and vascular events

Sign	Spearman r	T (N-2)	p
Nonfatal myocardial infarction	0,13	1,36	0,18
Non-fatal brain stroke	-0,08	-0,77	0,44
Fatal myocardial infarction	-0,07	-0,71	0,48
Fatal brain stroke	0,22	2,36	0,02
Fatal and nonfatal myocardial infarction	0,07	0,74	0,46
Fatal and non-fatal cerebral stroke	0,04	0,44	0,66
Fatal vascular events	0,06	0,57	0,57
Nonfatal vascular events	0,07	0,71	0,48
Fatal and non-fatal vascular events	0,09	0,98	0,33

Thus, renal factors in general and damage to the renal parenchyma during coronary angiography, in particular, are associated with the development of vascular events in the distant period, which justifies the transition to analysis of the role of contrast-induced nephropathy in predicting the course of cardiovascular pathology in this group of patients .

A Kaplan-Meier survival analysis showed a statistically significant effect of contrast-induced nephropathy on the risk of death from myocardial infarction (Figure 1).

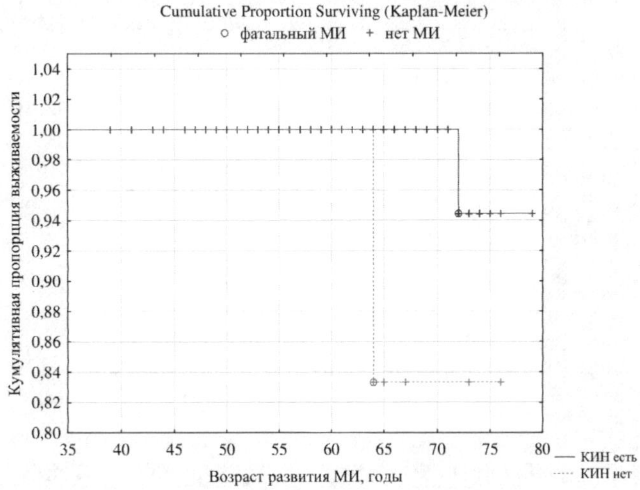

Figure 1. Schedule of cardiovascular survival of patients (development of fatal cerebral stroke) in the presence or absence of CIN. Gehan's Wilcoxon Test WW = 39,0, Sum = 2542,0 Var = 220,09, Test statistic = 2,59, p = 0,009

When conducting a logistic regression analysis, it was shown that contrast-induced nephropathy increased the risk of further development of fatal and non-fatal infarction by 20% (table 7).

IOP Conf. Series: Materials Science and Engineering **680** (2019) 012011 doi:10.1088/1757-899X/680/1/012011

Table 7.-Logistic regression analysis of the effect of contrast-induced nephropathy on the risk of developing vascular events

Sign	Constanta B0 Est	Constanta B0 OR	Estimate	OR (unit ch)	OR (range)	χ^2	df	p
Nonfatal myocardial infarction	2,17	8,8	24,1	$3{,}0*10^{10}$	$3{,}0*10^{10}$	2,24	1	0,13
Non-fatal brain stroke	2,48	12,0	24,1	$2{,}9*10^{10}$	$2{,}9*10^{10}$	1,67	1	0,19
Fatal myocardial infarction	2,49	12,2	23,0	$9{,}9*10^{9}$	$9{,}9*10^{9}$	1,65	1	0,19
Fatal brain stroke	4,64	104	-2,34	0,09	0,09	2,21	1	0,14
Fatal and nonfatal myocardial infarction	1,57	4,83	25,9	$1{,}8*10^{11}$	$1{,}8*10^{11}$	3,91	1	0,047
Fatal and non-fatal cerebral stroke	-2,37	0,09	0,06	1,06	1,06	0,003	1	0,95
Fatal vascular events	-2,37	0,09	0,06	1,06	1,06	0,003	1	0,95
Nonfatal vascular events	-1,6	0,19	-25,4	$1*10^{-5}$	$1*10^{-5}$	3,36	1	0,07
Fatal and non-fatal vascular events	-1,12	0,33	-1,07	0,34	0,34	1,27	1	0,26

When analyzing factors that could influence the formation of contrast-induced nephropathy, an interesting circumstance was that during the survival analysis it was shown that a more severe myocardial damage in the form of acute myocardial infarction contributed to the development of contrast-induced nephropathy during coronary angiography in compared with a less severe lesion in the form of unstable angina (Figure 2). One gets the impression of the presence of a mutually aggravating effect of CIN and coronary pathology on each other.

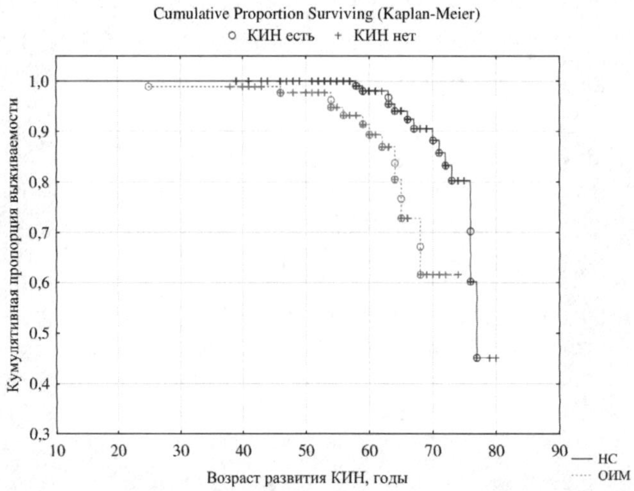

Figure 2. Survival chart of patients (development of contrast-induced nephropathy) with the development of unstable angina pectoris or myocardial infarction. Gehan's Wilcoxon Test WW = 1095,0, Sum = $4238*10^2$, Var = 96921, Test statistic = 3,52, p = 0,0004

4. Discussion

According to the literature, the likelihood of developing kidney dysfunction in patients with cardiovascular disease is significantly higher than in the general population. Thus, according to NHANES III (Third National Health and Nutrition Examination Survey), a combination of any two risk factors for developing cardiovascular diseases leads to a decrease in glomerular filtration rate of less than 60 ml / min / 1.73 m2 by a factor of 3.7, than with conserved kidney function. In addition, according to the results of a meta-analysis performed in 2002 by J. Suwaidi et al., In acute coronary syndrome, a slight decrease in renal function was accompanied by a higher mortality rate and frequency of repeated myocardial infarction during 6 months of follow-up [3].

According to the results of numerous clinical studies, the role of renal dysfunction as a factor worsening the prognosis in acute coronary syndrome has been proved. According to the Canadian register GRACE (Global Registry of Acute Coronary Events), in patients with myocardial infarction with ST segment elevation, a decrease in GFR to 30-60 ml / min increases the risk of death by 2.09 times; with GFR <30 ml / min, the probability of an adverse outcome increased by 4 times.

In addition, it has been proven that even moderate renal dysfunction is an independent predictor of myocardial infarction and death [4-6]. A retrospective analysis in 16,248 patients who underwent administration of a contrast medium showed that even a slight decrease in renal function can lead to a huge increase in mortality, regardless of other risk factors [7–9].

During the first year after administration of the contrast agent, the mortality rate in patients with chronic kidney disease that existed before the contrast remains very high. In particular, in the course of our work, it was demonstrated that the decrease in GFR after coronary angiography was associated with the development of subsequent non-fatal myocardial infarction, as well as non-fatal vascular events.

During the first year after administration of a contrast medium, the mortality rate in patients with chronic kidney disease, which existed before contrast, remains very high.

Mortality reaches 45.2% in the group of patients on hemodialysis, is 35.4% in patients with impaired renal function and 19.4% in patients with intact renal function. According to the Mayo Clinic register, mortality within 1 year after PCI is directly correlated with creatinine clearance, making up 1.5% among patients with creatinine clearance of more than 70 ml / min and 18.3% among patients with creatinine clearance of less than 30 ml / min [10].

According to our Kaplan-Meier survival analysis, we showed a statistically significant effect of contrast-induced nephropathy on the risk of long-term death from myocardial infarction in patients with acute coronary syndrome.

5. Conclusion

Thus, using a rank correlation analysis, a preliminary analysis of the relationship between the increase / decrease in blood creatinine, blood urea, GFR (Cockcroft-Gault) and GFR (CKD-EPI) after coronary angiography with vascular events was performed.A Kaplan-Meier survival analysis demonstrated the statistically significant effect of contrast-induced nephropathy on the risk of death from myocardial infarction in patients with long-term acute coronary syndrome.

Acknowledgements

The study had no sponsorship.

References

[1] StatSoft Russia. STATISTICA. The Official Guide, 2007, vol. 3, 783 p.

[2] Zharkov A. V. Statisticheskiye svoistva funktsii Gompertsa (Statistical properties of the Gompertz function). Obozrenie prikladnoy i promyshlennoy matematiki, 2010, vol. 17, no. 3, pp. 405–406.

[3] *Eggers P.W.* Has the incidence of end-stage renal disease in the USA and other countries stabilized? Curr Opin Nephrol Hyper- tens 2011; 20: 241—245.

[4] *Tasanarong A., Hutayanon P., Maric C. et al.* Urinary Neutrophil Gelatinase-Associated Lipocalin predicts the severity of contrast-induced acute kidney injury in chronic kidney disease patients undergo- ing elective coronary procedures Nephrology 2013; 14: 1471—2369.

[5] *Reaven G.M.* Insulin resistance: the link between obesity and cardiovascular disease. Med Clin North Am 2011; 95: 875—892.

[6] Rajendran P. The Vascular Endothelium and Human Diseases / Rajendran P, Rengarajan T, Thangavel J, Nishigaki Y, Sakthisekaran D, Sethi G, Nishigaki I. // Int J Biol Sci. – 2013.– № 9 (10) –P.1057-1069.

[7] Ponikowski P., Voors A.D., Anker S, Bueno H.J. Coats A.S., Falk V., Gonzalez-Juanatey J., Harjola V., A. Jankowska E., Jessup M., et.al.2016 ESC Guidelines for the diagnosis and treatment of acute and chronic heart failure.2016 // Russian Journal of Cardiology. 2017. № 1. 7-81.

[8] Кобзева Н. Терентьев В. Влияние стеноза почечных артерий в комплексе с другими факторами на риск развития фатальных и нефатальных сосудистых событий у пациентов ОКС // Терапия. – М., 2018. – No 3. – 98–110.

[9] Roffi M, Patrono C., Collet J.P., Mueller C., Valgimigli M, Andreotti F. et al. 2015 ESC Guidelines for the management of acute coronary syndromes in patients presenting without persistent ST-segment elevation // Eur Heart J. 2016. Vol. 37, № 3. P. 267-315

[10] Windecker S., Kolh P., Alfonso F., Collet J.P., Cremer J., Falk V. et al. 2014 ESC/EACTS Guidelines on myocardial revascularization // Eur Heart J. 2014. Vol. 35, № 37. P. 2541-2619.

IOP Conf. Series: Materials Science and Engineering **680** (2019) 012012 doi:10.1088/1757-899X/680/1/012012

Determination of the practical accuracy of the operation of the safety friction clutch with a variable gain installed in the forging equipment ensuring the absence of breakdowns

K O Kobzev, E S Bozhko, A V Mozgovoi, A G Iliev, A A Kovaleva

Don State Technical University, Gagarin's square 1, Rostov-on-Don 344000, Russia

E-mail: 5976765@mail.ru

Abstract. The created version of SFC with a variable value of the feedback gain theoretically has the accuracy of operation of the "ideal" SFC when it is installed in forging and stamping equipment. A variable value of the gain is obtained by changing the pressure angle of the rolling element control device depending on the transmitted load.
To obtain a variable value, the gain in the SFC was changed in the design of the control device.
The experimental studies of the full-scale SFC sample with a variable gain, however, showed that the coupling does not have the accuracy of operation inherent in the "ideal" SFC. Since the gain was experimentally studied in the static loading mode on the installation with a shortened kinematic chain, i.e., with a smooth increase in the load (the time of the full overload, which leads to the operation of the SFC, is longer than the period of natural vibrations of the system), the effect of unaccounted dynamic loads is practically excluded.

1. Introduction

The created version of the first-generation SFC with a variable value feedback gain theoretically has the accuracy of operation of the "ideal" SFC. A variable value of the gain is obtained by changing the pressure angle of the rolling element control device depending on the transmitted load.

To obtain a variable value, the gain in the SFC was changed in the design of the control device. The experimental studies of the full-scale SFC sample with a variable gain, however, showed that the coupling does not have the accuracy of operation inherent in the "ideal" SFC. Since the gain was studied experimentally in a static loading mode on a setup with a shortened kinematic chain, i.e., with a smooth increase in load (the time of full overload, which leads to SFC operation, is longer than the period of natural vibrations of the system), the effect of unaccounted dynamic loads was practically eliminated .

2. Methods

The equation of axial equilibrium of the support sleeve 6, according to the diagram of forces depicted in Fig. 1 has the form:

$$F_p - F_n - F_{tr} = 0, \qquad (1)$$

F_p – current coefficient reinforcing spacer force between rolling elements 4 and supporting sleeve 6; F_n – coefficient of reinforcing tension force spring 9; F_{tr} – friction force between the pressure plate

IOP Conf. Series: Materials Science and Engineering **680** (2019) 012012 doi:10.1088/1757-899X/680/1/012012

7 and the guide key 8 [1-3].

Unlike the first three members of equation (1), the force F_{tr} adopted linearly as a function of torque SFC. The value of the coefficient of friction between the pressure plate and the guide key, according to the work, is considered constant.

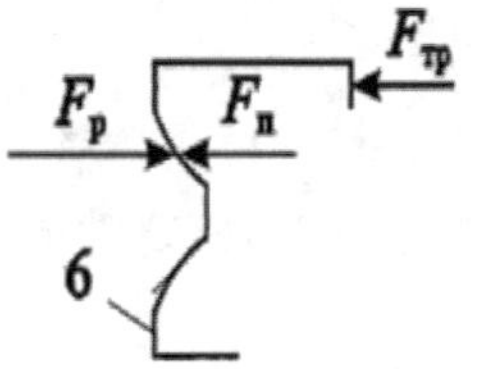

Figure 1. The design scheme

In equation (1), the terms can be represented as

$$F_p = \frac{T_{nom}}{r} \text{tg}\alpha_i ; \; F_{tr} = \frac{2T_{nom}}{d} f_1 \tag{2}$$

T_{nom} – rated torque SFC; α_i – current coefficient reinforcing pressure angle between the rolling body 4 and the socket wall; r – radius of the circle on which the rolling bodies are located 4; c_0 – axial spring stiffness 9; x_i – current coefficient reinforcing abscissa of the contact point between the rolling body 4 and the wall of the socket; f_1 – coefficient of friction between the pressure plate 7 and the guide key 8; d – diameter of the central hole of the pressure plate 7.

Substituting the right sides of formulas (2) into equation (1), we obtain:

$$\text{tg}\alpha_i = \frac{c_0 x_i r}{T_{nom}} + \frac{2rf_1}{d}. \tag{3}$$

Integrating function (3), we find:

$$y_i = \frac{c_0 r}{2T_{nom}} x_i^2 + \frac{2rf_1}{d} x_i + C \tag{4}$$

C – integration constant determined from the initial conditions: $y = y_0$, $x = 0$. $C = y_0$, y_0 – the initial ordinate of the generatrix of the nest, which is determined according to the scheme depicted in Fig. 2. According to the scheme we get:

$$y_0 = \frac{D}{2}\cos\alpha_{min} = \frac{D}{2\sqrt{1+tg^2\alpha_{min}}}. \tag{5}$$

Then relation (5) can be written in the following form:

$$y_i = \frac{c_0 r}{2T_{nom}} x_i^2 + \frac{2rf_1}{d} x_i + \frac{D}{2\sqrt{1+tg^2\alpha_{min}}}. \tag{6}$$

The rated torque SFC is calculated, according to the work, with a minimum value of the coefficient of friction, according to the following formula:

$$T_{nom} = zF_n R_{cp} \frac{f_{min}}{1 + zC_{min} f_{min}}, \tag{7}$$

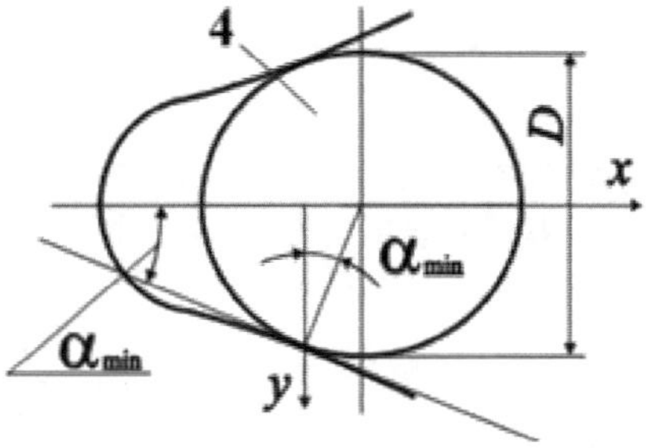

Figure 2. The design scheme

To establish dependence $C\left(f_i\right)$ we use the condition according to which the current coefficient is the amplifying value of the torque SFC at any value f_i remains constant and must be equal to the nominal torque of the coupling [4-9].

Equating to each other the right parts of the relation and relations (7), we obtain an equation whose solution has the following form:

$$C_i = \frac{f_i - f_{\min} + zC_{\min} f_i f_{\min}}{zf_{\min} f_i}. \tag{8}$$

On the other hand $C_i = \frac{R_{cp}}{r}\mathrm{tg}\alpha_i$, therefore, from expression (8) we obtain:

$$\mathrm{tg}\alpha_i = \frac{r(f_i - f_{\min} + zC_{\min} f_i f_{\min})}{R_{cp} zf_{\min} f_i}. \tag{9}$$

From relations (2) and (8) we find:

$$x_i = \frac{T_{\text{nom}}}{c_0}\left[\frac{f_i - f_{\min} + zC_{\min} f_i f_{\min}}{R_{cp} zf_{\min} f_i} - \frac{2f_1}{d}\right]. \tag{10}$$

Substituting (10) in (8), we obtain:

$$y_i = \frac{rT_{\text{nom}}}{2c_0}\left[\frac{f_i - f_{\min} + zC_{\min} f_i f_{\min}}{R_{cp} zf_{\min} f_i} - \frac{2f_1}{d}\right]^2 + \frac{2rT_{\text{nom}} f_1}{dc_0}\left[\frac{f_i - f_{\min} + zC_{\min} f_i f_{\min}}{R_{cp} zf_{\min} f_i} - \frac{2f_1}{d}\right] + \tag{11}$$

$$+\frac{D}{2\sqrt{1 + tg^2\alpha_{\min}}}.$$

3. Results

Consider the circuit shown in Fig. 3. The diagram shows the process of moving the rolling body 4 with automatic control [10-15].

Position I, II of the rolling body corresponds to the operation of the SFC with the minimum and maximum values of the coefficient of friction. Abscissa axis x passes through the center of the rolling body, the ordinate axis y − through the point of amplification of the contact of the side wall of the socket and the rolling body.

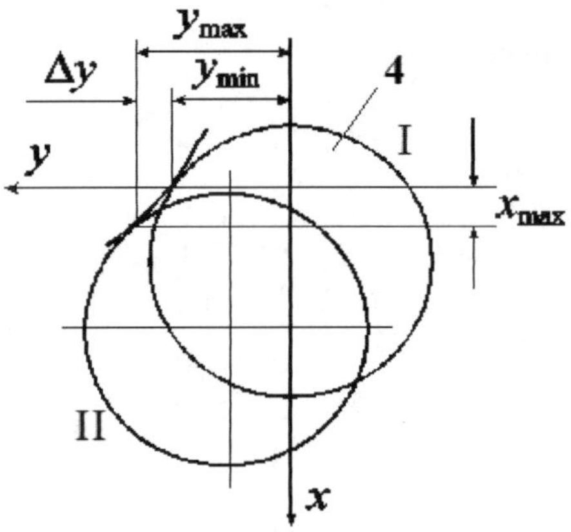

Figure 3. The movement pattern of the rolling body

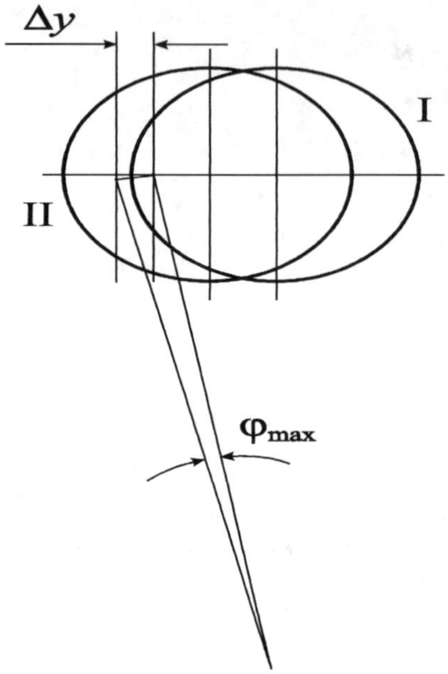

Figure 4. The design scheme

In the indicated position 1, the initial ordinate of the contact point is $y_{\min}$, and position 2 of the rolling body corresponds to the ordinate of the contact point $y_{\max}$.

$$\Delta y = y_{\max} - y_{\min}. \tag{12}$$

Substituting relations (10) and (9) into equality (12), we obtain:

$$\Delta y = \frac{rT_{\text{nom}}}{2c_0}\left[\frac{f_{\max} - f_{\min} + zC_{\min} f_{\max} f_{\min}}{R_{cp} z f_{\min} f_{\max}} - \frac{2f_1}{d}\right]^2 + \\ + \frac{2rT_{\text{nom}} f_1}{dc_0}\left[\frac{f_{\max} - f_{\min} + zC_{\min} f_{\max} f_{\min}}{R_{cp} z f_{\min} f_{\max}} - \frac{2f_1}{d}\right].$$

Consider the circuit shown in Fig. 4. The diagram corresponds to the view shown by the arrow in fig. 3. Positions I and II in Fig. 2.11 correspond to similar positions of the rolling body in Fig. 3.

Moving the rolling body from position I to position II (Fig. 3) leads to relative twisting of the SFC half-coupling by an angle $\varphi_{\max}$ (fig. 4). By producing geometric constructions in fig. 4, we find

$$\frac{\Delta y}{2r} = \sin\frac{\varphi_{\max}}{2}. \tag{13}$$

$$\varphi_{\max} = 2\arcsin\frac{\Delta y}{2r},$$

$$\Rightarrow \varphi_{\max} = 2\arcsin\left(\begin{array}{l} \frac{T_{nom}}{4c_0}\left[\frac{f_{\max} - f_{\min} + zC_{\min} f_{\max} f_{\min})}{R_{cp} z f_{\min} f_{\max}} - \frac{2f_1}{d}\right]^2 + \\ + \frac{T_{\text{nom}} f_1}{dc_0}\left[\frac{f_{\max} - f_{\min} + zC_{\min} f_{\max} f_{\min}}{R_{cp} z f_{\min} f_{\max}} - \frac{2f_1}{d}\right] \end{array}\right). \tag{14}$$

Using the main idea of the energy balance method, we write:

$$A_t = P_{np} + A_{tp}.$$

A_t – torque work, A_{tr} – work against friction, P_{np} – potential energy of the spring 9.

In equation (15), the terms can be represented as:

$$A_t = \frac{\Delta T}{2}\varphi_{\max}, \quad (15)$$

$$A_{tp} = 2F_{\text{тр}}x_{\max} = \frac{2\Delta T}{d} f_1 x_{\max}, \quad (16)$$

ΔT – magnitude of overload, i.e. maximum increase in external load [16-19].

Using the relations, we find the expression for calculating the overload ΔT, associated with an increase in the coefficient of friction and the onset of general overload in the drive of the machine:

$$\Delta T = \frac{4dc_0 x_{\max}^2}{d.\varphi_{\max} - 4f_1.x_{\max}}. \quad (17)$$

The practical accuracy factor SFC can be calculated by the formula:

$$K_t = \frac{T_{\max}}{T_{\min}} = \frac{T_{nom} + \Delta T}{T_{nom}} = 1 + \frac{\Delta T}{T_{nom}} \quad (18)$$

Obviously, the quantity K_t will approach unity if the torque decreases ΔT.

4. Discussion

Since the gain is the magnitude of the torque T_{nom} during the calculation and design, the SFC is the initial one and is defined as a function of various operational factors (the rated drive power of the machine, the location of the SFC in the kinematic circuit of the drive, etc.), it is considered constant.

Consider the possibility of decreasing ΔT by changing the value $x_{\max}$. We turn to relation (16). Decrease value x_i, $x_{\max}$ possibly due to an increase in axial stiffness c_0. In relation (19) we obtain that indicates a decrease in torque ΔT with increasing c_0.

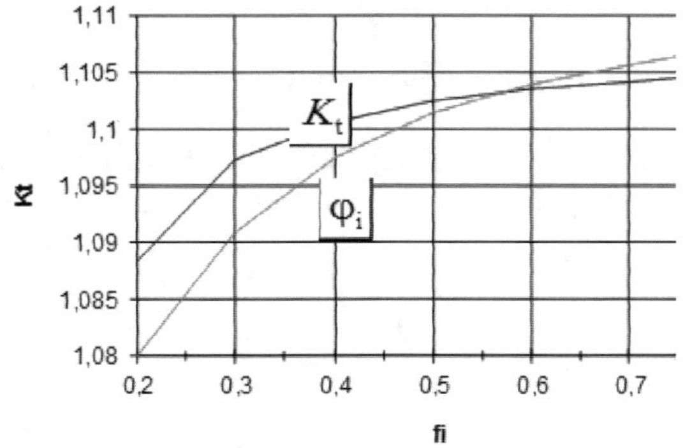

Figure 5. Areas of solution.

Dependency Function Charts $K_{\text{т}}(f_i), \varphi_i(f_i)$ are shown in fig. 5. When calculating $\varphi_{\max}$ relation (16) is used, in which, in turn, the quantity $y_{\max}$ was calculated by relation (15), without taking into account the last term in it. In this case, the following initial data are accepted: r = 0,03m, D = 0,016m, z = 6, $R_{\text{ср}}$ = 0,1 m, $f_{\min}$ = 0,2 , $f_{\max}$ = 0,8, $T_{\text{ном}}$ =16 Nm и $F_{\text{п}}$ =366 N.

The graph in fig. 5, the curve is constructed using relation (18). Curve $\varphi_i(f_i)$ constructed using relation (18). Angle change φ_i directly proportional to the change f_i.

Thus, a kinematic scheme for the occurrence of overloads during the operation of the first generation SFC with a variable gain is developed.

The cause of overloads during operation of the SFC with a variable gain is the additional energy costs necessary to move the rolling bodies relative to the side walls of the sockets of the control device.

With an increase in the coefficient of friction f_i (0.2 to 0.8) overload value ΔT rises from 0 to the maximum value.

The maximum overload during operation of the SFC occurs at the maximum value of the coefficient of friction, if before the increase in the coefficient of friction, the coupling worked at the minimum value of the coefficient of friction.

The SFC accuracy factor increases from 1 to 1.105 as the coefficient of friction increases from minimum to maximum.

Acknowledgements
The study had no sponsorship.

References

[1] Zhivov L.I., Ovchinnikov A.G., Skladchikov E.H. Forging and stamping equipment: Textbook for universities / Ed. L.I. Zhivova. - M.: Publishing House of MSTU. N.E. Bauman, 2015 .-- 560 f .: ill.

[2] Zhivov L.I., Chumakov B.N., Drozdov N.G. Features of the dynamics of the hot stamping crank press for stamping low forgings. “University News. Mechanical Engineering ”, 2012, No. 1, p. 155-159

[3] Zalessky V.I. Equipment forging shops. Ed. 2nd, overwork, and add. Textbook for high schools. M., "Higher School", 2009. 632 p.

[4] Crank forging machines / Ed. Vlasova V.I. - M.: Mechanical Engineering, 2012 .-- 424 p.

[5] Noskov G.P., Rodov G.M., Vyatkin V.P. An experimental study of the loads in the crank press drive during a technological operation. Forging and stamping production, 2016, No. 5, p. 30 - 32

[6] Svistunov V.E. Forging and stamping equipment. Crank presses: Study Guide. - M.: MGIU, 2016 .-- 704 p.

[7] Svistunov V.E. The results of mathematical modeling of crank presses with compact actuators. Forging and stamping production, 2015, No. 10, 24 - 27

[8] Sokov V.I. Experimental determination of the friction moment when the clutch is engaged and braking of the hot-stamping crank presses. Forging and stamping production. Metal forming. 2015, No.10, 29-35

[9] Truskovsky V.I. The dynamics of forging machines. - Chelyabinsk: Publishing house of SUSU, 2015.-79c.

[10] Fedorkevich V.F. About the rigidity of modern crank hot stamping presses / Forging and stamping. Metal forming. 2013, No. 5, p. 23 - 25

[11] Hoopfer P. Dynamic loads in crank presses. “Forging and stamping production”, 2011, No. 2, p. 28 - 31

[12] Chubukov V.A., Gartvig A.A. Investigation of the influence of the design and parameters of hot-stamping crank presses on the nomenclature and accuracy of stamped products. Collection of scientific reports of the VI International Conference "The participation of young scientists, engineers and educators in the development and implementation of innovative technologies." - M.: MGIU, 2016, p. 134-138

[13]. Scheglov V.F., Maksimov L.Yu., Linz V.P. Forging machines. - 2nd ed., Revised. and add. - M.: Mechanical Engineering, 2010 .-- 304 e., Ill.

[14] Patent RU 2427466. A method for protecting crank presses from overloads by force on a slider. Svistunov V.E., Chubukov V.A., Matveev A.G. Publ. 08/27/2011
[15] Schumann K. Methode zur rechnerischen Untersuchung der technologischen Stobbeansprucheng mechanischer Pressen. Umformtechnik, 24 (2012), No. 2, p. 29 - 35
[16] Hiraishi Kenji, Kagawa Toshiaki. Sumitomo jukikaigiho. Sumitomo Heavy Ind. Techn. Rev. 2010, No. 164, p. 9
[17] Schnellaufer-Press produziert Platinen. Maschinenmarkt. 2009, No. 41, p. 36
[18] Stanzoder Umformautomat: Application 1867469 EPO, IPC B 30 V 15/04 (2006.01), B30 B 15/00 (2006.01). Haulick + Roos GmbH, Siegel Andreas (Hoeger, Stellrecht & Partner Patentanwälte Uhlandstrasse 14 with 70182 Stuttgart): No. 06012074.8; Claim 06/12/2008; Publ. 12/19/2009.
[19] Stanzautomat selbst erganzen. Blesh InForm. 2009, No. 4, p. 87

IOP Conf. Series: Materials Science and Engineering **680** (2019) 012013 doi:10.1088/1757-899X/680/1/012013

The study of the use of multi-disc safety friction clutches in the working bodies of crank presses

K O Kobzev, E S Bozhko, A V Mozgovoi, E I Kostromina, L G Babenko

Don State Technical University, Gagarin's square 1, Rostov-on-Don 344000, Russia

E-mail: 5976765@mail.ru

Abstract. Studies of safety couplings have shown that when the friction forces in splined joints of friction disks and half couplings are taken into account, the moment of friction of the clutch increases compared to the moment calculated without taking these forces into account. Among the safety clutches, friction clutches occupy a special place in use in crank presses due to the increased accuracy of limiting the transmitted load. This allows you to increase the efficiency of protection of parts and drive units of machines, as well as reduce their weight and dimensions.

Friction disks are pressed against each other with the force created by the spring, and the moment of friction arising between them ensures the connection of the leading parts with the followers. The friction force in the slots significantly reduces the transmitted force of pressing on subsequent disks, but its influence is not taken into account in existing studies.

1. Introduction

Studies of safety couplings have shown that when the friction forces in splined joints of friction disks and half couplings are taken into account, the moment of friction of the clutch increases compared to the moment calculated without taking these forces into account.

Among the safety clutches, SFC friction clutches occupy a special place due to the increased accuracy of limiting the transmitted load. This allows you to increase the efficiency of protection of parts and drive units of machines, as well as reduce their weight and dimensions.

Friction disks are pressed against each other with the force created by the spring, and the moment of friction arising between them ensures the connection of the leading parts with the followers. The friction force in the slots significantly reduces the transmitted force of pressing on subsequent disks, but its influence is not taken into account in existing studies [1-5].

Previous studies have shown that when the friction forces in splined joints of friction disks and half couplings are taken into account, the moment of friction of the clutch increases compared to the moment calculated without taking these forces into account.

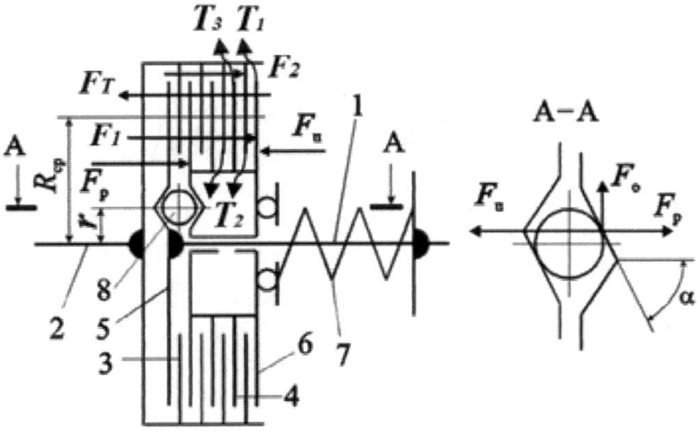

Figure 1. Multi – disk chart SFC

Consider the multi-disc chart SFC presented in Figure 1.

The coupling consists of coupling halves 1 and 2, a package of friction disks 3 and 4, a thrust disk 5, a pressure disk 6, a control device and a pressure spring 7. The disks 3 are connected to the drum of the coupling half 2, and the disks 4 to the hub of the pressure disk, which is mounted on the hub of the coupling half 1 with the possibility of rotation relative to it. The thrust disk 5 is rigidly mounted on the hub of the coupling half 1.

The control device consists of rolling elements 8, which are placed with a gap in the sockets (Fig. 1, section AA) of the thrust and pressure plates.

When SFC is operating, press forces 6 act on F_n – spring pull force 7, F_o – circumferential force and F_p – spacer (squeezing) force (Fig. 1, section AA). Force action F_p, as follows from fig. 1, reduces the force of pressing against each other by friction pairs 3-4. As the friction pairs move away from the support flange of the pressure plate, the action of the spacer force decreases due to the influence of friction forces F_t , F_{t1} .

2. Methods

Consider two cases:

1. Friction in the elements of the connection of the disks with the leading and driven parts is not taken into account.

The moment of friction forces SFC is determined by the formula:

$$T = zF_n R_{cp} \frac{f}{1+(z-1)Cf}. \quad (1)$$

2. Subject to the consideration of friction in splined joints.

Let us consider successively the transfer of the friction moment from the leading disks to the driven disks, shown in Fig. 1.

The moment of friction forces of the friction group of the clutch is equal to:

$$T = zR_{cp} f(F_n - F_p), \quad (2)$$

3. Results

Spacer force is determined by the formula:

$$F_p = \frac{T_1}{r} \text{tg}\alpha, \quad (3)$$

T_1 – torque transmitted by friction pairs that are associated with the pressure plate 6:

IOP Conf. Series: Materials Science and Engineering **680** (2019) 012013 doi:10.1088/1757-899X/680/1/012013

$$T_1 = \frac{z-1}{z} T, \tag{4}$$

Substituting expressions (1) and (2) into equation (3), we obtain:

$$F_p = \frac{(z-1)F_n Cf}{1+(z-1)Cf}, \tag{5}$$

Torque T_1 can be imagined as

$$T_1 = F_1 R_{cp} f,$$

F_1 – normal force acting on the first friction working surface from the support flange of the pressure plate:

$$F_1 = F_n - F_{t1},$$

F_{t1} – friction force in spline joints of the pressure plate:

$$F_{t1} = Ff_1 =_1 \frac{T_1}{R_H} f_1,$$

f_1 – coefficient of friction between the working surfaces of the spline connection; R_H – outer radius of friction disk 4; F – circumferential force transmitted by the friction disk:

$$F = (F_n - F_p) f. \tag{6}$$

Given relation (6), we obtain [6-9]:

$$F_1 = F - f_1 \frac{T_1}{R_H} = F - f_1 f \frac{R_{cp}}{R_H} F_1,$$

$$F_1 = \frac{F}{1 + f_1 f \frac{R_{cp}}{R_H}}. \tag{7}$$

The moment of friction, perceived by the second disk (the first driven disk)

$$T_2 = f_1 R_{cp} (F_1 + F_2).$$

F_2, acting on the second pair is determined by the formula

$$F_2 = F_1 - F_{т2} = F_1 - f \frac{T_2}{R_B} = F_1 - f_1 f (F_1 + F_2) \frac{R_{cp}}{R_B},$$

$$F_2 = F_1 \cdot \frac{1 - f_1 f \frac{R_{cp}}{R_B}}{1 + f_1 f \frac{R_{cp}}{R_B}} = F \frac{1 - f_1 f \frac{R_{cp}}{R_B}}{\left(1 + f_1 f \frac{R_{cp}}{R_H}\right)\left(1 + f_1 f \frac{R_{cp}}{R_B}\right)} \tag{8}$$

R_B – internal radius of the friction disk.

The moment of friction transmitted by the third disk: $T_3 = (F_2 + F_3) f R_{cp}$.

F_3, acting on the third pair is determined by the formula [10-14]

$$F_3 = F_2 - F_{t3} = F_2 - f_1 \frac{T_3}{R_H} = F_2 - f_1 f (F_2 + F_3) \frac{R_{cp}}{R_H},$$

$$F_3 = F\frac{\left(1-f_1 f\frac{R_{cp}}{R_B}\right)\left(1-f_1 f\frac{R_{cp}}{R_H}\right)}{\left(1+f_1 f\frac{R_{cp}}{R_B}\right)\left(1+f_1 f\frac{R_{cp}}{R_H}\right)^2}. \tag{9}$$

Using expressions (8) and (9), we write the spacer force F_{z_1} for an arbitrary odd pair z_1:

$$F_{z_1} = F\frac{\left[1-f_1 f\frac{R_{cp}}{R_B}\right]^{\frac{z_1-1}{2}}\left[1-f_1 f\frac{R_{cp}}{R_H}\right]^{\frac{z_1-1}{2}}}{\left[1+f_1 f\frac{R_{cp}}{R_B}\right]^{\frac{z_1-1}{2}}\left[1+f_1 f\frac{R_{cp}}{R_H}\right]^{\frac{z_1+1}{2}}}. \tag{10}$$

Accordingly, for an arbitrary even pair z_2 spacer force F_{z_2} defined as

$$F_{z_2} = F\frac{\left[1-f_1 f\frac{R_{cp}}{R_B}\right]^{\frac{z_2}{2}}\left[1-f_1 f\frac{R_{cp}}{R_H}\right]^{\frac{z_2}{2}-1}}{\left[1+f_1 f\frac{R_{cp}}{R_B}\right]^{\frac{z_2}{2}}\left[1+f_1 f\frac{R_{cp}}{R_H}\right]^{\frac{z_2}{2}}}. \tag{11}$$

The total moment of friction forces SFC is

$$T_f = \sum_1^z T_z = \sum_1^z R_{cp} f (F_n - F_z), \tag{12}$$

T_z – moment of friction forces of one friction pair.

We write expression (12) in the following form:

$$T_f = R_{cp} f[(F_n - F_1) + (F_n - F_2) + (F_n - F_3) + \ldots + (F_n - F_z)],$$

$$T_f = R_{cp} f\left[zF_n - (F_1 + F_2 + F_3 + \ldots + F_z)\right] = R_{cp} f\left[zF_n - \left(\sum_1^{z_1} F_{z_1} + \sum_2^{z_2} F_{z_2}\right)\right]. \tag{13}$$

In view of relations (11) and (12), expression (13) can be written as

$$T_f = R_{cp} f\left[zF_n - F\left(\sum_1^{z_1}\frac{\left[1-f_1 f\frac{R_{cp}}{R_B}\right]^{\frac{z_1-1}{2}}\left[1-f_1 f\frac{R_{cp}}{R_H}\right]^{\frac{z_1-1}{2}}}{\left[1+f_1 f\frac{R_{cp}}{R_B}\right]^{\frac{z_1-1}{2}}\left[1+f_1 f\frac{R_{cp}}{R_H}\right]^{\frac{z_1+1}{2}}} + \sum_2^{z_2}\frac{\left[1-f_1 f\frac{R_{cp}}{R_H}\right]^{\frac{z_2}{2}}\left[1-f_1 f\frac{R_{cp}}{R_H}\right]^{\frac{z_2}{2}-1}}{\left[1+f_1 f\frac{R_{cp}}{R_B}\right]^{\frac{z_2}{2}}\left[1+f_1 f\frac{R_{cp}}{R_H}\right]^{\frac{z_2}{2}}}\right)\right]. \tag{14}$$

We introduce the following notation:

$$A = \frac{1 - ff_1 \dfrac{R_{cp}}{R_B}}{1 + ff_1 \dfrac{R_{cp}}{R_B}} \; ; B = \frac{1 - ff_1 \dfrac{R_{cp}}{R_H}}{1 + ff_1 \dfrac{R_{cp}}{R_H}} .$$

Then expression (12), taking into account formula (11), can be written as

$$T_f = R_{cp} f \left[zF_n - \frac{f(F_n - F_p)}{1 + ff_1 \dfrac{R_{cp}}{R_H}} \left(1 + A + AB + A^2B + \ldots + A^{Z_2/2} B^{\frac{Z_2}{2}-1} \right) \right] =$$

$$= F_n R_{cp} f \left[z - \frac{f}{\left(1 + (z-1)Cf\right)\left(1 + ff_1 \dfrac{R_{cp}}{R_H}\right)} \times \frac{(1+A)\left[1 - (AB)^{z_2/2}\right]}{1 - AB} \right]. \quad (15)$$

$K_т$ calculated by the formula:

$$K_t = \frac{T_{\text{fmax}}}{T_{\text{fmin}}},$$

T_{fmin}, T_{fmax} – respectively, the minimum and maximum SFC torques, taking into account friction in splined joints, or, taking into account expression (15) [15-19]:

$$K_t = \frac{f_{\max}}{f_{\min}} \times \frac{\left[z - \dfrac{f_{\max}}{\left(1 + (z-1)Cf_{\max}\right)\left(1 + f_1 f_{\max} \dfrac{R_{cp}}{R_H}\right)} \times \dfrac{(1+A_{\max})\left[1 - (A_{\max}B_{\max})^{z_2/2}\right]}{1 - A_{\max}B_{\max}} \right]}{\left[z - \dfrac{f_{\min}}{\left(1 + (z-1)Cf_{\min}\right)\left(1 + f_1 f_{\min} \dfrac{R_{cp}}{R_H}\right)} \times \dfrac{(1+A_{\min})\left[1 - (A_{\min}B_{\min})^{z_2/2}\right]}{1 - A_{\min}B_{\min}} \right]} \quad (16)$$

To plot functions $T_f(z)$, $K_t(z)$ taking into account friction in spline joints, the following initial data were adopted: F_n=1000 N; $f_{\min}$=0,1; $f_{\max}$=0,8; f_1=0,1; f=0,1; C=1,25; R_{cp}=100 mm; $z=10$.

The geometric relationships of friction discs are found based on the following reasoning. The outer, middle and inner radii of the friction disk are interconnected by the ratio:

$$\nu = \frac{2(R_H - R_B)}{R_{cp}}, \quad (17)$$

ν – disk width ratio.

The second ratio, connecting the outer, middle and inner radii of the friction disk used in the calculations, has the form:

$$R_{cp} = \frac{R_H + R_B}{2}. \quad (18)$$

Solving the system of equations (1) and (4), we find:

$$R_H = \frac{R_{cp}(4+\nu)}{4}, \qquad (19)$$

$$R_B = \frac{R_{cp}(4-\nu)}{4}. \qquad (20)$$

Usually accepted ν =0,20, then, taking into account relations (18) and (19), we find:

R_H =0,106 м; R_B =0,094 м.

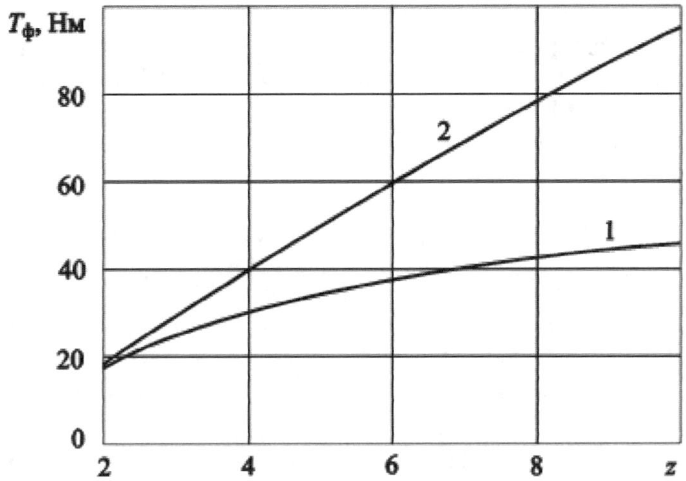

Figure 2. Graphs of the dependence of the torque SFC on the number of friction pairs

4. Discussion

Graphs are constructed using relations (3), (16) and (18). The accuracy factor SFC without taking into account friction in splined joints was calculated by the formula: $K_t = \frac{f_{max}[1+(z-1)Cf_{min}]}{f_{min}[1+(z-1)Cf_{max}]}$.

Dependency graphs $T_f(z)$ are shown in fig. 2. Curve 1 reflects the dependence of the SFC torque excluding friction in splined joints, curve 2 taking into account friction..

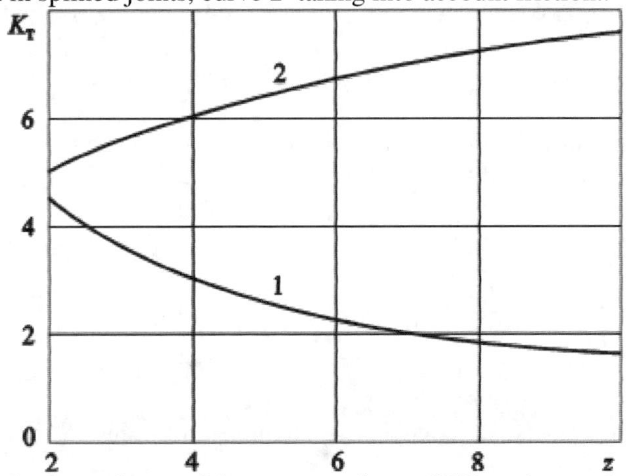

Figure 3. Graphs of the dependence of the accuracy factor SFC on the number of friction pairs

In fig. 3 curve 1 shows the dependence $K_t(z)$ excluding friction in the splines, curve 2 - taking into account friction.

Analysis of the graphs shows the following:

- SFC torque with regard to friction in splined joints at = 10 is approximately 2 times higher than without friction (Fig. 2);

Accuracies the accuracy coefficient SFC, taking into account friction in splined joints, increases with the number of friction pairs of the friction group, asymptotically approaching the value of the accuracy coefficient of the safety friction clutch without the control device (Fig. 3, curve 2);

- with a minimum number of friction pairs (z =2) indicators of load capacity and SFC response accuracy taking into account friction in spline joints and without taking into account friction are close to each other, which is explained by the minimal influence of friction forces in spline joints.

Thus, the study found that taking into account the frictional forces in the spline joints of friction disks with half couplings in multi-disc SFCs allows one to obtain refined dependences of the torque and accuracy coefficient on the number of friction pairs.

With a large number of friction pairs, the torque of multi-disk SFCs, taking into account friction in spline joints, significantly (two or more times) exceeds the torque without taking into account friction in spline joints.

Accounting for friction in splined joints significantly (up to 4 times) increases the accuracy coefficient of multi-disk SFC with a large number of friction pairs.

Acknowledgements
The study had no sponsorship.

References

[1] Zhivov L.I., Ovchinnikov A.G., Skladchikov E.H. Forging and stamping equipment: Textbook for universities / Ed. L.I. Zhivova. - M.: Publishing House of MSTU. N.E. Bauman, 2015 .-- 560 f .: ill.
[2] Zhivov L.I., Chumakov B.N., Drozdov N.G. Features of the dynamics of the hot stamping crank press for stamping low forgings. “University News. Mechanical Engineering ”, 2012, No. 1, p. 155-159
[3] Zalessky V.I. Equipment forging shops. Ed. 2nd, overwork, and add. Textbook for high schools. M., "Higher School", 2009. 632 p.
[4] Crank forging machines / Ed. Vlasova V.I. - M.: Mechanical Engineering, 2012 .-- 424 p.
[5] Noskov G.P., Rodov G.M., Vyatkin V.P. An experimental study of the loads in the crank press drive during a technological operation. Forging and stamping production, 2016, No. 5, p. 30 - 32
[6] Svistunov V.E. Forging and stamping equipment. Crank presses: Study Guide. - M.: MGIU, 2016 .-- 704 p.
[7] Svistunov V.E. The results of mathematical modeling of crank presses with compact actuators. Forging and stamping production, 2015, No. 10, 24 - 27
[8] Sokov V.I. Experimental determination of the friction moment when the clutch is engaged and braking of the hot-stamping crank presses. Forging and stamping production. Metal forming. 2015, No.10, 29-35
[9] Truskovsky V.I. The dynamics of forging machines. - Chelyabinsk: Publishing house of SUSU, 2015.-79c.
[10] Fedorkevich V.F. About the rigidity of modern crank hot stamping presses / Forging and stamping. Metal forming. 2013, No. 5, p. 23 - 25
[11] Hoopfer P. Dynamic loads in crank presses. “Forging and stamping production”, 2011, No. 2, p. 28 - 31
[12] Chubukov V.A., Gartvig A.A. Investigation of the influence of the design and parameters of hot-stamping crank presses on the nomenclature and accuracy of stamped products. Collection of scientific reports of the VI International Conference "The participation of young scientists, engineers and educators in the development and implementation of innovative technologies." - M.: MGIU, 2016, p. 134-138

[13]. Scheglov V.F., Maksimov L.Yu., Linz V.P. Forging machines. - 2nd ed., Revised. and add. - M.: Mechanical Engineering, 2010 .-- 304 e., Ill.

[14] Patent RU 2427466. A method for protecting crank presses from overloads by force on a slider. Svistunov V.E., Chubukov V.A., Matveev A.G. Publ. 08/27/2011

[15] Schumann K. Methode zur rechnerischen Untersuchung der technologischen Stobbeansprucheng mechanischer Pressen. Umformtechnik, 24 (2012), No. 2, p. 29 - 35

[16] Hiraishi Kenji, Kagawa Toshiaki. Sumitomo jukikaigiho. Sumitomo Heavy Ind. Techn. Rev. 2010, No. 164, p. 9

[17] Schnellaufer-Press produziert Platinen. Maschinenmarkt. 2009, No. 41, p. 36

[18] Stanzoder Umformautomat: Application 1867469 EPO, IPC B 30 V 15/04 (2006.01), B30 B 15/00 (2006.01). Haulick + Roos GmbH, Siegel Andreas (Hoeger, Stellrecht & Partner Patentanwälte Uhlandstrasse 14 with 70182 Stuttgart): No. 06012074.8; Claim 06/12/2008; Publ. 12/19/2009.

[19] Stanzautomat selbst erganzen. Blesh InForm. 2009, No. 4, p. 87

Theoretical foundations of the use of single-circuit negative feedback in safety friction clutches with differentiated friction pairs installed in forging equipment

K O Kobzev, E S Bozhko, A V Mozgovoi, M D Molev, N I Stuzhenko

Don State Technical University, Gagarin's square 1, Rostov-on-Don 344000, Russia

E-mail: 5976765@mail.ru

Abstract. From studies aimed at summarizing the results of private surveys, conditions of high accuracy SFC with single-loop negative feedback are obtained. These conditions reflect the dependences of the magnitude of the coefficient of force and the spacer force generated by the feedback control device on the current coefficient of the gain value of the coefficient of friction, which theoretically ensure the limiting value of the torque transmitted by the clutch, which leads to failure-free operation of the crank presses.

1. Introduction

The safety friction clutch (SFC) installed in crank presses with single-circuit negative feedback is classified by the first generation couplings and is most widely used in practice among other types of couplings. The reason for this, despite the relatively low gain enhance the accuracy of the response, is the structural simplicity, extensive research on this type of couplings, proven tuning techniques and extensive operational experience, which allowed us to obtain the results necessary to optimize the design parameters of the couplings.

It should be noted that in relation to the SFC of this type, only private studies have been carried out relating to certain aspects of their functioning. So, the influence on the response accuracy of the so-called gain of the feedback control device, the influence of the gain value on the load capacity of the couplings is studied quite thoroughly, questions of tuning the SFC to improve the reliability of load transfer in the kinematic circuit of the drive, as well as stability issues of the movement of the drive with SFC .

From studies aimed at summarizing the results of private surveys, conditions of high accuracy SFC with single-loop negative feedback are obtained. These conditions reflect the dependences of the magnitude of the coefficient of force and the spacer force generated by the feedback control device on the current coefficient of the gain value of the coefficient of friction, which theoretically ensure the limiting value of the torque transmitted by the clutch, which leads to failure-free operation of the crank presses.

2. Methods

A schematic diagram of a first-generation SFC with single-circuit negative feedback having differentiated friction pairs of the type “leading pairs - driven pairs” is shown in Fig. 1 (the upper part of the figure relative to the axis of rotation of the coupling).

Two kinematic half-couplings 1 and 2 coaxial with each other are connected to each other in the circumferential direction of the friction group consisting of friction disks 3 and 4. The disks 3 are connected in the circumferential direction with the hub of the pressure disk 5, the disks 4 are connected to the half-coupling drum 2.

The pressure disk 5 is devoid of kinematic connection with the hub of the coupling half 1 in the circumferential direction, with the exception of slight friction between them, which is not considered further [1-3].

The control device is made in the form of rolling bodies 6, which are placed in beveled sockets made on the end surfaces of the pressure disk facing one another and rigidly fixed on the hub of the coupling half 1 of the thrust disk 7 (Fig. 1, section A-A).

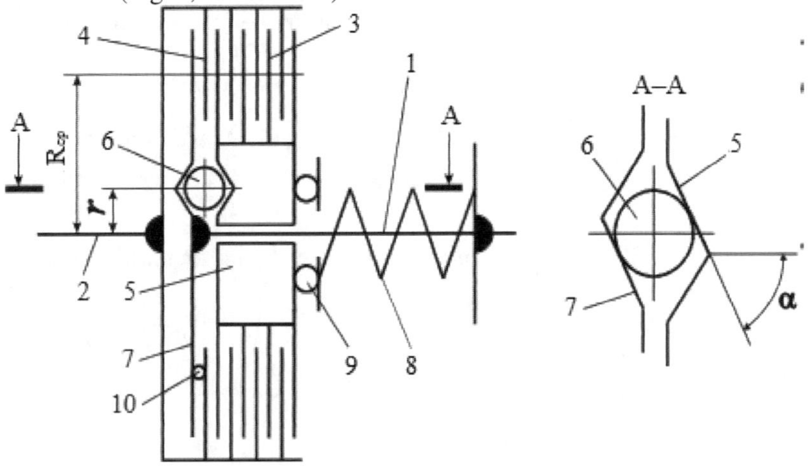

Figure 1. Schematic diagram of the first generation PFM with differentiated friction pairs.

The force closure of friction pairs is created by a spring 8, the force of which is transmitted to the pressure disk through a thrust bearing 9 to reduce friction between them[4-6].

Given the action of the spacer force in the SFC (see Fig. 1, section AA), we write the formula for the torque in the form:

$$T_p = zR\, f(F_p - F), \tag{1}$$

F – spacer force; other designations are looked above.

When the structural-layout scheme of the friction group, built according to the type of "leading friction pairs driven friction pairs", the spacer force is determined by the formula:

$$F = \frac{(z - z_1)T_p}{zr}\,\mathrm{tg}\alpha, \tag{2}$$

z_1 – the number of leading friction pairs of the friction group.

Putting the expression (2) in the formula (1), we obtain:

$$T_p = zF_p R\,\frac{f}{1 + (z - z_1)Cf}, \tag{3}$$

The greatest accuracy of the SFC operation will be when $T_p = \mathrm{const}$. In this regard, the question arises of choosing the required magnitude of the torque of the coupling.

The nominal value of the torque at which the SFC is adjusted is usually taken as the required value. SFC tuning to the rated torque is carried out according to the minimum coefficient of friction, which allows minimizing the clutch response when the coefficient of friction is reduced and increasing the reliability of load transfer without interrupting the course of the process performed by the machine.

3. Results

Based on the foregoing, we write the formula for the nominal torque of the SFC, using relation (3):

$$T_{\text{nmin}} = zF_p R \frac{f_{\min}}{1+(z-z_1)Cf_{\min}}, \tag{4}$$

$f_{\min}$ – minimum coefficient of friction.

Equating to each other the right-hand sides of relations (3) and (4), we find:

$$C_i = \frac{1}{z-z_1}\left(\frac{1}{f_{\min}} - \frac{1}{f_i}\right), \tag{5}$$

C_i – current gain gain value; f_i – current coefficient value of the coefficient of friction.

Expression (5) shows that for maximum accuracy of the SFC operation, the gain should be variable, and it functionally depends on the friction coefficient [7-10].

In addition, from the expression (5) it follows that for $f_i = f_{\min}$ should be $C_{\min} = 0$.

Finally, expression (5) shows that the value of the gain should increase with increasing f_i.

The graph of function (5) is shown in Fig. 2 Curve 1 is constructed from the following source data: z=6, z_1=1, $f_{\min}$=0,1.

Curve 1 shows that, with the indicated initial data, the absolute increase in the gain is 1.75 (for $C_{\min} = 0$).

Parameter Influence z_1 the magnitude of the gain is shown by curve 2, constructed at z_1=2: the absolute increase in gain in this case is 2.19.

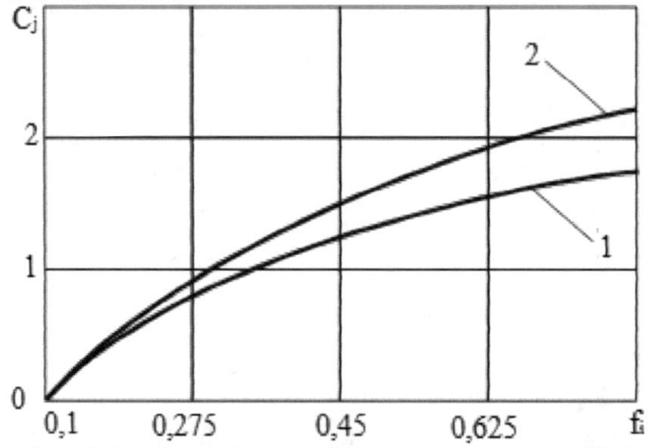

Figure 2. Graphs of the dependence of the gain on the coefficient of friction

Therefore, with an increase in the number of driven friction pairs, higher values of the gain are required to ensure the "ideal" SFC load characteristic. Physically, this is due to the fact that when $T_{\text{p.min}} = \text{const}$ a decrease in the number of driven friction pairs leads to a decrease in the torque transmitted by the pressure plate and rolling elements of the control device. To compensate for the decrease in spacer force, an increase in gain is required.

Second condition $C_{\min} = 0$ theoretically impossible, since the gain, according to the physical meaning of the gain (see above), must be equal to zero at least one of the two input parameters, R или α.

Value R cannot be equal to zero, since the gain is lost any meaning of the SFC functioning [11-15].

To achieve the condition $F_p = 0$ theoretically parameter α may be zero. However, from a practical point of view, this is not possible, since the gain at $f_i > f_{\min}$ SFC should work with negative feedback, i.e., with $F_p \neq 0$. Therefore, from the physico-mechanical point of view, in this case, the transition of the rolling elements of the control device from a zero pressure angle to a non-zero angle is necessary. At $\alpha_{min} = 0$ this seems impossible.

In the SFC of the first generation with differentiated friction pairs, there is a limit from above on the magnitude of the gain, which in this case has the form:

$$C_p \leq \frac{1}{z_1 f_{max}}, \tag{6}$$

f_{max} – maximum coefficient of friction.

Replacing the parameter in relation (5) f_i per parameter f_{max} and using the right-hand sides of relations (5) and (6) in the equality $C_i = C_p$, we get:

$$z_1 = \frac{z}{m}, \tag{7}$$

m – relative width of the range of the coefficient of friction $f_{min} \ldots f_{max}$:

$$m = f_{max} / f_{min}.$$

If accept z_1 =1, then the condition for equal values of gain C_i , C_p performed when

$$z = m. \tag{8}$$

Relation (7) shows that to ensure a minimum integer value z_1 =1 it is necessary that the total number of SFC friction pairs be at least equal to the coefficient m. Usually $m = 8\ldots10$, which corresponds to a multi-disc version of the clutch.

The same conclusion can be drawn from the analysis of relation (8).

Therefore, the implementation of the limitation of the magnitude of the gain at the maximum value of the coefficient of friction is possible only in the multi-disk version of the SFC with differentiated pairs of friction. The number of leading friction pairs should be minimal.

We are also investigating a variant of the structural-layout scheme of the friction group in which friction pairs are not differentiated into leading and driven, but all are driven. The indicated option is shown in fig. 1 (lower part relative to the axis of rotation of the coupling) [16-19].

The difference between the considered SFC variant is the introduction of a thrust bearing 10, installed between the leftmost (in Fig. 1) friction disk 4 and the thrust disk 7. The kinematic connection between them is broken, and the entire load of the coupling is transferred from the coupling half to the coupling half 2 (or vice versa) by means of rolling 6.

Applying the same method as for the SFC variant with differentiated friction pairs, the calculation gain method, we find the expression for calculating the coupling torque:

$$T_p = zF_pR\frac{f}{1+zCf}. \tag{9}$$

Relation (9) differs from relation (3), with the same parameters, by a lower value of the reduced coefficient of friction equal to:

$$f_p = \frac{f}{1+zCf},$$

and, accordingly, lower torque T_p.

This is explained by the greater magnitude of the spacer force acting on the friction pair, due to the fact that in this case the rolling elements of the control device transmit the full torque of the SFC, and not a certain part of it, as in the SFC with differentiated pairs of friction.

Using the mathematical calculations similar to the above, we find the dependence of the magnitude of the gain on the coefficient of friction:

$$C_i = \frac{1}{z}\left(\frac{1}{f_{min}} - \frac{1}{f_i}\right). \tag{10}$$

IOP Conf. Series: Materials Science and Engineering **680** (2019) 012014 doi:10.1088/1757-899X/680/1/012014

The relation (10) is similar to the relation (5): it gives the same general dependencies between the function and variable arguments z , f_i: The only difference is the smaller the current gain, the gain value at equal current gains, the values of the friction coefficient.

Since the gain, as noted above, at $f_{\min}$ should be $C_{\min} = 0$, introduce a new parameter $C'_{\min} \neq 0$. Using this parameter in the initial relation (4) and having performed the mathematical action described above, we find:

for the first version of the SFC layout scheme:

$$C'_i = \frac{1}{z - z_1}\left(\frac{1}{f_{\min}} - \frac{1}{f_i}\right) + C'_{\min}, \tag{11}$$

for the second version of the SFC layout scheme:

$$C''_i = \frac{1}{z - z_1}\left(\frac{1}{f_{\min}} - \frac{1}{f_i}\right) + C''_{\min}. \tag{12}$$

Formulas (11) and (12) differ from similar formulas (5) and (10) by the presence of terms $C'_{\min}$ $C''_{\min}$ respectively. In this case, when $f_i = f_{\min}$ will be $C_{\min} = C'_{\min} (C''_{\min})$. This means that with a minimum value of the coefficient of friction in the SFC, negative feedback will act.

The minimum and, at the same time, tuning (nominal) torque in this case should be calculated using relation (4) with the replacement of the parameter in it C per parameter $C'_{\min}$ (for the first version of the SFC).

For the second variant of SFC, formula (4) can also be used with the above replacement C and without taking into account the parameter z_1.

Consider the issue of parameter values $C'_{\min}$ $C''_{\min}$. We will solve this problem on the basis of the same nominal load capacity of the compared SFC options.

Using relations (3) and (9) with the replacement of parameters in each of them f ,C respectively on the parameters $f_{\min}$, $C'_{\min} (C''_{\min})$, which corresponds to the nominal torque of the SFC, we find:

$$C'_{\min} = \frac{1}{z - z_1}\left(\frac{zF_pR}{T_p} - \frac{1}{f_{\min}}\right), \tag{13}$$

$$C''_{\min} = \frac{1}{z}\left(\frac{zF_pR}{T_p} - \frac{1}{f_{\min}}\right), \tag{14}$$

An analysis of relations (13) and (14) shows that $C'_{\min} > C''_{\min}$. Therefore, given this result, we can conclude that, taking into account relations (11) and (12), $C'_i > C''_i$, moreover, the difference between these values increases in comparison with the values calculated by the relations (5) and (10).

The difference in the values of the gain for the first and second versions of the SFC, however, affects their nominal load capacity. Indeed, substituting relations (13) and (14) into expressions (3) and (9), respectively, leads to different results – $T'_{\text{п}\min} > T''_{\text{п}\min}$ (respectively, the nominal torques of the first and second variants of the SFC).

Moments $T'_{\text{п}\min}$, $T''_{\text{п}\min}$ obtained from expressions (3) and (9) by replacing the parameter in them f per parameter $f_{\min}$.

4. Discussion

For the first SFC variant, it is important to check the maximum value of the gain coefficient for its restriction due to the limitation of the gain above, for SFCs having differentiated friction pairs to reduce overload in crank presses. For SFC with all driven friction pairs of such a restriction, the gain in the stationary loading mode does not exist.

Using relation (6) for $z_1 = 1$ and expression (11), taking into account formula (13), in the equality $C'_{max} = C_п$, we find:

$$C'_{min} = \frac{1 + 1/m - 1/(z-1)(1-1/m)}{1 + (z-1) f_{min}}. \quad (15)$$

When deriving relation (15), expression (4) was used with a replacement in the last parameter C per parameter C'_{min}.

Relation (15) allows us to abstract from such SFC parameters as T_p, F_p, R, use when calculating C'_{min} minimum amount of source data.

As follows from relation (15), the value C'_{min} inversely proportional to the coefficient of friction f_{min}, if we consider the value of the coefficient of friction f_{max} variable, and $m = \text{const}$. Therefore, to minimize C'_{min} friction materials with friction coefficients as high as possible should be assigned f_{min}.

From relation (15) it also follows that the value C'_{min} decreases with increasing coefficient m. In this regard, it is advisable from the point of view of minimization C'_{min}, use friction materials of friction pairs with the greatest dispersion of the coefficient of friction.

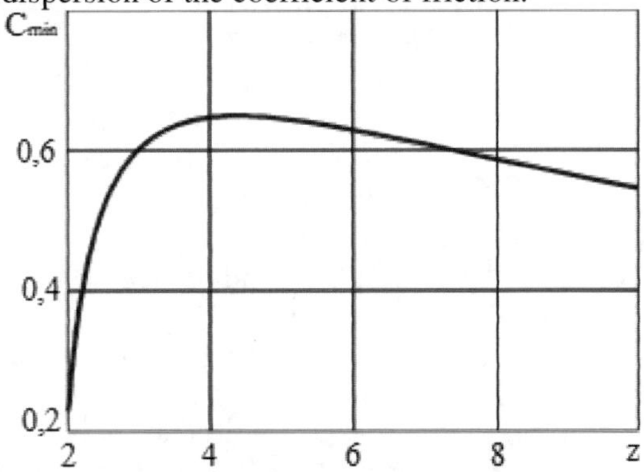

Figure 3. Graphs of the dependence of the minimum value of the gain on the number of friction pairs

Nature of function $C'_{min}(z)$ can be traced by the graph gain, constructed in accordance with relation (15). The graph is built at f_{min} =0,1 и m =8, which corresponds to real data.

The graph (Fig. 3) shows that C'_{min} with increasing increases in the range of 2 ... 4, then, having reached a maximum, decreases. With this minimum value z corresponds to C'_{min}, much smaller than with $z > 4$.

Therefore, with the minimum value of the parameter z better use is achieved from the point of view of minimizing the spacer force with a minimum value of the coefficient of friction and, thus, increasing the nominal load capacity of SFC.

Acknowledgements
The study had no sponsorship.

IOP Conf. Series: Materials Science and Engineering **680** (2019) 012014 doi:10.1088/1757-899X/680/1/012014

References

[1] Zhivov L.I., Ovchinnikov A.G., Skladchikov E.H. Forging and stamping equipment: Textbook for universities / Ed. L.I. Zhivova. - M.: Publishing House of MSTU. N.E. Bauman, 2015 .-- 560 f .: ill.

[2] Zhivov L.I., Chumakov B.N., Drozdov N.G. Features of the dynamics of the hot stamping crank press for stamping low forgings. "University News. Mechanical Engineering ", 2012, No. 1, p. 155-159

[3] Zalessky V.I. Equipment forging shops. Ed. 2nd, overwork, and add. Textbook for high schools. M., "Higher School", 2009. 632 p.

[4] Crank forging machines / Ed. Vlasova V.I. - M.: Mechanical Engineering, 2012 .-- 424 p.

[5] Noskov G.P., Rodov G.M., Vyatkin V.P. An experimental study of the loads in the crank press drive during a technological operation. Forging and stamping production, 2016, No. 5, p. 30 - 32

[6] Svistunov V.E. Forging and stamping equipment. Crank presses: Study Guide. - M.: MGIU, 2016 .-- 704 p.

[7] Svistunov V.E. The results of mathematical modeling of crank presses with compact actuators. Forging and stamping production, 2015, No. 10, 24 - 27

[8] Sokov V.I. Experimental determination of the friction moment when the clutch is engaged and braking of the hot-stamping crank presses. Forging and stamping production. Metal forming. 2015, No.10, 29-35

[9] Truskovsky V.I. The dynamics of forging machines. - Chelyabinsk: Publishing house of SUSU, 2015.-79c.

[10] Fedorkevich V.F. About the rigidity of modern crank hot stamping presses / Forging and stamping. Metal forming. 2013, No. 5, p. 23 - 25

[11] Hoopfer P. Dynamic loads in crank presses. "Forging and stamping production", 2011, No. 2, p. 28 - 31

[12] Chubukov V.A., Gartvig A.A. Investigation of the influence of the design and parameters of hot-stamping crank presses on the nomenclature and accuracy of stamped products. Collection of scientific reports of the VI International Conference "The participation of young scientists, engineers and educators in the development and implementation of innovative technologies." - M.: MGIU, 2016, p. 134-138

[13]. Scheglov V.F., Maksimov L.Yu., Linz V.P. Forging machines. - 2nd ed., Revised. and add. - M.: Mechanical Engineering, 2010 .-- 304 e., Ill.

[14] Patent RU 2427466. A method for protecting crank presses from overloads by force on a slider. Svistunov V.E., Chubukov V.A., Matveev A.G. Publ. 08/27/2011

[15] Schumann K. Methode zur rechnerischen Untersuchung der technologischen Stobbeansprucheng mechanischer Pressen. Umformtechnik, 24 (2012), No. 2, p. 29 - 35

[16] Hiraishi Kenji, Kagawa Toshiaki. Sumitomo jukikaigiho. Sumitomo Heavy Ind. Techn. Rev. 2010, No. 164, p. 9

[17] Schnellaufer-Press produziert Platinen. Maschinenmarkt. 2009, No. 41, p. 36

[18] Stanzoder Umformautomat: Application 1867469 EPO, IPC B 30 V 15/04 (2006.01), B30 B 15/00 (2006.01). Haulick + Roos GmbH, Siegel Andreas (Hoeger, Stellrecht & Partner Patentanwälte Uhlandstrasse 14 with 70182 Stuttgart): No. 06012074.8; Claim 06/12/2008; Publ. 12/19/2009.

[19] Stanzautomat selbst erganzen. Blesh InForm. 2009, No. 4, p. 87

Aspects of using single-circuit negative feedback in a safety friction clutch with all leading friction pairs, taking into account installation in crank presses

K O Kobzev, E S Bozhko, A V Mozgovoi, Y E Chertov, I A Zanina

Don State Technical University, Gagarin's square 1, Rostov-on-Don 344000, Russia

E-mail: 5976765@mail.ru

Abstract. A feature of the considered option is that, as an analog, a safety friction clutch (SFC) of the usual accuracy of operation in crank presses, in which there is no feedback control device, is adopted. Moreover, this case is applied only with a minimum value of the coefficient of friction, i.e., a constructive version of such a SFC should provide for the feedback control device to be turned off from operation under the indicated condition. Taking into account the structure of the ratio for calculating the magnitude of the gain, as well as the fact that for the "ideal" SFC, this value should be variable, it should be noted, taking into account the previous experience in calculating and constructing SFC of this type, that it is most convenient to vary the magnitude of the gain by changes in the angle (pressure angle of the rolling element control device).

1. Introduction

A feature of the considered option is that, as an analogue, a safety friction clutch (SFC) of the usual accuracy of operation in crank presses, in which there is no feedback control device, and therefore, $C = 0$. Moreover, this case is applied only with a minimum value of the coefficient of friction, i.e., a constructive version of such a SFC should provide for the feedback control device to be turned off from operation under the indicated condition. Taking into account the structure of the ratio for calculating the magnitude of the gain, as well as the fact that for the "ideal" SFC, this value should be variable, it should be noted, taking into account the previous experience in calculating and constructing SFC of this type, that it is most convenient to vary the magnitude of the gain by changes in the angle (pressure angle of the rolling element control device) [1-4]. Obviously, in this case, when $f_{\min}$, $C_{\min} = 0$, $\alpha_{\min} = 0$ ($\alpha_{\min}$ - minimum pressure angle).

2. Methods

It is technically impossible to fulfill this condition, since there will be no spacer force at zero angle of inclination of the side walls of the nests F_{p}, and at any additional external load, the rolling elements will not transition to the profile of the side wall of the socket with a non-zero angle α. Consequently, the principle of automatic SFC regulation will be lost.

IOP Conf. Series: Materials Science and Engineering **680** (2019) 012015 doi:10.1088/1757-899X/680/1/012015

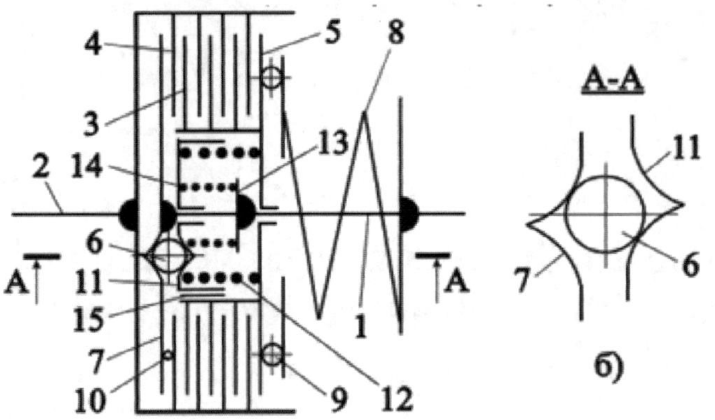

Figure 1. Schematic diagram of the AFM (first modified version)

The indicated principle can be implemented, for example, as shown in Fig. 1. Here is a schematic diagram of a SFC with a control device that conditionally implements a zero angle α , $f_{\min}$. This is achieved due to the fact that the connection between the rolling bodies 6 and the pressure plate 5 in the axial direction is not directly carried out, as in the basic version of the SFC (Fig. 1), but by means of the support sleeve 11, which is installed in the central hole of the pressure plate.

The support sleeve is spring-loaded with respect to the pressure disk 5 by the main spring 12, and with respect to the stop 13 rigidly fixed on the coupling half 1 - with the help of an additional spring 14 (Fig. 1). To transmit torque between the pressure disk 5 and the supporting sleeve 11, a key 15 is installed. The seats for the rolling elements 6 made in the thrust disk 7 and in the supporting sleeve 11 are profiled by curved lines convex in the direction of the rolling bodies to reduce the angle α from the bottom of the nest to its periphery [5-8].

According to the condition of functioning of the control device and the task, the tension force of the spring 12 should be zero at a minimum value of the coefficient of friction. This is due to the lack of a spacer force acting on the pressure plate 5 under the specified condition. The tension force of the spring 14 is not equal to zero, since the gain is not a zero angle. Therefore, the tension force of the spring 14 should compensate for the spacer force, i.e.

$$F_{\text{p.min}} = F_{\text{p14}}, \tag{1}$$

$F_{\text{p.min}}$ – minimum spacing; F_{p14} – spring tension force 14 at $f_{\min}$. $F_{\text{p.min}}$ can be calculated using the previously obtained ratio.

3. Results

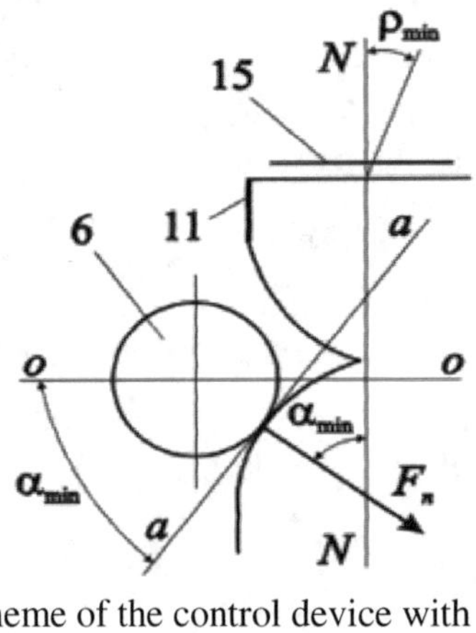

Figure 2. The scheme of the control device with a variable angle.

$$F_{p.\min} = (z - z_1) F_p C_{\min} f_{\min}, \quad (2)$$

Using the result contained in the right-hand side of equality (1) in expression (2), we can find the minimum spring tension $F_{п14}$.

Equality (2) contains the term $C_{\min}$, to determine which it is necessary to fulfill a condition that excludes self-braking when moving to the right (in Fig. 1) of the support sleeve 11 relative to the guide key 15.

The ability to carry out the movement of the support sleeve 11 to the right as a result of the influence of a spacer force on it is determined by the ratio of the values of the friction forces between the support sleeve and the guide key and the spacer force feedback control device. This ratio is expressed by the following equilibrium of the support sleeve

$$F_{p.\min} - F_{tr} - F_{p.14} = 0, \quad (3)$$

F_{tr} – friction force between the support sleeve and the key.

The thrust force is calculated by the formula:

$$F_{p.\min} = \frac{T_p (z - z_1)}{zr} tg\alpha_{\min}, \quad (4)$$

$\alpha_{\min}$ – the minimum angle between the tangent at the contact point of the rolling body 6 with the socket of the support sleeve 11 and the axial line of the socket[9-13].

Given the previously obtained expression, we can write expression (4) in the form:

$$F_{p.\min} = (z - z_1) F_p R_{cr} f_{\min} \frac{tg\alpha_{\min}}{r}, \quad (5)$$

The friction force between the support sleeve and the guide key is calculated by the formula:

$$F_{tr} = \frac{2(z - z_1) T_p f_1}{zd_0}, \quad (6)$$

f_1 – coefficient of friction between the support sleeve and the key 15; d_0 – the diameter of the Central hole of the support sleeve 11.

Angle $\alpha_{\min}$, as shown above depends on the coefficient of friction f_1. To establish the relationship between them, we turn to the circuit shown in Fig. 2. The diagram shows a fragment of the nest of the support sleeve 11 and the rolling body 6 located in the nest. Minimum angle $\alpha_{\min}$ between the tangent A-A to the profile of the wall of the nest at the point of contact between it and the rolling body and the axial line *o-o* nests corresponds to the position of the rolling body with a minimum value of the coefficient of friction. Normal $N - N$ held to the guide key 15 and forms normal pressure forces with the vector F_p angle equal to angle $\alpha_{\min}$. The friction angle between the support sleeve and the guide key is indicated as $\rho_{\min}$.

The condition for the motion of the bodies of the frictional contact "body 11 – body 15" is written as:

$$\alpha_{\min} \geq \rho_{\min}. \quad (7)$$

After substituting the right-hand sides of expressions (6) and (7) into equation (5), taking into account condition (3) (in the form of equality, as a limiting case of the relative motion of bodies 11 and 15), we obtain:

$$F_{p14} = (z - z_1) F_p R_{cr} f_{\min} f_1 \left(\frac{1}{r} - \frac{2}{d_0} \right). \quad (8)$$

In the expression (8), the replacement $tg\alpha_{\min} = f_1$. The second case relates to the operation of the SFC in adaptive mode with a minimum value of the coefficient of friction. Obviously, in this case, the gain should not be zero. Then the nominal torque of the coupling will be equal, in accordance with the expression:

$$T_p = zF_pR_{cr}\frac{f_{\min}}{1+(z-z_1)C_{\min}f_{\min}}. \tag{9}$$

Equating to each other the right parts of the previously found expression and relation (9) and replacing in the first f, f_i, C, C_i, we get the dependence C_i, f_i based on the solution of the equation:

$$C_i = \frac{1}{z-z_1}\left[\frac{1}{f_{\min}} - \frac{1}{f_i} + (z-z_1)C_{\min}\right]. \tag{10}$$

Comparing the previously obtained dependence and dependence (10), we see that the current coefficient is the gain value, the gain coefficient C_i in the second case, an increase of $C_{\min}$, than in the first case. In addition, as follows from expression (10), the nominal SFC torque is less than $1+(z-z_1)C_{\min}f_{\min}$ times than the SFC torque with differentiated friction pairs[14-19].

Using the previously obtained relation, we find the restrictions on the values for the parameters z, z_1. Solving quadratic inequality with respect to the unknown z_1:

$$z_1 \in \left[-\infty; \frac{f_{\max}(1+zC_{\min}f_{\min}) - \sqrt{f_{\max}^2(1+zC_{\min}f_{\min}) - 4zC_{\min}f_{\min}^2 f_{\max}}}{2C_{\min}f_{\min}f_{\max}}\right] \cup \tag{11}$$
$$\left[\frac{f_{\max}(1+zC_{\min}f_{\min}) + \sqrt{f_{\max}^2(1+zC_{\min}f_{\min}) - 4zC_{\min}f_{\min}^2 f_{\max}}}{2C_{\min}f_{\min}f_{\max}}; +\infty\right].$$

Many solutions of (11) were obtained taking into account the positive term $C_{\min}f_{\min}f_{\max}$ with an unknown second degree in the square inequality, as well as the fact that $D > 0$ (D – discriminant of square inequality). Really, $D = f_{\max}^2 + 2zC_{\min}f_{\min}f_{\max}^2 + z^2C_{\min}^2 f_{\min}^2 f_{\max}^2 - 4zC_{\min}f_{\min}^2 f_{\max}$. (12)

In expression (12) we compare the second and fourth terms

$$\Delta = 2zC_{\min}f_{\min}f_{\max}\left(f_{\max} - 2f_{\min}\right). \tag{13}$$

From relation (13) it follows that $\Delta > 0$, since the gain is the difference in brackets is positive (for known combinations of modern friction materials used in SFC).

4. Discussion

Graphically, the set of solutions (13) is shown in Fig. 3. Line 1 reflects the right boundary of the left half-region of solutions (13), line 2 - the left border of the right half-region. Parameter Values z_1, satisfying the necessary constraint C above (see above), are located in the regions of solutions located respectively below line 1 and above line 2 (see Fig. 3).

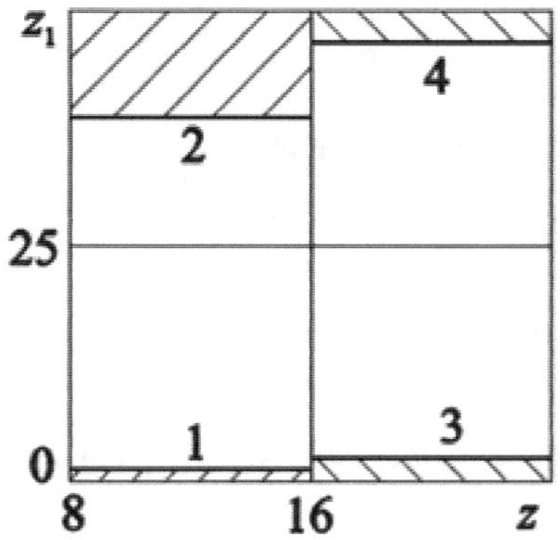

Figure 3. Solution Areas.

The graphs shown in Fig. 3, are constructed according to the following initial data: $f_{max} = 0.8$, $f_{min} = 0.1$, $C_{min} = 0{,}3$, $z = 8$. It follows from solutions (13) that the parameter z_1 increases with increasing z: slower for the right boundary of the left half-region of solutions, faster for the left border of the right half-region of solutions (lines 3 and 4, respectively, constructed with the same initial data, except for the parameter $z = 16$).

At the same time, it follows from Fig. 3 that the left boundary of the right half-region of solutions (13) reflects the values of the parameter z_1, which exceed the value of the parameter z, taken as the initial one when plotting charts. This contradicts the physical picture, according to which the parameters z, z_1 ($z > z_1$). Therefore, the right half-domain of solutions (13) cannot be taken into account, since the gain leads to results that really cannot be put into practice.

In addition, the graph in Fig. 3 (line 1) shows that the value of the parameter z_1, gain necessary to achieve a sufficient upper limit of the value C_{max}, does not meet the set condition, since the gain is less than unity. Parameter increase only z up to sixteen allows you to implement the option SFC with parameter value $z_1 = 1$ (line 3).

Thus, if all the SFC friction pairs are leading, then the current gain is the feedback gain less than when the SFC is made with differentiated friction pairs.

A schematic diagram of a feedback control device with switching off the action of the latter with a minimum value of the coefficient of friction is proposed and calculated, which allows the construction of a SFC with increased load capacity for crank presses.

Acknowledgements

The study had no sponsorship.

References

[1] Zhivov L.I., Ovchinnikov A.G., Skladchikov E.H. Forging and stamping equipment: Textbook for universities / Ed. L.I. Zhivova. - M.: Publishing House of MSTU. N.E. Bauman, 2015 .-- 560 f .: ill.

[2] Zhivov L.I., Chumakov B.N., Drozdov N.G. Features of the dynamics of the hot stamping crank press for stamping low forgings. "University News. Mechanical Engineering ", 2012, No. 1, p. 155-159

[3] Zalessky V.I. Equipment forging shops. Ed. 2nd, overwork, and add. Textbook for high schools. M., "Higher School", 2009. 632 p.

[4] Crank forging machines / Ed. Vlasova V.I. - M.: Mechanical Engineering, 2012 .-- 424 p.

IOP Conf. Series: Materials Science and Engineering **680** (2019) 012015 doi:10.1088/1757-899X/680/1/012015

[5] Noskov G.P., Rodov G.M., Vyatkin V.P. An experimental study of the loads in the crank press drive during a technological operation. Forging and stamping production, 2016, No. 5, p. 30 - 32
[6] Svistunov V.E. Forging and stamping equipment. Crank presses: Study Guide. - M.: MGIU, 2016 .-- 704 p.
[7] Svistunov V.E. The results of mathematical modeling of crank presses with compact actuators. Forging and stamping production, 2015, No. 10, 24 - 27
[8] Sokov V.I. Experimental determination of the friction moment when the clutch is engaged and braking of the hot-stamping crank presses. Forging and stamping production. Metal forming. 2015, No.10, 29-35
[9] Truskovsky V.I. The dynamics of forging machines. - Chelyabinsk: Publishing house of SUSU, 2015.-79c.
[10] Fedorkevich V.F. About the rigidity of modern crank hot stamping presses / Forging and stamping. Metal forming. 2013, No. 5, p. 23 - 25
[11] Hoopfer P. Dynamic loads in crank presses. “Forging and stamping production”, 2011, No. 2, p. 28 - 31
[12] Chubukov V.A., Gartvig A.A. Investigation of the influence of the design and parameters of hot-stamping crank presses on the nomenclature and accuracy of stamped products. Collection of scientific reports of the VI International Conference "The participation of young scientists, engineers and educators in the development and implementation of innovative technologies." - M.: MGIU, 2016, p. 134-138
[13]. Scheglov V.F., Maksimov L.Yu., Linz V.P. Forging machines. - 2nd ed., Revised. and add. - M.: Mechanical Engineering, 2010 .-- 304 e., Ill.
[14] Patent RU 2427466. A method for protecting crank presses from overloads by force on a slider. Svistunov V.E., Chubukov V.A., Matveev A.G. Publ. 08/27/2011
[15] Schumann K. Methode zur rechnerischen Untersuchung der technologischen Stobbeanspruchеng mechanischer Pressen. Umformtechnik, 24 (2012), No. 2, p. 29 - 35
[16] Hiraishi Kenji, Kagawa Toshiaki. Sumitomo jukikaigiho. Sumitomo Heavy Ind. Techn. Rev. 2010, No. 164, p. 9
[17] Schnellaufer-Press produziert Platinen. Maschinenmarkt. 2009, No. 41, p. 36
[18] Stanzoder Umformautomat: Application 1867469 EPO, IPC B 30 V 15/04 (2006.01), B30 B 15/00 (2006.01). Haulick + Roos GmbH, Siegel Andreas (Hoeger, Stellrecht & Partner Patentanwälte Uhlandstrasse 14 with 70182 Stuttgart): No. 06012074.8; Claim 06/12/2008; Publ. 12/19/2009.
[19] Stanzautomat selbst erganzen. Blesh InForm. 2009, No. 4, p. 87

Objectives of increase of capacity and lifetime of municipal solid waste dump according to density index study

D Larin[1], E Staseva[2] , E Khetsuriani [1]

[1] Platov South-Russian State Polytechnic University (NPI), str. Prosvesheniya 132, Novocherkassk 346428, Russia Federation
[2] Don State Technical University, Gagarin Square 1, Rostov-on-Don 344000, Russia Federation

E-mail: elena_staseva@mail.ru

Abstract. The problem of the optimal waste management of industrial and household waste is extremely important due to large amount of accumulated and annually produced waste. One of the crucial aims to improve ecological quality and environmental safety and compliance is to improve efficiency of waste management. The vast majority of solid domestic waste landfills are used up, but design, construction and place new territories into operation for receiving and waste disposal will take 3-5 years. In this work, issues of definition of the residual capacity of landfill are considered , aiming to expand the capacity and usage period, based on the solid landfill body density index research .Study object is solid household waste landfills in Rostov region, study subject is anthropogenic filling layers of the landfill. In order to support business objectives, there was conducted engineering and geological research, wells were drilled, laboratory studies were undertaken and soil tests were analyzed. Experimental results show the possibility to continue waste disposal works in the current landfill by means of increase of waste density index and free geometric scope.

1. Introduction

Nowadays, the issue of health protection and environment is world-widely topical. The amount of wastes is incrementally increasing as a result of industrial production and footprint.

More than 80 billion tons of various types of waste, including environmentally hazardous [1, 2], were disposed in Russia.

Waste treatment is implemented in different countries, according to economic capabilities, economic policy and legislation. The simplest and relatively cheap method of solid household waste (SHW) treatment is storage at sanitary landfills. The method of solid household waste disposal in landfills is one of the most widely used in Russia [3].

Rostov region is one of the leading regions because of its strategic position and economic and resource potential.

The source of the waste is population pressures, factories and organizations. There are 719 solid household waste disposal facilities, which annually take 4,5 million tons of waste, in Rostov region [4, 5]. The problem of the optimal waste management of industrial and household waste is extremely important due to large amount of accumulated and annually produced waste [6-7]. One of the crucial aims to improve ecological quality and environmental safety and compliance is to improve efficiency of waste management [3, 5].

2. Problem urgency

The majority of landfills in Rostov region were built in the end of 90's of the 20th century. Their project involvement will be used up in the near future. However, they are free and underutilized according to height project index within the scope of given locations. The service age of a landfill is accounted according to the waste disposal capacity and process engineering solutions made in the project [3,6].

Process engineering solutions in the project development were based on the process of waste collection, transportation and disposal, relevant at that time, based on the usage of domestic car equipment and relevant characteristics in the majority of waste transportation and collection companies.

Current project landfills have been designed and used for 10-15 years, over this time there have been following significant changes, resulting in free capacity in the landfill within the scope of allotment:

- there has been high quality technical retooling of companies operating landfills, due to acquisition of equipment, not only for city sanitation, but also for works in a landfill;
- in accordance with rise in living standards, increase in supermarket chains which use almost completely plastic packages, resulting in changes in morphological composition of population and commercial city centers waste;
- Landfill performance characteristics have increased, resulting in the increasing density in waste working maps;
- total amount of waste in landfills has decreased because of waste sorting implementation, in other words development of waste management technologies;
-over the years, the height of waste storage has increased, which leads to vertical deformation and consolidation of underlying layers by overlying level.

Current landfills of Rostov-on-Don, as well as other cities of Russian Federation, are almost used up, but design, construction and place new territories into operation for receiving and waste disposal will take 3-5 years.[6].

Nowadays, objectives of calculating volume of actual waste and residual capacity is crucial.

The major contribution to research of solid household waste landfills, biochemical and physical processes and their influence on the environment was made by: Academy of community facility n.a. K.D. Pamfilov, Moscow State University, Institute of Biochemistry of Russian Academy of Sciences, Research Institute of Ecological Safety of Russian Academy of Sciences, St. Petersburg State University, St. Petersburg Polytechnic State University, Russian State University of Technology, Russian Federation Ministry of Natural Resources, State Sanitary & Epidemiological Surveillance Agency, Federal Service of Russia for Hydrometeorology and Monitoring of the Environment, Institute of geology and geochronology of Precambrian of Russian Academy of Sciences and others, and also domestic and foreign experts: Abramov N.F., Vaisman Ya. I., Gribanova L.P., Bartsev I.A., Maksimova S.V., Likhachev U.M., Nozhevnikova A.N., Pantskhava E.S., Christensen T., Shtengmann P. and others. [7-18]. Nowadays, there is insufficient information about density waste in landfills, and also, the subject of expansion of currrent landfills capacity and usage period has been little explored.

3. Discussion

The aim of the work is to research and provide data for identifying indicators of soil density and make decisions on study of residual capacity and usage period of SHW landfill. The investigation is concerned with solid household waste landfills in Rostov region, study subject is anthropogenous filling layers of the landfill, analysis and research of mechanical and physical properties of soil.

IOP Conf. Series: Materials Science and Engineering **680** (2019) 012016 doi:10.1088/1757-899X/680/1/012016

In order to support business objectives, there has been conducted the research in current landfill in Rostov region. The results of geophysical and geochemical observations show the active anaerobic destruction process in loaded landfill maps and vertical deformation.

Vertical deformations of working substance begin during operational time within the process of SHW daily subsidence and continue after the closure under its own weight and developing biochemical processes and waste destruction [10-12].

Deformation rate and scope are of heterogeneous nature and can be determined by morphological composition of incoming waste; storage height and compaction degree during stowing; time, after stowing; landfill capacity; climate and geographic conditions and type of soil used for daily filling; type and characteristics of final covering.

Process of subsidence at different stages of a landfill life cycle is demonstrated in Table 1 [13].

Table 1 - Scheme of subsidence at different stages of a landfill lifecycle

Lifecycle stage	Mechanism of subsidence	Phase of subsidence	Duration
Operational	Deformation of waste inerts; compression	Starting compression	During operational time
Recultivating Intense methane phase	Filtration compaction; Secondary consolidation; biochemical destruction	First compression	From 2 months to 2-5 years after the closure
Post-recultivating (stable methanogenetic phase)	Biochemical destruction	Second compression	Until the end of decay process

The processes of mechanical compression and filtering compaction prevail at the early stages of landfill life cycle. These vertical deformations are called landfill sludge. Biochemical processes of waste destruction with the biogas production prevail at the stages of remediation and post-remediation. These deformations are defined as subsiding. All the above factors influence the formation of the current landfill and increase the possibility to use its capacity by further filling. [14-15].

Also, high quality technical retooling of waste collection companies has had a high impact on the process of waste compaction: purchase of waste collection vehicles with Faun and Zoeller equipment, with higher density levels in the process of loading from container staging area into a vehicle. Also, the index of density at these vehicles is from 5,0 to 10,0, in comparison with previously used domestic waste collection vehicles with density index from 1,5 to 2,5.

Similar companies of Rostov-on-Don are also switching to vehicles with the back loading, with density index up to 6,0. However, nowadays, domestic vehicles are used to access to some districts, because of the type of urban planning, Waste collection with the help of the vehicles with small capacity and low density index has two stages: waste collected by these vehicles are loaded to special containers of large capacity at the stations of waste and compaction reloading, also the density index rises to 8,0-12,0.

Waste collection vehicles move in special ladle of reloading station and unload its content there. The ladle can get load of several vehicles. After loading, the ladle is brought into the position of use and empty its contain out in the receiving unit.

In the receiving, waste is pressed into the large capacity container with compaction at the same time.

The large capacity container is connected with the receiving unit, with the help of oleo-dynamic latch mechanism. After loading, large container is loaded on the traction engine multi-lift and taken out to the landfill. Under these circumstances, the research of density index of current SHW landfill in Rostov region was conducted with the aim of increase its capacity and usage period.

4. Methods

In order to support business objectives, engineering and geological research was conducted, wells were drilled, laboratory studies were undertaken and soil tests were analyzed. The drilling was made by 146 mm diameter, by a churn-line method, by the driller УГБ-1ВС (UGB-1VS). There were drilled 6 technical strings up to 24 m down. To identify physical indicators of soil, there were taken undisturbed structure soil samples, according to the technology "in one stroke". The interval of soil samples was 2,0 m.

The laboratory works resulted in the identifying of natural water content, density, plastic limit of soil.

Research is conducted in accordance to the requirements of current environmental regulations. The determination of physical characteristics was conducted according to State All-Union standard 5180-2015 [19]. The determination of soil density was conducted by the method of weighing in water, according to State All-Union standard 5180-2015. Statistical processing of soil data was according to State All-Union standard 20522-2012 [20].

In accordance to State All-Union standard 20522-2012, normative and physical soil characteristics were estimated from selected layer. Statistical estimates according to abovementioned State All-Union standard were performed for this. In this regard, measures of variability of soil index were taken into consideration. As a criterion of level of variability, used coefficient of variation was according to State All-Union standard 25022-2012.

5. Results

In a SHW landill there were opened quaternary anthropogen soil formations, according to State All-union standard 25100-2011 [21], related to the category of man-made soils, with mechanic and hydrogenous colloidal structural bonds. The folding is monoclinal. According to research based on State All-union standard 20522-2012, there are the following geological layers in geological lithological section:

Layer-1 (tQIV) from 0,0 to 24,0m – to 1,0m: heterogeneous loam soil with the construction and household waste up to 40%. Lower: household waste, anthropogenic domestic and productive activity waste, with loamy layers, with capacity 20-3-cm, class of hazard – IV-V. Layer capacity – 24,0m. The layer was drilled by all wells. The drilling to the depth 24,0m №№ 1-6 did not open up groundwater.

Physical and mechanical soil properties of the current landfill were analyzed according to performance-test data, the results are listed in the Table 2.

Table 2 – Physical mechanical soil properties

Index	Moisture load, %	Density, g/cm^3			Poriness, %	Porosity fractions , unit fraction
		Soil particle	Soil	Dry soil		
Quantity	62	9	72	72	9	9
Characteristic value	16,3	2,70	1,35	1,15	56,8	1,32
Minimum	11,6	2,69	1,25	1,07	55,9	1,27
Maximum	21,30	2,70	1,46	1,33	58,1	1,39
Average deviation	2,40	0,01	0,06	0,06	0,72	0,04
Variability index	0,10	0,00	0,04	0,04	0,01	0,03
Estimated value:						
- under probability 0,85			1,34	1,14		
- under probability 0,95			1,34	1,14		

Geotechnical and topographical surveys of monolith selection by a churn-line method in 6 wells, were carried out in order to confirm by experiment.

As a result of selected monolith weighing, according to the volumetric data, the average value of density is 1,34 g/cm3

According to t.2.6 "Design, operation and re-cultivation of SHW landfills guideline» [22], after double SHW compaction by bulldozer, density characteristic value is 570-670 kg/m3 (0,57-0,67 g/cm3), after quadruple – 670-800 kg/m3 (0,67-0,8 g/cm3). Soil density data, at natural moisture, determined by practical consideration, is from 1,25 to 1,46 and it is 1,34g/cm3. Comparison of SHW density data with *-guideline, t.2.6 [22] , leads to the conclusion that the level of compaction is high according to studied landfill expansion and depth up to 24m.

The method of compaction is mechanized, by quadruple bulldozer compaction REM-25 (РЭМ-25), with the operating weight 26 t., with the maximum of crushing force more than 127кН. Through this process, 1m3 of SHW transform into 0,25 m3 of consolidated waste. Geometrically, 4m3 can be stored on 1 m3. According to the engineering survey, calculation of residual geometrical capacity of the landfill is 212400 m3. As a result, taking into consideration the quadruple compaction, 849600 m3 of SHW can be stored in the landfill. Total capacity of the landfill will be 1138464 tons, at the waste density of 1,34 t/m3, At the rate of 150000 tons of annual intake the usage time of the landfill will be 7,5 years.

6. Conclusions

The implementation of the aforementioned modern technologies for waste collection and waste disposal decreased physical amount of SHW disposal by increasing density index from calculated 0,8-0,85 t/m3 to really measured 1,34 t/m3, which is confirmed by research. In connection with the above, the current landfill has the capacity to continue works by increasing of density.

In view of this, to measure waste density and estimate residual capacity to increase usage period, it is necessary to do extra research in current landfills.

The solution of this problem will improve efficiency of current landfills, which is important not only for Rostov region, but also for Rostov-on-Don.

Acknowledgements

The study had no sponsorship.

References

[1] Gosudarstvenny`j doklad «O sostoyanii i ob oxrane okruzhayushhej sredy` Rossijskoj Federacii v 2017 godu» [E`lektronny`j resurs].– URL:http://www.mnr.gov.ru/docs/

[2] Prikaz Minprirody` Rossii ot 14.08.2013 N 298 «Ob utverzhdenii kompleksnoj strategii obrashheniya s tverdy`mi kommunal`ny`mi (by`tovy`mi) otxodami v Rossijskoj Federacii» [E`lektronny`j resurs]. – URL: http://docs.cntd.ru/document/499041934. (data obrashheniya 20.06.2019)

[3] E`kologicheskie problemy` antropogenny`x lanshaftov Rostovskoj oblasti / V.V. Privalenko, O.S. Bezuglova. T.1: E`kologiya goroda Rostova-na-Donu. Rostov-na-Donu, 2003.

[4] Upravlenie otxodami. Poligonny`e texnologii zaxoroneniya tverdy`x by`tovy`x otxodov. Rekul`tivaciya i poste`kspluatacionnoe obsluzhivanie poligona: monografiya / Ya.I. Vajsman [i dr.]; pod red. Ya.I. Vajsmana. – Perm` izd-vo Prem. nacz. issled. Politexn. un-ta, 2012. – 244 s.

[5] Omelchenko E.V., Trushkova E.A., Sidelnikov M.V., Pushenko S.L., Staseva E.V. Algorithm research exposure dust emissions enterprises of building production on the environment. // IOP Conference Series: Earth and Environmental Science Current Problems and Solutions. Ser. "Ecology and Safety in the Technosphere: Current Problems and Solutions" 2017. S. 012018.

[6] Rumyanceva A.V., Berezyuk M.V., Rumyanceva E.I. E`kologoe`konomicheskoe obosnovanie proekta po pererabotke tverdy`x kommunal`ny`x otxodov na osnove sovremenny`x texnologij // Vestnik VGTU. Seriya E`konomika i upravlenie. № 3, 2017. S.31-38.

[7] Lassini, P. Reclamation of old and new landfills and their integration with the environment / P. Lassini, G. Sala, F. Sartori / 6th, International waste management and landfill symposium. Sardinia. –1999. – V. 4. – P. 415 – 422 [E`lektronny`j resurs]. Rezhim dostupa: http://www.gbv.de/dms/tib-ub-hannover/319191907.pdf. (data obrashheniya 12.06.2019)
[8] R.P.Stearns, T.D.Wright, B.A.Stirrat. Landfill gas recovery and utilization at Industry Hills, California // Waste Management & Research. – 1984. - №2. – P. 153-161 [E`lektronny`j resurs]. – URL: https://www.sciencedirect.com/science/article/pii/0734242X8490137X. (data obrashheniya 19.05.2019)
[9] Utilizaciya i pererabotka tvyordy`x by`tovy`x otxodov : uchebnoe posobie / A. S. Klinkov, P. S. Belyaev, V. G. Odnol`ko, M. V. Sokolov, P. V. Makeev, I. V. Shashkov. – Tambov : Izd-vo FGBOU VPO «TGTU», 2015. – 100 e`kz. – 188 s.
[10] Armisheva, G.T. Rekuperaciya resursov pri zaxoronenii tverdy`x by`tovy`x otxodov / Avtoreferat dissertacii na soiskanie uchenoj stepeni kandidata texnicheskix nauk, g. Perm` – 2008. [E`lektronny`j resurs]. – URL: http://earthpapers.net/rekuperatsiya-resursov-pri-zahoronenii-tverdyh-bytovyh-othodov#ixzz5J8Dp3eZH. (data obrashheniya 22.06.2019)
[11] Vajsman, Ya.I. Razrabotka metodologicheskix principov sozdaniya i optimizacii ucheta dvizheniya otxodov s cel`yu povy`sheniya e`kologo-e`konomiko-social`noj e`ffektivnosti upravleniya ix obrashheniem / Ya.I. Vajsman , O.A. Tagilova, E.L. Sadoxina. [E`lektronny`j resurs]. – URL: http://www.komeco.ru/pressa/ecip_2013_12.pdf.
[12 Barcev, I.A. Xarakteristika sistemy` upravleniya pererabotkoj TBO: struktura i sostavny`e e`lementy` / I.A. Barcev, I.S. Docenko // UE`kS, 10/2013. [E`lektronny`j resurs]. – URL: http://uecs.ru/marketing/item/2434-2013-10-17-07-25-35.
[13] Maksimova Svetlana Valentinovna. E`kologicheskie osnovy` osvoeniya territorij zakry`ty`x svalok i poligonov zaxoroneniya tverdy`x by`tovy`x otxodov : Dis. ... d-ra texn. nauk : 03.00.16 : Perm`, 2004 285 c. RGB OD, 71:05-5/357 [E`lektronny`j resurs]. – URL: http://www.dslib.net/ekologia/jekologicheskie-osnovy-osvoenija-territorij-zakrytyhsvalok-i-poligonov-zahoronenija.html. (data obrashheniya 24.05.2019)
[14] Dobroserdova, E.A. Organizaciya i obrashhenie s tverdy`mi by`tovy`mi otxodami : Uchebnoe posobie. – /E.A. Dobroserdova.– Kazan` : Izd-vo KGASU, 2015.– 65 s.
[15] Berezyuk M.V., Rumyanceva A.V. Novaya sistema upravleniya TKO: innovacionny`j podxod // Innovacionnoe razvitie e`konomiki: nauchno-prakticheskij i teoreticheskij zhurnal № 5 (35). 2016. S.19-29/
[16] Slyusar`, N.N. Izuchenie frakcionnogo i morfologicheskogo sostava otxodov stary`x svalok i poligonov / N.N. Slyusar`, Yu.M. Zagorskaya, G.V. Il`iny`x // Vestnik PNIPU. Prikladnaya e`kologiya. Urbanistika. – 2014. – № 3. – s. 77-85.
[17] Slyusar`, N.N. Principy` upravleniya poligonom zaxoroneniya tverdy`x kommunal`ny`x otxodov na razny`x e`tapax zhiznennogo cikla / N.N. Slyusar`, A.Yu. Puxnyuk // Vestnik PNIPU. Prikladnaya e`kologiya. Urbanistika. – 2016. – № 2. – s. 148-164.
[18] Methodological tool «Tool to determine methane emissions avoided from disposal of waste at a solid waste disposal site» (Version 06.0.1) // UNFCCC CDM Executive Board, EB 66 Report, Annex 46. – 2012. – 16 p.
[19] GOST 5180-2015 Grunty`. Metody` laboratornogo opredeleniya fizicheskix xarakteristik [Tekst]. / — M. - 2015
[20] GOST 20522-2012 Grunty`. Metody` statisticheskoj obrabotki rezul`tatov ispy`tanij [Tekst] / / — M. - 2012.
[21] GOST 25100-2011 Grunty`. Klassifikaciya (s popravkoj). – Vzamen GOST 25100-95; Vved. 01.01.2013.
[22] Instrukciya po proektirovaniyu, e`kspluatacii i rekul`tivacii poligonov dlya tverdy`x by`tovy`x otxodov / Utverzhdena Ministerstvom stroitel`stva Rossijskoj Federacii 2 noyabrya 1996 goda.

IOP Conf. Series: Materials Science and Engineering **680** (2019) 012017 doi:10.1088/1757-899X/680/1/012017

Reconstruction of electromechanical device for blocking dust emissions

M S Minkin, D N Kuimov

Department of Automation of Production Processes, Don State Technical University, Gagarin's square 1, Rostov-on-Don 344000, Russia

E-mail: m.s.minkin@mail.ru

Abstract. As a result of the work, a new configuration of the electrode system was investigated based on the application of the particle behavior prediction method. The new configuration of the electrode system allows for a lower supply voltage to increase the bulk density of the electric field charge in the interelectrode space, to increase the time of the dust particle in the interelectrode space, and also reduces the risk of electrical discharges on the corona electrodes. Calculation on the developed special software product was made for different geometry of the interelectrode window (120,150,200 microns). Computer simulation of physical processes in the interelectrode space, analysis of the dynamics of the charge set by the dust particle, as well as calculation of the forces acting on the dust particle confirmed the efficiency of the chosen geometry of the electrode system.

1. Introduction

Lack of many devices which are earlier developed and applying to the display corona discharging category to blocking of dust fractions consists in the insufficient area of to the display corona discharging electrodes and small time spent of a particle in a zone of to the display corona discharging electrode.

Technical task of the offered device consists in increase in effectiveness of suppression of dust emissions for decrease in environmental of various technological objects, at decrease energy consumption and mass-dimensional indicators of the device.

The possibility of the solution of an objective is reached by change of geometry of an electrode system. In difference from earlier studied configuration of the electromechanical device of suppression of dust fractions of needle type, the offered option of a design provides increase in the area of influence of to the display corona discharging electrode, increase in the time spent of a dust particle in a scope of an electric field and also indirect mechanical suppression of driving of dustair mix.

2. Main part

At the expense of that, to the display corona discharging electrode is executed from the separate plates bent at an angle connected to the negative electrode of the power supply an apparent density of a charge in interelectrode space increases at the expense of additional streamer streams and by that receiving time dust particles of the limit charge decreases, the time spent of particles in the field of a corona discharge increases and also the effort from an electric field increases.

IOP Conf. Series: Materials Science and Engineering **680** (2019) 012017 doi:10.1088/1757-899X/680/1/012017

In turn the form-building electrode is made in the form of a grid of unmagnetized material of the steam-and-gas mixture tense perpendicular to a stream and connected to a positive pole of a source of high-voltage tension.

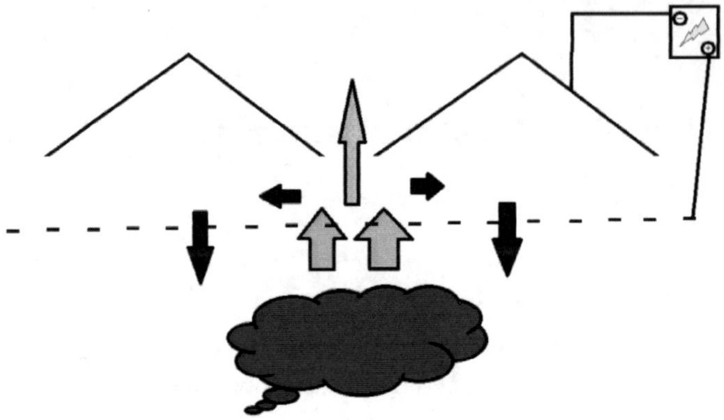

Figure 1. The scheme of driving of dust streams in the electromechanical device of suppression of dust fractions of blinds type

We will practically not distinguish the principle of action of this design from early configurations of electrode systems. Dustair mix, after a cyclone moves in the bunker in the perpendicular direction to the form-building and to display corona discharging electrodes. In the course of passing through interelectrode space each particle of dustair mix carries out accumulation of a charge. With increase in a charge in process of passing of a way from a form-building electrode to the display corona discharging electrode the braking effect amplifies. Additional effect of application of this configuration is the design of to the display corona discharging electrode which is providing the larger area of a covering and mechanically stopping movements of dust particles. However It should be noted that this effect isn't the basic. By results of model operation it is possible to tell that the most efficient will be that geometry of an electrode system at which the stop and redirection of driving of dust particles will be carried out before contact of particles with to the display corona discharging electrode that will allow to reduce negative effects of possible categories. Achievement of this effect will allow to apply installation in more adverse conditions with the explosive environment, for example, in case of application of installation in mines to purification of air.

Within realization of the real work various options of designs of the electromechanical device of suppression of dust emissions are investigated. Input datas, applicable to all researches of a chapter 3, were divided into the following groups.

- the unchangeable specifications determined by technological process.
- the unchangeable specifications determined by parameters of a power line and design features of the considered device: - a delivery is carried out from single-phase network by tension U = 220 V / 30 kV with f frequency = 50 Hz / 0 Hz (reverse polarity); cleaning of working parts – air.

In the considered options of devices of blocking of dust emissions a number of the physical processes having direct impact on overall performance of this type of devices proceeds.

For determination of design data of the device at which its effective functioning is ensured it is necessary to use mathematical model operation of physical processes in the device.

Mathematical model of electric processes of the dust fractions arising in the fissile region of the electromechanical device of suppression:

$$\frac{\partial^2 \phi}{\partial x^2} + \frac{\partial^2 \phi}{\partial y^2} = -\frac{\rho}{\varepsilon_0}, \tag{1}$$

$$\rho = \frac{j}{kE} \quad (2)$$

$$\overline{E} = -grand\varphi \quad (3)$$

$$F_1 = mg \quad (4)$$

$$\overline{F}2 = q(t)\overline{E} \quad (5)$$

$$F_3 = pS \quad (6)$$

$$F_4 = p_i \frac{\pi r^2}{4} \quad (7)$$

$$m\frac{d\overline{V}}{dt} = \Sigma \overline{F} \quad (8)$$

$$q(t) = q_m \frac{en_0kt}{4e_0 + en_0kt} \quad (9)$$

$$q_m = 4\pi\varepsilon_0(1 + 2\frac{\varepsilon - 1}{\varepsilon + 2})r^2 E \quad (10)$$

The approach based on use of a finite element method was applied to calculations.

In the course of calculations the finite element method, for a single particle of dust fraction carried out calculation of the charge acquired by it at stay in a particular point of space with coordinates (x, y). For each point of space taking into account the acquired charge the particle, carried out calculation of all accompanying efforts and the balance of forces was checked.

Realization of the above described calculations allowed to receive characteristics of the physical processes arising in the fissile region of an electrode system of the electromechanical device of suppression of dust fractions and also to estimate behavior of particles of the dust fractions which are carrying out immediate driving in an action field of a corona discharge. The main difference from earlier developed and applied techniques is the possibility of accounting of "ionic" wind.

Within this project model operation of physical processes and assessment of behavior of particles of fine fraction in the fissile region of an electrode system of the electromechanical device of suppression of dust fractions for the available design of the experimental exemplar "needle plane" was carried out that in subsequent allowed to confirm adequacy of the chosen method of assessment of behavior of fine fraction.

Calculation of dependences was performed with use of a finite element method, for a trajectory of driving of a particle of fine fraction. For exact realization of the functions, calculation was carried out for the worst provision of a dusty particle in the fissile region of an electrode system in which it is affected by minimum effort from an electric field.

In figure 2 the electrode system of the electromechanical device of suppression of dust fractions is represented. The electrode system is executed from two types of electrodes: to display corona discharging, presented by plates with sawtooth edges and form-building in the grid form. At power connection to an electrode system, the negative to the display corona discharging electrode and positive to a form-building electrode, in a zone of interelectrode space volume electric charge is created. As a result the electrode system creates two multidirectional efforts that is caused by a configuration and arrangement of electrodes. The dominance of an axial component of electric forces leads to an expulsion of particles of dust from a gap between in parallel the located precipitation electrodes which are bound to a positive pole of a source of high tension. Because to the display corona discharging electrode consists of plates the time spent of dust particles in an area of coverage of to the display corona discharging category increases that leads to increase in effectiveness of a lock.

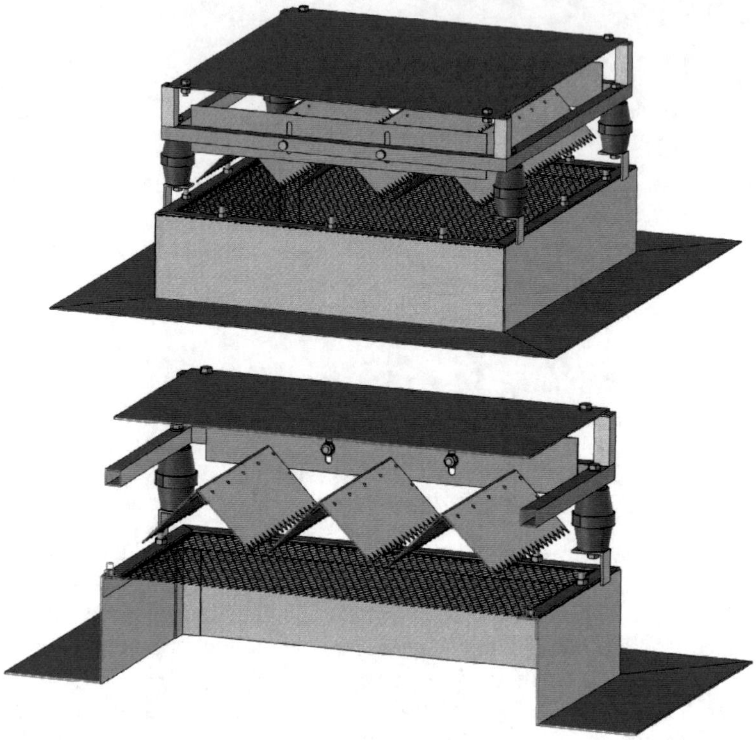

Figure 2. Three-dimensional graphic model of the electromechanical device of suppression of dust emissions

Thus, the electromechanical device of suppression of dust fractions with the improved characteristics blocks driving dust parts at the free exit of scrubbed gas that provides prevention a diversity of dust and keeping of dust particles in a dusting source zone.

For a lock of this type, in view of construction features, it makes sense to consider processes only in an interelectrode window between to the display corona discharging electrodes. In this research model operation of physical processes for a new design of the device was carried out. Results of the carried-out calculations of dynamics of charging of the particles and efforts operating in this electrode system at change are given below:

- interelectrode space of to the display corona discharging electrodes relatively each other.
- sizes of a particle of fine fraction.

Table 1. Dynamics of enrollment of a charge a particle of fine dust fraction depending on the sizes of an interelectrode window

Entrance to the fissile region	Interelectrode window of 150 mm	Interelectrode window of 200 mm	Interelectrode window of 120 mm
50 mm	3,57752E-13	3,08703E-13	3,11245E-13
100 mm	7,11909E-13	6,11325E-13	6,19361E-13
150 mm	1,057E-12	9,02139E-13	9,1959E-13
Guaranteed stop line	1,3818E-12	1,17343E-12	1,20216E-12

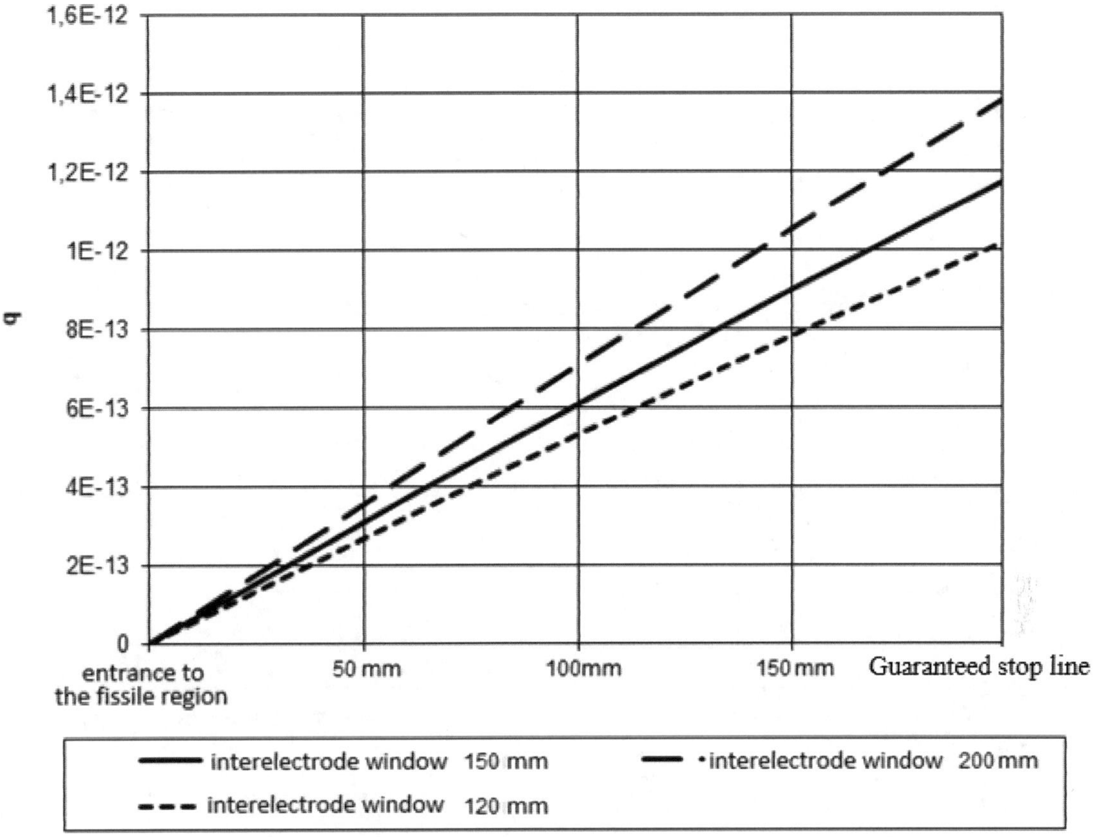

Figure 3. Dynamics of enrollment by a dust particle of a charge

Table 2. Calculation of effort from an electric field, acting on a particle of fine fraction, depending on geometry of an interelectrode window.

Entrance to the fissile region	Interelectrode window of 150 mm	Interelectrode window of 200 mm	Interelectrode window of 120 mm	Pressure force from a stream
50 mm	2,12E-09	1,58E-09	1,85E-09	2,51E-09
100 mm	4,14E-09	3,01E-09	3,6E-09	2,51E-09
150 mm	5,73E-09	4E-09	4,99E-09	2,51E-09
Guaranteed stop line	6,05E-09	4,14E-09	5,26E-09	2,51E-09

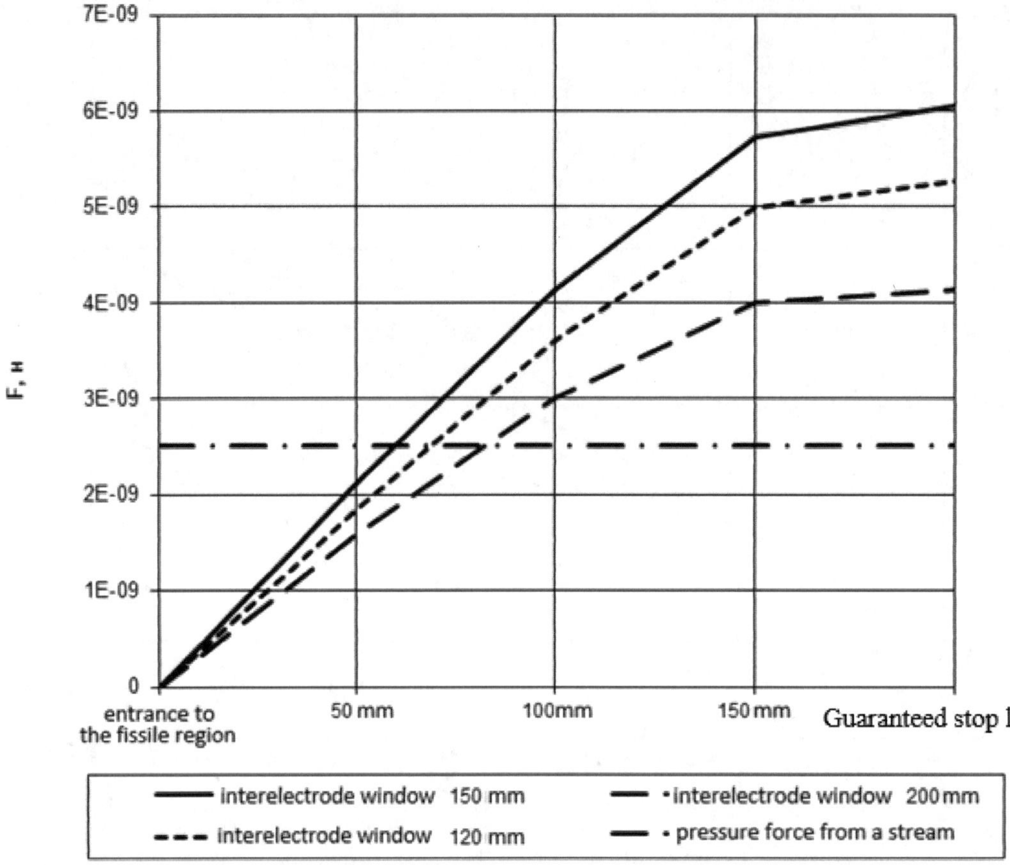

Figure 4. The effort from an electric field operating on a particle of fine fraction depending on geometry of an interelectrode window

For the purpose of confirmation of adequacy of a mathematical apparatus of the developed model and correctness of its assumptions the experimental check of theoretical results of model operation was executed.

Table 3. Observed datas on the designed installation of blocking of dust emissions.

Parameters	Unit of measure	The device is switched-off	The device is switched-on
Air temperature in the place of measurement	°C	38	38
Air speed on a pollution	m/s	1.38	1.02
Production time of tests	min	3	20
Quantity of an aspiring air on emission	m^3/c	370	320
Concentration of dust	mg/m^3	469.2	8.1
Amount of dust in a stream	g/s	0.046	0.0009
Extent of cleaning	%		98.27

IOP Conf. Series: Materials Science and Engineering **680** (2019) 012017 doi:10.1088/1757-899X/680/1/012017

Research and development of the new power supply will become an important factor of increase in overall performance of the electromechanical device of suppression of dust fractions. The analysis of scientific publications showed that success in creation of highly effective devices achieved in recent years with extent of purification of dustair mixes to 98 – 99% is inseparably linked with perfecting of units of a delivery and an intensification of the modes of their work.

Polarity given on to the display corona discharging tension electrodes the negative. It is bound to the fact that at the negative polarity of value of breakdown voltage of a working interval of the device is much higher, than at positive.

Regulation of tension of the unit is made, as a rule, in indirect parameters, such as average tension, frequency of scintillations, power and so on.

The overall performance of the electromechanical device of suppression of dust fractions as practice shows, in essential degree depends on a regularity of the choice of a diet of the unit: the form of power voltage caused by a rectification circuit, a way of regulation of tension, level of electric parameters, the scheme of turning on of the unit on loading. The diet in turn is defined by type of the dust caught in the device, properties of the dust environment, a design of the electromechanical device of suppression of dust fractions, features of operation of units in the mode of breakdowns. As show the pilot studies on the electromechanical device of suppression of dust fractions of needle type, at achievement of a defined critical value of tension extent of cleaning practically doesn't change at increase in current and tension or can even decrease, at the same time insignificant breakdowns were observed. This effect can affect negatively at application in rooms with the explosive environment, for example in mines.

References

[1] Stel'makh S.A., Shcherban' E.M. 2018 Electroactivation of foam concrete for buildings and structures with improved constructive and energy efficient characteristics *IOP Conf. Ser.: Mater. Sci. Eng.* vol 463

[2] Shcherban E.M., Tkachenko G.A., Goltsov Yu.I., Stelmakh S.A. 2012 About the impact of processing of foam-concrete mix a variation electric field on properties of foam concrete *The Modern problems of science and education* No. 1

[3] Shcherban E.M., Goltsov Yu.I., Tkachenko G.A., Stelmakh S.A. 2012 Prescription technology factors and their role in formation of properties of the foam concretes received from the mixes processed by a variation electric field *The Engineering bulletin of Don* No. 3

[4] Minkin M.S., Lukyanov A.D., Kuimov D.N. 2016 Development of the energy-saving air regeneration system in production rooms *Materials Science Forum* pp 625-629

[5] Minkin M.S., Kuimov D.N. 2018 Development of the principles of mathematical modelling and management of behaviour of fine dust fractions of various nature in non-uniform electric field *MATEC Web of Conferences* pp 20-29

[6] Minkin M.S., Kuimov D.N., Mozgovoj A.V. 2017 Simulation of physical processes in devices with non-magnetic, milk-dispersed secondary part *MATEC Web of Conferences T. Abbasian Najafabadi, I. Sevostianov, K. Yeghiazaryan, A. Nguyen Dong and V. Mladenovic*

[7] Minkin M.S., Lukyanov A.D., Kuimov D.N. 2016 Development of the energy-saving air regeneration system in production rooms *Procedia engineering* pp 1353-1358

[8] Minkin M.S., Lukyanov A.D., Kuimov D.N. 2016 Separation of fine particles with the set dimensional and physical characteristics by method of the crown separation *Materials Science Forum* T. 870 pp 625-629

[9] Pavlov A.N., Goltsov Yu.I., Stelmakh S.A., Shcherban E.M. 2015 Foam concrete durability at influence of a variation electric field *The Scientific review.* No. 10-1 pp 147-150

IOP Conf. Series: Materials Science and Engineering **680** (2019) 012018 doi:10.1088/1757-899X/680/1/012018

Modeling the mechanical characteristics of tribotechnical composites

T V Suvorova[1], O A Belyak[1], P G Ivanochkin[2] and S A Danilchenko[2]

[1]Department of Higher Mathematics, Rostov State Transport University,2 Rostovskogo Strelkovogo Polka Narodnogo Opolchenia Sq., Rostov –on- Don, 344038, Russia
[2]Department of Theoretical Mechanics, Rostov State Transport University,2 Rostovskogo Strelkovogo Polka Narodnogo Opolchenia Sq., Rostov –on- Don, 344038, Russia

E-mail: o_bels@mail.ru

Abstract. Some aspects of creating composite materials with a modified matrix based on aromatic polyamide phenylone FS-2 are considered. The effective characteristics of the composites are calculated on the basis of the self-consistency method and compared with experimental data obtained by the nanoindentation method. The stress-strain state of the composite during tribological contact was investigated on the basis of the contact problem of the motion

1. Introduction

Recently, oil-filled nanocomposites of the "oil-bearing" class have been actively used for assemblies and parts for tribotechnical applications on polymer matrices [1-3]. Obtaining a synergistic effect due to the simultaneous use of nanofillers and components with a lubricating effect, allows to achieve significant (10...25%) savings in the base material of the polymer matrix [4]. This fact significantly reduces the cost of production. Modified composite materials have a wide range of performance properties. Such materials compete with traditional metal, including ceramic, alloys. For the development of the country's economy, this increases the attractiveness of this area of materials science. In general, materials of this class can be considered as a three-component system, which consists of a polymer binder, plasticizer and multifunctional filler additives. The use of oil-filled composites in friction units requires a study of the features of contact interaction mechanics for materials of this class. In this paper, we propose one of the possible approaches to the formulation of contact problems for these materials. The theoretical basis for creating polymer composite materials that have adaptive functions in friction units is the possibility of forming adaptive principles for reducing wear and friction directly in the process of friction interaction. Theoretical approaches are based on optimizing the properties of the starting components. Moreover, the achievement of threshold values of external influence leads to a violation of the principles of compatibility of components and the formation of a certain free volume of one of the components involved in the reduction of frictional bond forces formed in the contact zone of materials. Mechanical compatibility is achieved provided that the deformation characteristics, thermal expansion coefficients, and other properties of the components correspond to each other. In this case, the integrity of the system during operation is not violated. Common to these materials is the realization of the "self-lubrication" effect based on plasticization of polymers. It is quite obvious that the issue of creating composite oil-filled nanomaterials with adaptive operational properties requires the simultaneous consideration of a number of interrelated tasks. The solution of such problems often turns out to be contradictory, since the antifriction indices can be in conflict with the indices of the bearing ability during friction or the indices of frictional heat resistance. In this context, a plasticizer may be a substance that has pronounced lubricating properties. Such a tribological system is thermodynamically and kinetically unstable. This fact makes it possible to lay adaptive functions into it, depending on the conditions of frictional interaction at the design stage. An increase in the wear resistance of composites is achieved not only by increasing the rigidity of the polymer matrix, but also by the formation of easy slip planes in the friction transfer film due to the introduction of volumes of a lubricant with high chemical affinity for it into the polymer composition. In particular, with increasing temperature, the mobility of the macromolecular chain of the polymer matrix increases. This causes the desorption (sweating) of the plasticizer substance located in the interpackage space of the macromolecular chain outward - into the friction zone, characterized by the highest gradient of deformation and temperature. This leads to a decrease in the coefficient of friction due to a decrease in the number of physical bonds between the macromolecules of the polymer matrix and the friction surface, an increase in the flexibility and mobility of its interstitial fragments, and, correspondingly, a decrease in the temperature of the contact zone, which induces the sweating of the plasticizer. All these facts slow down the process of sweating out the substance of the plasticizer and perform the function of automatically adjusting the speed of its feed into the contact zone, equivalent to the process of wear of the polymer matrix. This means that conditions are being created for the adaptive adaptation of the composite material to friction conditions. The energy effect activates the structure of such materials, creating conditions for opening channels of energy dissipation, aimed at "healing" the resulting defects, reducing operational damage. The more intense the external energy effect is, the greater the reaction the polymer system forms, performing the function of auto-regulation of the friction characteristics of the friction unit. The use of minerals as fillers, which have the ability of metal cladding by creating a layer of inorganic polymer, as well as high absorption characteristics with respect to hydrocarbon media, the degree of which is regulated by the form of preliminary thermochemical treatment and increased dispersion (with the transition to the nanodispersed state) of minerals allows to achieve a synergistic effect of increasing antifriction characteristics of self-lubricating composites.

2. Study of the mechanical properties of composites

The authors used the aromatic polyamide phenylone FS-2 as a polymer matrix, which is designed to produce composites by injection molding. The manufacture of parts based on phenylone requires temperatures above 310-325°C. This polymer matrix is characterized by high physicomechanical properties in comparison with other polymers, which allows for the processing of parts such as grinding and polishing. The filler was magnesium aluminate spinel, ground on a NETZSCH bead laboratory mill, type MicroCer, to a size of 40-50 nm. At the stage of laboratory testing, PFMS (polymethylphenylsiloxane), VGO (non-hydrotreated vacuum gasol), bitumen, C-52 (cylinder oil) I-40 mineral oil in the amount of 5-10% were studied as plasticizer samples. Studies have established that up to external heating temperatures of 80...100°C, the coefficient of friction of oil-filled nanocomposite samples was 0.22, provided that the coefficient of friction of unmodified samples was 0.42. When the external heating temperature exceeds 100°C, the friction coefficient of the unmodified sample increased to 0.46, which led to the destruction of the sample on the friction path of 1020 m. The friction coefficient of the modified sample remained unchanged when the preheating temperature of the friction zone reached 100 ° C, which confirms the structural adaptability of the material developed composite to a change in the external conditions of friction. Further determination of the mechanical characteristics of the studied composites was carried out by nanoindentation. For nanoindentation, pre-prepared samples (cylinders with a diameter of 10 mm and a thickness of 1.5 mm) from the studied materials were used. The billets were ground on a Sapphir 550 single-disk grinding and polishing machine in order to reduce the effect of the surface roughness of the test sample on the final indentation results. The mechanical characteristics of the studied materials were determined using the NanoTest 600 complex. In this work, we considered a composite in which C-52 cylinder oil was used as a plasticizer. Oil-filled composites were molded with a ratio of components: 7% oil C-52 + 3% magnesium aluminate spinel (composition is indicated in the table). Below are the results for samples of different moldings.

Table 1. The values of the mechanical characteristics of the studied materials.

Material	Microhardness *H*, GPa	Elastic modulus *E*, GPa	*H/E*	H^3/E^2
FS-2	0.395	5.950	0.066	0.001741
FS-2 + 3% spinel	0.409	5.967	0.069	0.001935
FS-2 + 10% composition (v.1)	0.27	4.931	0.055	0.000816
FS-2 + 10% composition (v.2)	0.255	3.843	0.068	0.001231
FS-2 + 10% composition (v.3)	0.3	4.752	0.063	0.001219

As can be seen from the data in table 1, the modification of phenylone with aluminum-magnesium spinel leads to an increase in hardness and Young's modulus compared to phenylone [5]. Adding C-52 cylinder oil to the modified matrix reduces the hardness and Young's modulus, however, during tribological tests [2, 6] it was shown that the friction coefficient of such a composite material is significantly reduced (0.42..0.18) compared to phenylone.

3. Modeling the mechanical properties of composites

A very important problem when creating composite materials is the directed modeling of their effective elastic modules. In the present work, the calculation of the effective elastic constants of the composite is based on the self-consistency method. Consider an elementary representative volume of a composite medium. At each point in the medium with a radius vector $\tilde{r}$ at the micro level, the generalized Hooke law:

$$\tilde{\sigma}(\tilde{r}) = [C^m(\tilde{r}) + h(\tilde{r})(C^*(\tilde{r}) - C^m(\tilde{r}))] \cdot\cdot\, \tilde{\varepsilon}(\tilde{r}), \quad (1)$$

$C^m(\tilde{r})$, $C^*(\tilde{r})$ – tensors of elastic modules of the fourth rank of the matrix of the composite medium and inclusions, respectively, $h(\tilde{r})$ – step function equal to unity inside the inclusion and zero in the matrix of a heterogeneous medium. Let us average the expression (1) over an elementary representative volume under the assumption that within its limits the tensor of the elastic modules of the matrix and inclusions is constant. As a result, we obtain the following relationship:

$$\sigma(r) = C^m(r) \cdot\cdot\, \varepsilon(r) + \varphi(C^*(r) - C^m(r)) \cdot\cdot\, \varepsilon^*(r), \quad (2)$$

φ – volume fraction of inclusions, ε^* – strain tensor inside inclusions. Following the ideas of [7], in the absence of interaction of inclusions, the study reduces to the problem of a single inclusion in an infinite homogeneous medium. According to this approach, expression (2) can be represented as:

$$\sigma(r) = C^{eff}(r) \cdot\cdot\, \varepsilon(r),$$
$$C^{eff}(r) = C^m(r) + \phi\big(C^*(r) - C^m(r)\big) \cdot\cdot\, A(r), \quad (3)$$
$$A(r) = (I + S(r) \cdot\cdot\, (C^m(r))^{-1} \cdot\cdot\, (C^*(r) - C^m(r)))^{-1}.$$

$S(r)$– asymmetric Eshelby tensor [8, 9] of the fourth rank, I– unit tensor. The calculation of nonzero components of the elastic modulus tensor C^{eff} of the composite was carried out under the assumption that the composite matrix is isotropic and the inclusions in the matrix are spherical [10].

In this case, the nonzero components of the Eshelby tensor are known and have the form [9]. It should be noted that with this form of inclusions distributed in an isotropic matrix, the desired composite inherits the isotropy of its matrix, which means that the task of determining the effective elastic moduli of the composite reduces to calculating two independent constants: $C_{1111}^{eff}, C_{1122}^{eff}$ by which the desired values were determined *E, υ*. It should be noted that the considered approach to the determination of effective modules gives a result that agrees quite well with experiment only for small values of the volume fraction of inclusions (up to 15%) [11]. A series of numerical experiments was carried out to calculate the effective elastic modules of composites based on formula (3). Tables 2, 3 present the Young's modulus and Poisson's ratio for composite materials, composition 2 and 3 (table 1). In this case, for the composite composition v1, v2, v3 (table 1), we first calculated the effective elastic modules of the matrix modified by magnesium aluminate spinel in a mass fraction of spinel equal to 3% (table 2), and then the obtained values of the Young's modulus and Poisson's ratio were input data when calculating the effective elastic moduli of the desired composite depending on the volume fraction of cylinder oil (table 3). In the calculations, the following initial data were taken: Young's modulus E =5.783 [GPa], Poisson's ratio ν =0.34 of the phenelone, which are determined experimentally [6], Young's modulus and Poisson's ratio of the magnesium aluminate spinel E =271 [GPa], ν =0.26 [12], bulk modulus of the oil K =1450 [Pa].

Table 2. The dependence of the composite modules E, ν on the volume fraction of magnesium aluminate spinel φ.

φ (%)	E (GPa)	ν
0	5.7830	0.3400
1	5.8982	0.3391
2	6.0132	0.3382
3	6.1283	0.3373

Table 3. The dependence of the composite modules E, ν on the volume fraction of oil φ.

φ (%)	E (GPa)	ν
0	6.1283	0.3373
3	5.7599	0.3316
7	5.2655	0.3222
10	4.8913	0.3130

Note that the obtained simulation results of effective elastic moduli of the composite were compared with the experimental data obtained on the basis of nanoindentation, given in Section 2. For example, the relative error between the experimental (table 1) and calculated Young's modulus for composite (FS- 2 + 3% spinel) did not exceed 3%, for composite (FS-2 + 3% spinel + 7% cylinder oil, v.1) did not exceed 6%.

4. The stress-strain state of the composite material with tribological contact

In order to study the stress-strain state of a composite material under tribological contact, the contact problem of the motion of a punch with a flat base of width $2a$ at a speed V in the positive direction of the axis Ox_1 along the boundary $x_2 = 0$of an elastic isotropic half-space under the action of a force $P = \{P_1, P_2, 0\}$ was considered. The half-space is modeled by a two-phase medium consisting of an elastic isotropic matrix and spherical inclusions, the effective modules of which are determined on the basis of the self-consistency method in Section 3. The punch moves uniformly, is in a state of ultimate equilibrium under the action of force P. There is full contact with the surface of the half-space. We assume that under the action of the normal component of the force P applied to the punch with an eccentricity, only the punch settles, parallel to the axis Ox_2. Contact stresses are related by the Coulomb-Amonton law. The boundary of the half-plane outside the contact area is free of stresses. The method of applying external force P allows us to consider the problem in a two-dimensional formulation:

$$\sigma_{ij,j} = \rho \ddot{u}_i, \quad i,j = 1,2$$
$$\sigma_{ij} = C_{ijkl}^{eff} u_{k,l}$$
$$\sigma_{12} = \sigma_{22} = 0, \quad if \quad |x_1 - Vt| > a$$
$$\sigma_{12} = k\sigma_{22}, \quad \sigma_{22} = -q(x_1), \quad u_2 = -\delta, \quad if \quad |x_1 - Vt| \leq a, \qquad (4)$$

k - coefficient of friction. We assume that the speed of the punch V is less than the speed of the Rayleigh wave.

Next, we move to the moving coordinate system with the origin in the center of the punch [13, 14]. Relations (4) will be considered in a moving coordinate system in a dimensionless form, with linear dimensions assigned to the half-width of the punch, and stresses to the shear modulus G of the elastic matrix. It is required to determine the distribution of contact pressure $q(x_1)$ under the punch. The described problem is numerically solved by the finite element method in ANSYS. Calculation of normal and tangential contact stresses for a two-component composite material, with a matrix based on FS-2, modified with magnesium aluminate spinel (3%) and oil C-52 (7%) (table 3) was carried out for the following values: ρ_f=1200 [кг/м^3], ρ_m=930[кг/м^3], speed of the punch V/V_R=0.1, V_R- speed of the Rayleigh wave. Based on numerical experiments, it was found that normal contact stresses are less sensitive to changes in the coefficient of friction than tangential contact stresses. The stress distribution over the contact area is asymmetric, which is typical for contact problems with friction [12], with a decrease in the coefficient of friction, the normal and tangential contact stresses decrease, the friction unit operates in a more favorable mode.

5. Conclusions

Modeling of the effective elastic moduli of the composite material gives an acceptable error, compared with the experimental data, which allows one to directionally simulate the mechanical properties of the composite. The numerical solution of the contact problem allows one to predict the friction force arising in the tribosystem when using the composite, and an increase in the content of the liquid oil fraction in the composite reduces the friction coefficient.

Acknowledgments

The reported study was funded by RFBR, project number № 18-08-00260-a.

References

[1] Derlugyan P D and Levintsev V A 2014 *University news. North-Caucasian region. Technical sciences series* **2** 9-15
[2] Ivanochkin P G, Danilchenko S A and Novikov E S 2016 *Procedia Engineering* **150** 520–26
[3] Derlugyan P D, Loginov V T, Sukhov A S and Derlugyan I D 1987 *University news. North-Caucasian region. Technical sciences series* **3** 61-7
[4] Sviridenok A.I., Kravcevich A.V. and Mikulich S.I. 2013 *Perspective materials and technologies* **6** 116-133
[5] Kolesnikov I V 2017 *System analysis and synthesis of processes occurring in metal-polymer friction units of friction and antifriction purposes* (Moscow: AUSTIC RAS)
[6] Ivanochkin P G, Suvorova T V, Danilchenko S A, Novikov E S and Belyak O A 2018 Vestnik RGUPS **4** 54–62
[7] Kristensen R 1982 *Introduction to the mechanics of composites* (Moscow: Mir)
[8] Eshelby J D 1957 *Proceedings of the Royal Society A* **241** 376-96
[9] Tsvelodub I Y 2010 *Journal of Applied Mathematics and Mechanics* **74** 249-252

IOP Conf. Series: Materials Science and Engineering **680** (2019) 012018 doi:10.1088/1757-899X/680/1/012018

[10] Giraud A, Huynh Q V, Hoxha D and Kondo D 2007 *Mechanics of Materials* **39** 1006-24
[11] Belhouideg S 2018 *European Mechanical Science* **2** (2) 68-75
[12] Golieva E V et al. 2016 *Journal of Optical Technology* **83** (2) 67-72
[13] Suvorova T V, Belyak O A and Usoshin S A 2011 Vestnik RGUPS **4** 155-61
[14] Belyak O A and Suvorova T V 2018 *Ecological Bulletin of Scientific Centers of Black Sea Economic Cooperation* **3** 25-31

IOP Conf. Series: Materials Science and Engineering **680** (2019) 012019 doi:10.1088/1757-899X/680/1/012019

The role of local plastic deformation in the formation of structure and properties of materials under extreme heating

A V Brover, G I Brover and O B Moysova

Don State Technical University, Gagarin Sq., 1, Rostov-on-Don 344000, Russia

E-mail: brover@mail.ru

Abstract. Based on the physical processes analysis in the areas of pulsed laser applied on metals, the research reveals the peculiarities in self-organization of the structure, which leads to the formation of non-trivial properties of surface layers. Structural features of the hardening process during high-speed laser processing depend on the resulting level of strains in the processing zones, that is, on the ratio of plastic deformation processes during heating under the influence of thermal and phase strains and energy dissipation by polygonization and recrystallization. As a result, the effect of material strengthening is achieved not only by martensitic transformation, microchemical heterogeneity, but also by high-temperature riveting, increasing the density of crystal structure defects, plastic shifts under the strain of different nature.

1. Introduction

The study of phenomena occurring on the surface of metal alloys when they are processed by concentrated energy flows (CEF), is technically the most promising in a number of modern scientific areas. Such processing allows to get results unattainable with traditional technology.

Under the influence of CEF, laser radiation particularly, complex structural and phase transformations occur in steel within thousands of a second. Such transformations lead to significant hardening. The mechanical properties of steels depend on many factors: the degree of martensitic transformation, the initial chemical contents, the amount and dispersion of carbide phase. Experimental studies proved the maximum hardening after laser processing on steels with more than 0,6-0,8% carbon weight [1, 2]. High-speed processes typical to laser method of hardening, increase the hardness of steel by 20-30% compared to traditional bulk hardening.

Better mechanical properties in the steel surface zones processed by CEF is due to the high speeds of heating and cooling, a significant temperature gradient in the areas, which, along with $\alpha \leftrightarrow \gamma$ recrystallization lead to the appearance of the thermal strain (due to the difference of specific volume and linear extension coefficient of coexisting phases), relaxing due to the formation, movement and interaction of dislocations.

Thus, the hardening of alloys observed under the laser pulse radiation is determined by:

- martensite transformation;
- the influence of high-temperature local plastic deformation, leading to a special dislocation structure with the increased dispersion of blocks as during the irradiation process, as in the following fixation of non-equilibrium structures;
- acceleration of the process of atom mass transfer by "drift" mechanism.

The leading role in the organization of finite structures belongs to internal strain fields, as well as to incompleteness of homogenization processes and the nature of distribution of impurities and alloying elements in the solution before cooling. In this case, the hardness of the final structures will depend not only on the level of total saturation in solid solutions, but also on the degree of preserved microchemical heterogeneity around the dissolution sources. Localization of relaxation processes as a result of the beginning and development of deformation flows along the interfacial boundaries, next to the inclusions of carbides, nonmetallic inclusions and in the adjacent areas of the steel matrix causes the formation of long-range strain fields during laser processing, accompanied by localization of the origin of dissipative structures and indicating the processes of self-organization in the irradiated material.

The practical use of laser irradiation effects is hindered by the lack of a unified concept of the nontrivial structure formation mechanism and the physical processes in the areas of laser radiation used on metals, taking into account the role of the structural heredity phenomenon, the occurrence of significant temperature gradients that lead to local plastic deformation, as well as the degree of relaxation of the riveting consequences.

2. Research methods

The studied materials are: carbon and alloy steels, as well as "model" materials – copper, nickel, steel 12X18H9T and 08X13.

Pulsed laser surface processing was carried out at laser equipment “Kvant-16” with a change in the radiation power density in the range of 80-150 MW/m^2. The degree of beam defocus (3-6 mm) and the duration of radiation ($3 \cdot 10^{-3}$ s - $6 \cdot 10^{-3}$ s) allowed us to vary the density of radiation power within wide limits.

Surface processing was applied on more than 7 samples of 10×10×15 mm for each hardening. Measuring different values, their average measures, standard deviations and trust intervals were determined at the level of reliability P=0,95 (Student's criteria was 2,447).

Microstructures demonstration were made on cross and lengthwise sections on the microscope Neophot-21. Studies of thin structure of the processed layers were carried out on the electron microscope EMMA-4 at magnifications of 15000-20000 by the foil method, which is based on the image formation by diffraction contrast.

Measurements of microhardness were carried out on PTM-3 with a load of 0,49 N. We stick to the standard rules of placement of prints, according to which the minimum distance between the centers of neighboring prints was 30 μm, from the center of the print to the end of the sample – 20 μm.

The "lines" ("steps") of shear, which occurred during plastic deformation, and their measurements were carried out using the Linnik microinterferometer MII-4M.

Identification of phase compounds in the areas of thermoinfluence was made by diffraction x-ray depth analysis, after electrolytic polishing of samples and metallographic control of the studied surface structure. The study was carried out on defractor DRON-0,5 in filtered FeK_{α} - a radiation with registration of the intensity of scintillators. X-ray lines profiles scanning was a nonstop recording at a speed of 1 deg/min.

3. Results and discussion

Experimentally proved [3] that due to uneven distribution of energy within the laser beam cross section, the irradiated metal zones have a heterogeneous structure. The maximum amount of energy is absorbed in the center of the spot, the maximum heating temperature is reached, which decreases to the periphery of the irradiated area.

It should be noted that the reduction of laser-hardened structure can be a result of a dynamic polygonization [4-6], as a result of relaxation of thermal strains, with the formation of the smallest subgrains (nanostructural effect) with a developed surface of the boundaries having low energy.

Hardened spot zones are of different morphology. In the case of laser processing with surface melting of the sample, there are generally three zones in the spot. The central, melted zone, of 5-20 μm

IOP Conf. Series: Materials Science and Engineering **680** (2019) 012019 doi:10.1088/1757-899X/680/1/012019

deep, has a dendritic-cellular structure and contains α-, γ-phases and some insoluble carbides. Nevertheless, γ-phase reaches 30-90% depending on the chemical compound of the irradiated steel.

In the cross section, columnar crystals look like equally axed grains consisting of separate blocks within each grain in a certain way, that is, the central part of the spot has a mosaic substructure.

The research found that in the areas of laser quenching due to temperature and concentration gradients there is a high degree of non-equilibrium of the liquid phase, bifurcation instability of the melt and the transformation of laminar fluid flow to turbulent. The vortices in result provide energy dissipation, abnormal mass transfer flows and dispersion of growing crystals at different scale levels.

The peculiarity of the molten zone structure is the curved grain boundaries in the form of serration, which, apparently, is a consequence of the boundaries migration due to the significant strain and plastic deformation in the micro volumes under laser processing. It should be noted that in the case of metal surface laser melting, that is, when a layer of solidified melt appears in the center of the spot, the main source of deformation is the thermal strains during crystallization, as well as strains at the boundary of the melt zone and the solid metal during crystallization.

Thus, during laser processing with melting of the surface, after rapid absorption of the laser pulse energy follows the expansion of the heated area surrounded by a large mass of cold starting metal. Thus, with a large temperature gradient there is an uneven thermal expansion. This process with local volume changes in polymorph $\alpha\leftrightarrow\gamma$ transformations leads to the appearance of the next (second) zone of thermal strain irradiated spot, relaxing due to the formation, motion and interaction of dislocations.

Here are some typical results of the study of local plastic deformation effects in the area surrounding laser quenching zone in liquid state.

Tracing plastic deformation during irradiation is a difficult task. Plastic deformation absolute value is, obviously, small due to the short-term light pulse. Moreover, normal deformation effects (changing the shape of grains, the formation of twins and deformation bands) on most materials are not found. However, in the study of "model" alloys with single-phase structure or a structure with a small amount of the second phase it is possible to prove the existence of the plastic deformation effect.

Copper and nickel alloys, as well as technical iron containing little carbon were chosen for the experiment in this paper. The samples had a polished surface to study the traces of slip and rotation of the grains. It is found that in copper and nickel alloys next to the melted zone (in the zone of thermal influence) there is a deformation relief and grain rotation.

These phenomena are connected with the summ of micro-displacements of a large number of dislocations in their slip planes, which leads to the formation of macro-displacements that appear on the surface of the polished samples in the form of slip lines or bands and relief grain boundaries. Moreover, it is possible to find the traces of sliding of several systems, the nature of which indicates the high-speed plastic deformation [7-9].

Thus, by selecting materials and irradiation conditions, the existence of local plastic deformation in the processing of metals by concentrated energy flows is proved. In the areas of laser irradiation of alloys, a complex strain occurs, accompanied by relaxation processes reducing localized stresses and mass transfer. Complex deformed dislocation dissipative structures are formed as a result.

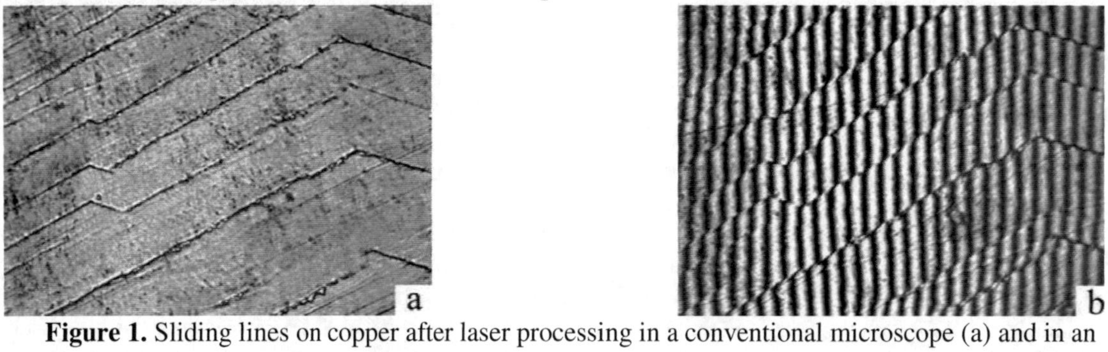

Figure 1. Sliding lines on copper after laser processing in a conventional microscope (a) and in an interference microscope (b) (×700).

Deformation is possible due to a vivid transverse slip in the grain (figure 1, a). In a large slip band (length L=10^{-2} mm), the number of dislocations accumulating on the ends (at the edges) is rather high about $n \approx 1000$.

Around such horizontal dislocation groups there will be a field of high strains directed against the source and caused stoppings in the neighboring sliding planes. Due to the large clusters of dislocations, it is possible to bend and rotate the slip bands, which also leads to the stops of moving dislocations and hardening.

Interference microscopy with the use of microinterferometer MII-4M was used to study the slip traces. Using the results obtained on the "model" alloys, an attempt is made to find the shear strain causing plastic deformation of the metal in the laser processing zones.

Along the grain boundaries, v-shaped protrusions are formed on the surface associated with the need for accommodation of different deformation in neighboring grains. The development of a strong surface relief caused by sliding and discharges inside the grains is a result of structural changes in the local plastic deformation due to the impact of CEF. By measuring the displacement of interference fringes of equal thickness we can judge about the magnitude of the shear strains that cause local plastic deformation on the surface of the irradiated spot.

As can be seen in figure 1, b, the displacement of interference lines appears on the slip lines. The measurements of these displacements (Δ) were made, then the height of the step (h) formed on the sliding line was calculated by:

$$h = \frac{\lambda}{2} \cdot \frac{\Delta}{l}, \quad (1)$$

λ - wavelength of monochromatic radiation;

Δ - displacement;

l - the distance between the interference bands.

The height of the step h is proportional to the number of dislocations coming from the current dislocation source nb (n is the amount of dislocations that came to the surface, b is the Burgers vector). The probability that dislocations output was perpendicular to the observed plane is very small. Taking the angle of the output equal to 45°, we get

$$nb = h_l = \frac{2h}{\sqrt{2}} \quad (2)$$

The elastic deformation preceding the slip in the crystal is τ/G, with τ - the applied strain, G - the shear module. When sliding starts, the elastic deformation relaxes in the area with a diameter of $2L$, with L being the length of the dislocation sliding lines [7].

Taking full conversion of total elastic deformation ($2L\tau/G$) in plastic, equal to distribution of n dislocation loops size nb, we get

$$nb = 2L\tau/G \quad (3)$$

Thus, using the expressions (2) and (3), the shear stress (τ) causing plastic deformation can be calculated (with L=10^{-2} mm).

The results of the calculations are given in table 1. The variety of calculations is associated with a large number of measurements. Despite this, the results show that the shear strain causing local plastic deformation on the surface of the irradiated spot is much higher than the yield strength of the alloys. This difference is due to stress relaxation by plastic deformation as the yield strength of the material is reached.

The role of inclusions of different compounds of the initial structure in the formation of the structure and properties of laser-hardened layers, especially on alloyed steels, should be singled out separately. Inclusions, being strain concentrators and contributors to localization of relaxation as a result of the beginning and development of deformation flows along the interfacial boundaries and in the neighboring parts of the steel matrix, cause the formation of long-range stress fields, accompanied by the localization of dissipative structures and, thus, proving the processes of self-organization in the inclusion-matrix system.

IOP Conf. Series: Materials Science and Engineering **680** (2019) 012019 doi:10.1088/1757-899X/680/1/012019

Table 1. Calculations of shear stress in laser-irradiated layers of "model" materials.

Material	Shear strain τ, MPa	Yield strength $\sigma_{0,2}$, MPa
Copper	320.9-423.8	40-80
Nickel	311.7-578.4	120-160
Technical iron	519.6-607.8	120
Steel 08X13	410.7-547.6	205
Steel 12X18H9T	386.5-485.3	195-205

During the metal-physical experiments, it was found that the laser beam energy was sufficient for melting refractory and melting fusible carbide inclusions in alloy steels, as well as for the development of mass transfer across the boundaries of the inclusion - matrix. The matrix zones next to the inclusions are filled with the components of inclusions, which are fixed under spontaneous cooling in solid solution. Apparently, relaxation microplastic deformation occur next to the inclusions, and during cooling process – relaxation of thermal strains.

It should be noted that the studies have revealed the occurrence in the irradiation zone of conditions for melting inclusions, some of which may harden in the form of thin amorphous films during melting and rapid cooling. This factor makes an additional contribution to the overall hardening of steel during laser processing.

Metallographic studies of steels have shown that since the temperature increases in the laser impact zone, recrystallization occur, which have a dynamic nature and are classified by the type of steel and its initial state. Here dynamical polygonization, primary, secondary and collective recrystallization are possible.

Particularly, it was found that in the areas of laser irradiation on samples of technical iron (low-carbon steel) there is a significant increase in hardness (2800-3200 MPa) compared to the original structure with a hardness of 1000 MPa. In this case, the metal structure near the melted zone is characterized by fineness – the average grain diameter of 3-20 μm (in the original structure – 80-100 μm) [9]. This effect is comparable with deformation hardening with high degrees of compression. Grain refinement and increase in hardness of irradiated technical iron is due to the phase riveting at $\alpha \leftrightarrow \gamma$ recrystallization, plastic deformation under the heat strains and further recrystallization [10].

Studies of the surface layers of steel 45 after CEF processing showed that the microhardness of martensite areas on the former pearlite grains is 8200-8500 MPa, and the former ferrite grains are 3200-3500 MPa at an initial hardness of 1000 MPa. This increase in the hardness of ferrite is explained by its riveting during phase recrystallization and deformation of ferrite grains due to volumetric changes in neighboring grains that have undergone martensitic transformation.

Of considerable interest are the results of a series of experiments to study the causes of achieving abnormally high hardness of steel 45 after double hardening. To do this, the samples were first subjected to ordinary quenching (volumetric heating in the furnace to the optimum temperature and cooling at a rate above critical), then local laser treatment. The results of measuring the surface hardness of the samples showed values up to 11 GPa. At high-speed laser heating of quenching martensite or low tempering martensite, the reverse martensite transformation occurs ($M \rightarrow A$). Thus, in steel there are 3 martensitic transformations – 2 direct and 1 reverse. And along with the local plastic deformation, the flowing phase riveting leads to immediate increase in the density of dislocations and other defects of crystal lattices, which pre-determines the increase in the hardness and wear resistance of steel.

4. Summary

As a result of the research, this paper presents a new approach to the study of the mechanism of structure formation in metal materials under hyper speed heating by laser radiation.

It is shown by metal-physical methods that complex structural changes occur during laser surface processing in milliseconds, which are imposed by plastic deformation caused by the appearance of internal strains during polymorphic transformation and thermal strains due to the huge temperature gradient between the heated and cold metal zone. A special role of these strains is to increase the free energy of the phases, to gain the driving force of the phase transitions, which is possible by increasing the rate of transformations and getting a special structural state of the metal, unattainable with bulk heating.

As a result, the effect of hardening of materials under the laser impact is achieved not only by martensitic transformation, dissolution of carbides, filling of matrix with their components, micro-chemical heterogeneity, but also by high temperature riveting, increasing density of defects of crystal structure, the plastic changes under the influence of strains of different nature.

Structural features of the hardening process during high-speed laser processing depend on the resulting level of strains in the processing zones, that is, on the ratio of plastic deformation processes during heating under the influence of thermal and phase strains and energy dissipation by polygonization and recrystallization.

References

[1] Mirkin L I 1975 *Physical Bases of Materials by Laser Beams Processing* (Moscow: Moscow State University)
[2] Grigoryants A G and Safonov A N 1987 *Methods of Surface Laser Processing* (Moscow: Higher school)
[3] Vedenov A A and Gladush G G 1985 *Physical Processes in Laser Processing of Materials* (Moscow: Enegroatomizdat)
[4] Tushinsky L I 1986 *Study of the Structure and Physical and Mechanical Properties of Coatings* (Moscow: Mashinostroenie)
[5] Brover A V 2005 Structural features of the process of surface hardening of steel with concentrated energy flows *Materialovedenie* **9** 18
[6] Bernstein M. L and Pustovoit V N 1987 *Heat Processing of Steel Products in a Magnetic Field* (Moscow: Mashinostroenie)
[7] Velikikh V S, Voronov I N and Goncharenko V P 1982 Radiographic study of residual stresses arising after pulse laser hardening of steels *Phys. and Chem. of Materials* **6** 138
[8] Lomaev G V and Kharanzhevsky E V 2002 Hardening surface treatment by high-speed laser recrystallization *Metal Sci. and Thermal Processing of Metals* **3** 27
[9] Brover A V 2006 The appeared effects of local plastic deformation in the surface layers of steel in the treatment of concentrated energy flows *Hardening Technol. and Coat.* **7** 27
[10] Brover A V and Dyachenko L D 2007 Self-organization of surface layers of metal materials in the processing of concentrated energy streams *Hardening Technol. and Coat.* **3** 8

IOP Conf. Series: Materials Science and Engineering **680** (2019) 012020 doi:10.1088/1757-899X/680/1/012020

The heating rate influence on the elastic modulus temperature dependence of a corrosion-resistant nickel alloy

A Yu Zhilyakov, S V Belikov, I B Polovov, A F Gibadullina, E V Elkina

Ural Federal University, 19 Mira Str., Ekaterinburg 620002, Russian Federation

E-mail: a.y.zhilyakov@urfu.ru

Abstract. The temperature dependence of the elastic modulus of nickel-chromium-molybdenum alloy C4 was investigated by the method of dynamic mechanical analysis. At all heating rates, an abnormal increase in the elastic modulus is observed upon heating in the temperature range of 200-300 °C. It is shown that an increase in the heating rate shifts the beginning of the observed effect to the region of higher temperatures. The temperature anomaly in the elastic modulus of the C4 alloy is explained from the standpoint of the formation of a short-range order in a solid solution. The thermodynamically equilibrium temperature of the "disorder – short-range order" transition in C4 alloy is determined to be 205 °C.

1. Introduction

Nickel-chromium-molybdenum alloys are one of the main structural materials of the chemical industry and nuclear energy. Their use is caused by high corrosion resistance both at room and at elevated (up to 400 °C) temperatures [1, 2]. It is also proposed to use these alloys in the environment of molten chlorides at elevated temperatures (350-650 °C) [3].

It is known that in the indicated temperature range, short-range and long-range effects are observed in alloys of the Ni-Cr-Mo system. The appearance and destruction of ordered structures is accompanied by a change in the forces of interatomic interaction in the alloy, which affects the physical properties of the alloy and its corrosion resistance. Properties such as specific heat, thermal diffusivity, Coefficient of Thermal Expansion (CTE), and electrical resistivity are fairly well studied for nickel-chromium and nickel-chromium-molybdenum alloys of different chemical compositions. The nonlinear nature of the change in these properties with temperature is general. Basically, the excesses on the curves "temperature - property" are associated with the transitions "disorder - short-range order - long-range order - short-range order" [4–8]. However, literature data on changes in the elastic modulus of these alloys upon heating and cooling are rather scarce.

Thus, the nonlinear nature of the Young's modulus of pure nickel during heating and cooling was shown in [9] (Fig. 1). Below the Curie temperature, the measurements were carried out in a saturating magnetic field. The effect was attributed to the influence of residual stresses on magnetic ordering.

In [10], the authors investigated the bulk modulus of elasticity of Ni–Cr, Fe–Ni–Cr–Mo, and Mn–Fe alloys. In the latter case, an anomaly was observed in the curve at a temperature of 470 K. The researchers associated this effect (Fig. 2) in an alloy of the Mn-Fe system with antiferromagnetic ordering. Unfortunately, there was no explanation why in the Ni-Cr system (also a ferromagnet-antiferromagnet pair) such an anomaly was not found. In the educational literature [11] it is indicated that the formation of short-range order (known as the K-state) in this system leads to an increase in the elastic modulus by 1.5%. The existence of different explanations for the same effect requires additional study of this issue.

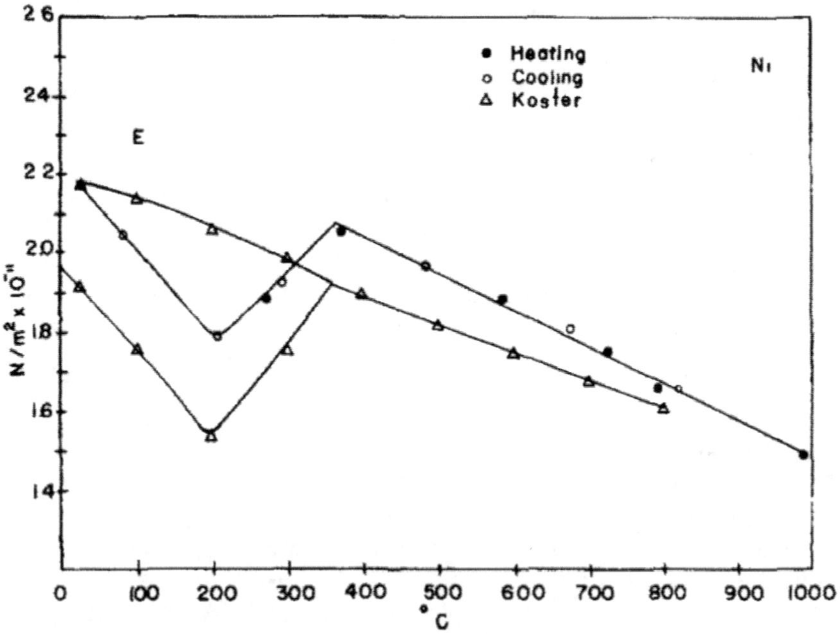

Figure 1. The temperature dependence of Young's modulus of nickel [9]

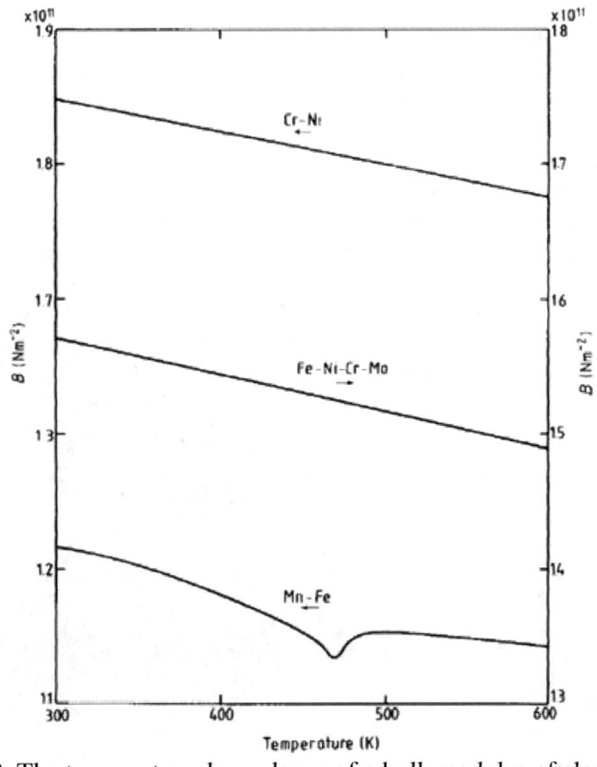

Figure 2. The temperature dependence of a bulk modulus of elasticity [10]

IOP Conf. Series: Materials Science and Engineering **680** (2019) 012020 doi:10.1088/1757-899X/680/1/012020

The aim of this work is to investigate and analyze the influence of the heating rate on the temperature dependence of the elastic modulus of a corrosion-resistant alloy of the Ni-Cr-Mo system.

2. Materials and methods

The research material was nickel-chromium-molybdenum alloy C4. The chemical composition, wt. %: Cr 14.5 ... 17.5%; Mo 14.0 ... 17.0%; Fe not more than 3.0%; P no more than 0,020%; Ti not more than 0.7%; Co not more than 2.0%; Si not more than 0.05%; S not more than 0.010%; C not more than 0.009%; Mn not more than 1.0%.

The determination of the elastic modulus of the alloy was carried out by the dynamic mechanical analysis method using Netzsch DMA 242 C instrument, and the samples were deformed according to a three-point bending scheme. In the experiment, a plate with a width of 4 ± 0.2 mm and a thickness of 1.5 ± 0.2 mm was loaded with a cyclic load of 5.3 N at a frequency of 1 Hz, leading to elastic deformation, and the temperature was raised to 500 ° C. During heating, the change in the elastic modulus of the sample was recorded and the dependence of the elastic modulus on the temperature function was obtained.

Metallographic analysis was performed using a Jeol JSM-6490LV scanning electron microscope in the backscattered electron mode (orientation-compositional contrast).

X-ray diffraction analysis (XRD) was carried out on a Bruker D8 Advance X-ray diffractometer using Kα Cu radiation (filtering using a Sol-X energy dispersive detector) in the range of reflection angles 2°=30-140° at voltage U = 40 kV, current tubes I = 40 mA using the incident beam Soller slots; the measuring diameter was 500 mm, the step was 0.02°, the pulse set time at each step was from 3 to 5 s. Qualitative phase analysis was performed in the DiffracPlus® EVA software package by selecting standards from the ICDD PDF2 database of X-ray diffraction spectra.

In the manufacture of thin sections, alloy samples were processed on sandpaper with a sequential decrease in the abrasive fraction of the paper to the minimum using the Struers LaboPol-5 installation, and then polished on diamond suspensions with a decrease in the fraction from 9 to 1 μm. To obtain orientation-compositional contrast in a SEM, it is necessary to remove the surface stresses that arose during mechanical processing, so the samples were subjected to final polishing on colloidal silicon.

3. Results and discussion

According to the data of X-ray phase analysis, the main phase in the C4 alloy is γ - a solid solution based on nickel (Figure 3). The metallographic images obtained by SEM show the grain structure of the C4 alloy (Figure 4). The grain size ranges from 100-200 microns. Inside the grains there are a large number of annealing twins, which is typical for nickel-chromium-molybdenum alloys in the quenched state.

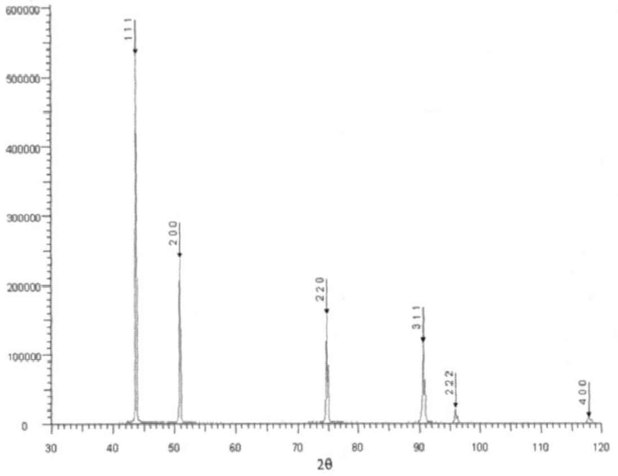

Figure 3. XRD of C4 alloy

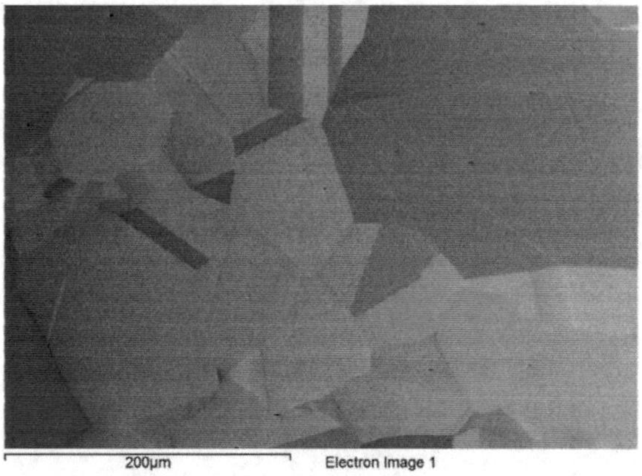

Figure 4. The structure of C4 alloy

Figures 3.4 and 3.5 show the results of an X-ray diffraction phase analysis during heating in a heat chamber. The kink at the temperature of 550 °C on line 222 (second) of the solid solution is clearly visible on the heating curve (Figure 5a). Such a kink indicates an increase in the growth rate of the sample volume above temperatures of 550 °C. This behavior can be explained by a decrease in the interatomic interaction forces caused by short-range order destruction, which, according to the literature, exists in Ni-Cr-Mo alloys in the temperature range from room temperature to 600 °C. During cooling, such inflection is also observed, but less pronounced, which indicates the restoration of short-range order in the temperature range of 550 °C (Figure 5b). No effects are observed on the thermal diffractogram from room temperature to 550 °C.

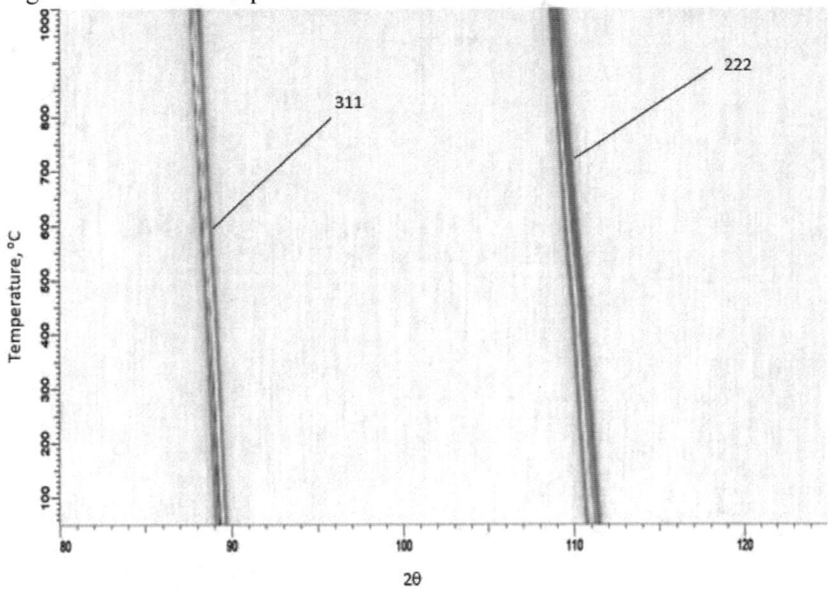

Figure 5a. Thermo-XRD of C4 alloy (heating)

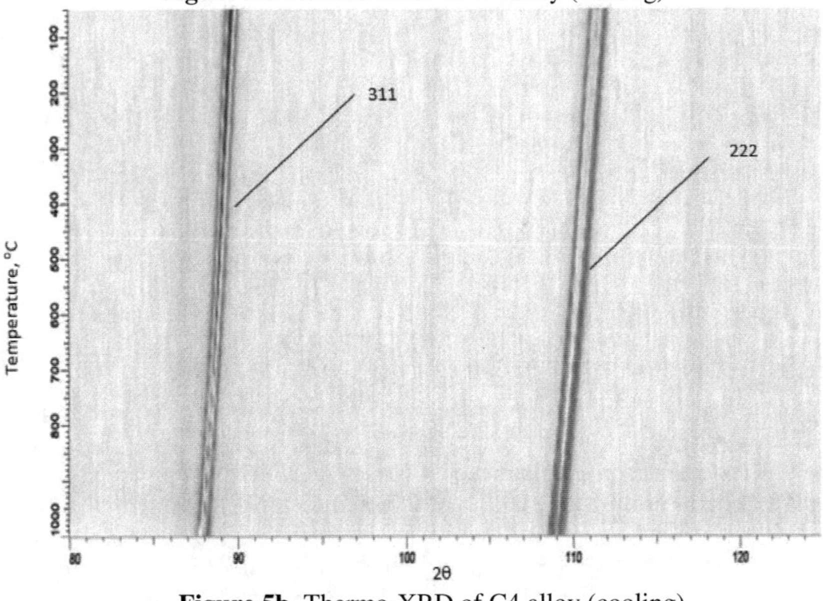

Figure 5b. Thermo-XRD of C4 alloy (cooling)

In this work, four measurements of the elastic modulus were carried out at different heating rates of the sample by dynamic mechanical analysis. The values obtained at room temperature fluctuate in the range of 210-215 GPa, which is in good agreement with the known data for this alloy. The relative change in the elastic modulus during heating for different heating rates is shown in Fig. 6.

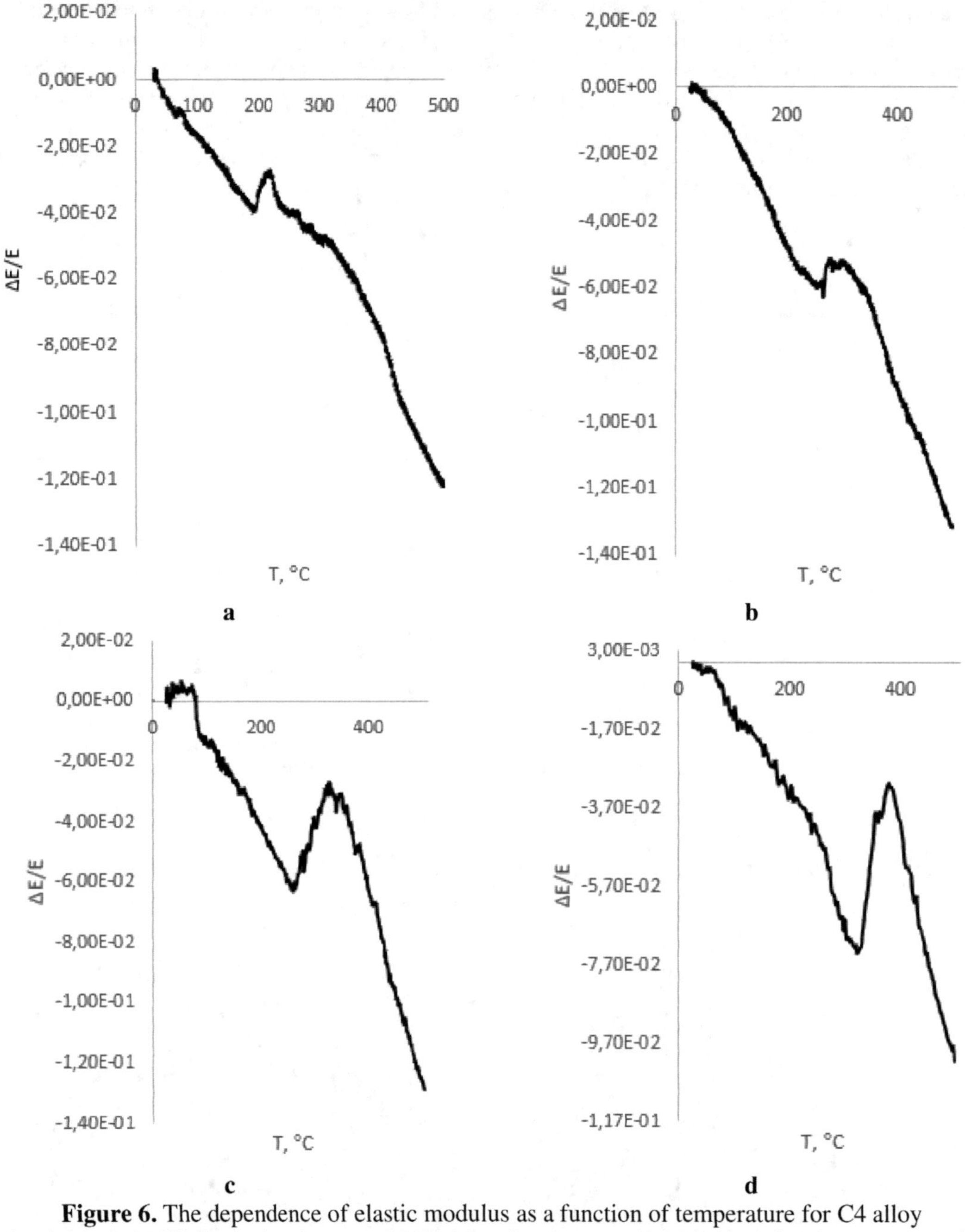

Figure 6. The dependence of elastic modulus as a function of temperature for C4 alloy
a- heating rate 1 °C/min; b- heating rate 5 °C/min; c- heating rate 10 °C/min; d- heating rate 20 °C/min

The inflection temperatures for all heating rates were determined graphically. They were determined to be 193 °C, 260 °C, 263 °C, 324 °C for heating rates of 1 °C/min, 5 °C/min, 10 °C/min, 20 °C/min, respectively. The shift of the inflection point to the region of higher temperatures with increasing heating rate is apparently associated with the leading role of diffusion in the process of formation of short-range order. Based on the obtained data, a dependence graph of the inflection temperature on the heating rate was constructed (Figure 7). The approximation shows a linear relationship with a correlation coefficient of R^2=0.8687. Extrapolation of this line to a zero heating rate gives an equilibrium inflection temperature, i.e. transition temperature "disorder - short range order". In our case, this temperature is 205 °C.

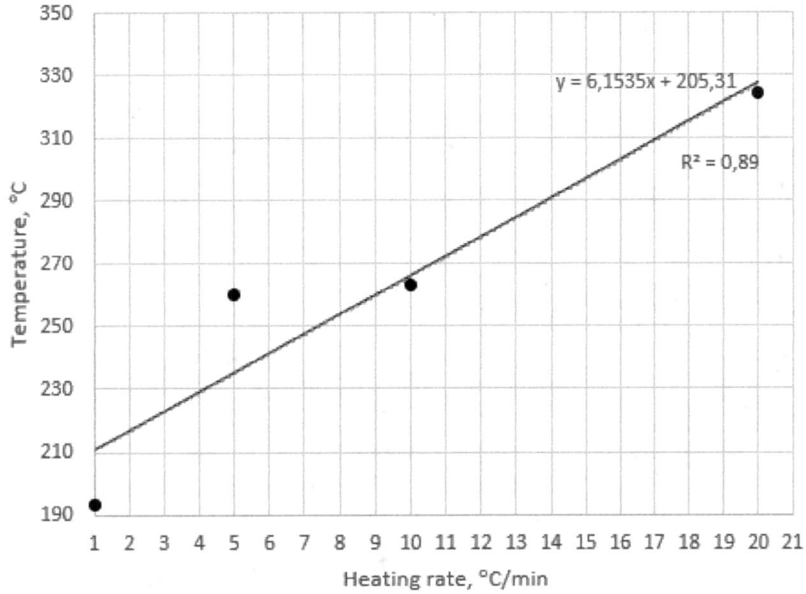

Figure 7. A dependence graph of the inflection temperature on the heating rate

The true causes of the anomaly in the elastic modulus have not been established by the methods used in this work. However, the fact that the temperature at which the anomaly occurs depends on the heating rate indicates the non-magnetic nature of this effect. The driving force of the process is diffusion, and the explanation of it by the formation of short-range order in our opinion is more correct. In addition, it is interesting that there is no reverse disorder – short-range transition during cooling (Figure 8). Herewith, the increase in the elastic modulus is about 15% relative to the initial value.

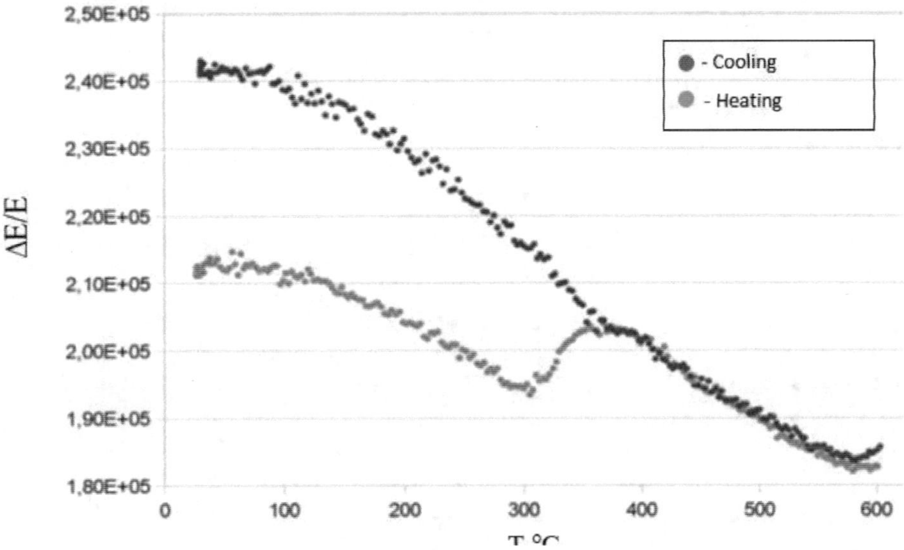

Figure 8. The dependence of elastic modulus as a function of temperature for C4 alloy

IOP Conf. Series: Materials Science and Engineering **680** (2019) 012020 doi:10.1088/1757-899X/680/1/012020

4. Conclusion

Using dynamic mechanical analysis, the dependence of the elastic modulus anomaly temperature of the C4 alloy on the heating rate was established. This indicates diffusion as the driving force of the process causing this effect. Comparing the obtained results with the published data, the cause of the temperature anomaly in the nickel-chromium-molybdenum alloy is the "disorder – short-range order" transition.

References

[1] Sims S.T., Stoloff N.S., Hagel W.S. Superalloys II. John Wiley & Sons, 1987
[2] Svistunova, T.V. Met Sci Heat Treat (2005) 47: 383
[3] Polovov I.B., Abramov A.V., Karpov V.V. et al. ECS Transactions (2017) 77, No.11: 753
[4] Lang E., Lupinc V., Marucco A. Materials Science and Engineering A (1989) 114: 147
[5] Tawancy H.M. Journal of Materials Science (1981) 16: 2883
[6] Verma A., Singh J.B., Sundararaman M. et al. Metallurgical and Materials Transactions A (2012) 43: 3078
[7] Dudova, N.R., Kaibyshev, R.O. & Valitov, V.A. Phys. Metals Metallogr. (2009) 108: 625
[8] Rabin B.H., Swank W.D., Wright R.N. Nuclear Engineering and Design (2013) 262: 72
[9] Farraro, R. & Mclellan, R.B. MTA (1977) 8: 1563
[10] Lenkkeri J. T. Journal of Physics F: Metal Physics. (1981) 11. No.10: 1991.
[11] Livshits, B.G .; Kraposhin, V.S .; Linetsky, Y.L. Physical properties of metals and alloys: M .: Metallurgy; 1980.320 s. [in Russian]

IOP Conf. Series: Materials Science and Engineering **680** (2019) 012021 doi:10.1088/1757-899X/680/1/012021

Conditions and mechanisms of the defects formation in vacuum ion-plasma coatings

V N Varavka, O V Kudryakov, I Yu Zabiyaka, O V Shvedchikova and E A Yadrets

Material Science Department, Don State Technical University, 1 Gagarin sq., Rostov-on-Don 344000, Russia

E-mail: kudryakov@mail.ru

Abstract. The results of studies of various structural features of vacuum ion-plasma PVD-coatings qualified by the authors as coating defects, are presented. Defects include a large volume of porosity, deformation of crystallites, exfoliation, etc. The systematization of defects is presented, depending on the nature of their nucleation (droplet, substructural, and growth defects) and on the character of their distribution in the coating (regular and stochastic). Coatings of various chemical composition were obtained by the method of ion-plasma deposition using both arc and magnetron evaporators. Experimental results are presented, the main array of which was obtained using a double-beam (ion / electron) electron microscope. A separate study on specific substructural defects was conducted. Statistical data on their geometrical dimensions and distribution in coatings is presented. Defects of this type have micron sizes (up to 5 μm in diameter) and a cylindrical shape with an axis perpendicular to the surface of the coating. A hypothesis about the dislocation mechanism of their nucleation on screw dislocations and the spiral mechanism of their growth has been put forward and substantiated.

1. Introduction

Obtaining of ion-plasma coatings is one of the most rapidly progressing scientific areas. Despite the high publication activity of scientists working in this field, we were not able to find deep fundamental research on the defects of ion-plasma coatings. The problem of defects is usually attributed to the category of technology and, if one wants to discuss it, not the substantive part of a scientific publication should be used, but a methodological one. Taking this consideration into account, for the present work devoted to the problem of defects in ion-plasma coatings, two aims were set: 1) to propose a classification basis for defects of this type of coating on the basis of a short review; 2) to conduct a study of one of the specific types of such defects, which is proposed to be called "defects of substructural origin".

To obtain coatings, vacuum-ion-plasma spraying units were used: the two-cathode system *PLATIT* π^{80}, equipped with two arc evaporators, as well as the modernized TINA-900 unit, equipped with magnetron evaporators. The operating modes of the coating were the following range of parameters: deposition temperature was 300...450°C; pressure in the vacuum chamber was (1.3...4.7) × 10^{-2} mbar; bias voltage was 100...150 V. In accordance with the above parameters of vacuum ion-plasma technology, the resulting coatings belong to the category of PVD-coatings. The coatings of various nitride and metal systems were studied. The main results on the study of substructural defects were

IOP Conf. Series: Materials Science and Engineering **680** (2019) 012021 doi:10.1088/1757-899X/680/1/012021

obtained on TiAlN coatings. The coatings hardness had exceeding the level of 12…15 GPa, which allowsed us to classify these coatings as wear-resistant.

Samples of steel 20Cr13, 08Cr18Ni10Ti, 38Cr2MoAl, 12Cr2Ni4 in a heat-treated state were used as substrates. The influence of the composition, structure, and properties of the substrate on the parameters of the coatings was not taken into account. Prior to coating deposition in the vacuum chamber, the surface of the samples was cleaned with a continuous stream of Ar ions for 5 min. To ensure high adhesion of the coatings, a sublayer of the metal was first applied to the sample surface cleaned by an ion beam.

To study the microstructure of the coatings and the relief of their surface with high resolution, we used ZEISS CrossBeam 340 scanning two-beam (electron/ion, SEM/FIB) microscope with an integrated X-Max EDAX energy-dispersive X-ray detector (Oxford Instruments) for microchemical surface analysis.

2. Classification signs of defects in ion-plasma coatings

One of the main conditions for defect-free coatings deposition using ion-plasma technology is the minimum branching of the surface relief of the substrate. For the practical implementation of the technology, the roughness with parameters not lower than $R_a \leq$ 0,12 μm; $R_z \leq$0,6 μm is considered a normative condition for the relief. Failure to comply with these conditions can lead to the formation of "regular growth defects" in the coating during deposition in the form of porosity, deformation of crystallites and lattices, internal stresses, etc. Separately located microroughness sites on a relatively smooth relief of the substrate lead to misorientation of the axes of the growing crystallites. It causes deformation of crystallites and forms incoherent intergranular boundaries with high porosity. A high density of microroughness sites leads to the formation of a large volume of porosity near the coating-substrate interface. A large macro-roughness of the surface of the substrate (for example, during rough grinding) forms high stresses in the coating. With weak cohesive bonds in the coating, it can lead to the coating delamination, and with weak adhesion, to the complete coating exfoliation. Coatings containing such defects generally do not meet even the most mild operating requirements in terms of protecting products from wear, corrosion, etc. impacts.

If the specified requirements for the surface roughness of the substrate are met and the coating is uniform in density and structure, this does not exclude the appearance of defects of a random nature in the coating. Such defects may include:

1) defects caused by the presence of a droplet phase;

2) defects of substructural origin.

Since the appearance of both types of defects in a specific microvolume of the coating is random in nature, unlike "regular growth defects", it is logical to combine them into a class of "stochastic defects". Among them, "droplet defects" are mainly formed when using powerful thermal evaporators of a vacuum installation, as well as during the deposition of low-melting coating elements. "Droplet defects" can be almost completely eliminated by magnetron evaporation. Examples of the droplet stochastic defects in ion-plasma coatings are shown in figure 1. On the surface, they have a characteristic shape with a flocculent configuration, which is indicated by dark arrows in figure 1 a. Getting on the coating surface during its application, the droplet phase disrupts the laminar dynamics of the normal growth of the coating, which manifests itself in the form of pores of irregular shape in the cross section of the coating (figure 1 b). In multilayer coatings, pores and structural inhomogeneities formed in the zones of the droplet phase can have an elongated shape. In this case, defects are located at the boundaries of neighboring layers saturated with a refractory and low-melting element (figure 1 c).

In general, all the coating defects considered above are of technological origin. Their appearance can be excluded or limited by optimization of the technological parameters of the ion-plasma method and equipment. As for defects of substructural origin, regulation or exclusion of their appearance in the coating seems unlikely, since they are caused by the nature of the real coating material. The results of the study of such defects are the main content of this work.

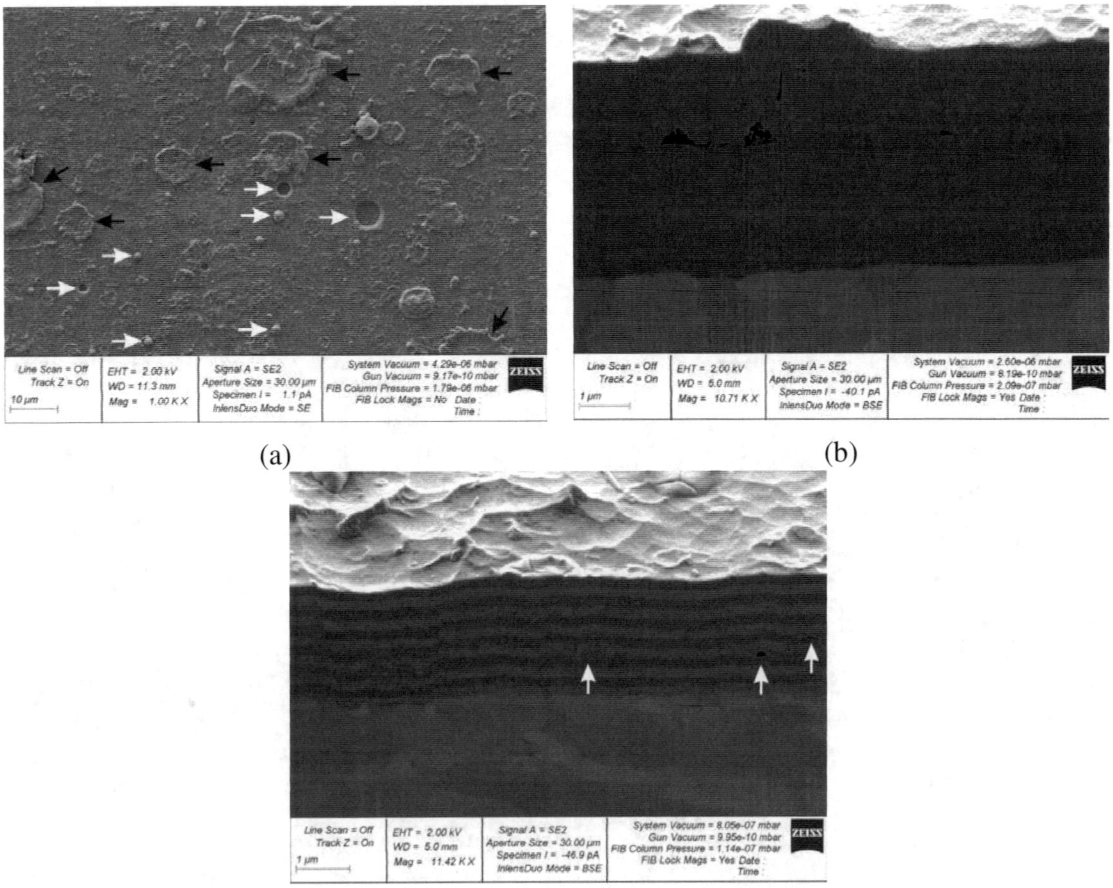

(a) (b)

(c)

Figure 1. Examples of droplet defects in TiAlN coatings, SEM: a - on the surface of the coating (shown by dark arrows); b - in the cross section of a monolayer coating; c - in the cross section of a multilayer coating (shown by bright arrows).

3. Defects of substructural origin

Examples of investigated defects of substructural origin are shown in the figure 1 a (shown by bright arrows) and in the figure 2. They have a characteristic geometrical shape: the cylindrical part of the defect is in the coating body, and the conical part protrudes above the coating surface. Under certain conditions, the defect that has formed is rejected (extruded, ejection) by the coating, leaving cylindrical recesses of the correct geometrical configuration in place of their localization. If the coating deposition process continues, the depression is filled with deposited ions of the coating material at a faster rate than the rest of the flat surface of the coating. Apparently, this is facilitated by the "edge effects" of the electromagnetic field generated by the bias voltage on the substrate. As a result, “healing” of cylindrical recesses occurs during the application process and the coating in this area is quite uniform. If the considered defects are extruded from the coating immediately before the end of the application process, the remaining cylindrical niches are clearly visible on the coating surface (figures 1 a and 2 a-d). Figure 2 shows a complete microscopic picture of the various stages of the life cycle of the substructural defects of the PVD coating of the TiAlN system.

Figure 2 a shows in a normal projection three defects of different sizes located at a distance of several microns from each other (marked by arrows). The conical shape of the protruding part of the localized defects and the cylindrical niche of the extruded defect are clearly visible. In figures 2 b-d these stages of the existence of defects are presented in a three-dimensional picture: the conical "heads" protruding from the coating are indicated by the number 1, cylindrical niches by the number 2, and the number 3 in figure 4 c indicates the place of the niche at the "healing" stage. A volumetric image was obtained by dissecting the cross-section FIB, perpendicular to the surface of the coating, and then tilting the sample towards the detector by 15-28 degrees. Figures 2 b, c show only the cross section of the coating and its surface, and figure 4 d shows the entire cross section as a whole. The preparation of cross-sections made it possible to obtain a cross section of the studied defects (figures 2 b, c). It is clearly seen that their lower base is flat, located inside the coating at various depths and is not morphologically related to the structure of the coating or to the relief of the substrate. Almost all defects visible in figure 2 have different diameters and heights, while the angle at the apex of the conical "head" (that is, the ratio of its diameter to height) varies slightly. This gives reason to believe that defects grow as the coating grows: they nucleate at some point in time on the surface of an existing growing coating and subsequently increase their diameter and height over time; the growth of the coating and the defect occurs simultaneously and at the same rate, except for the outstripping growth of the "head" of the defect, since the "head" must always be above the level of the surface of the coating until extrusion.

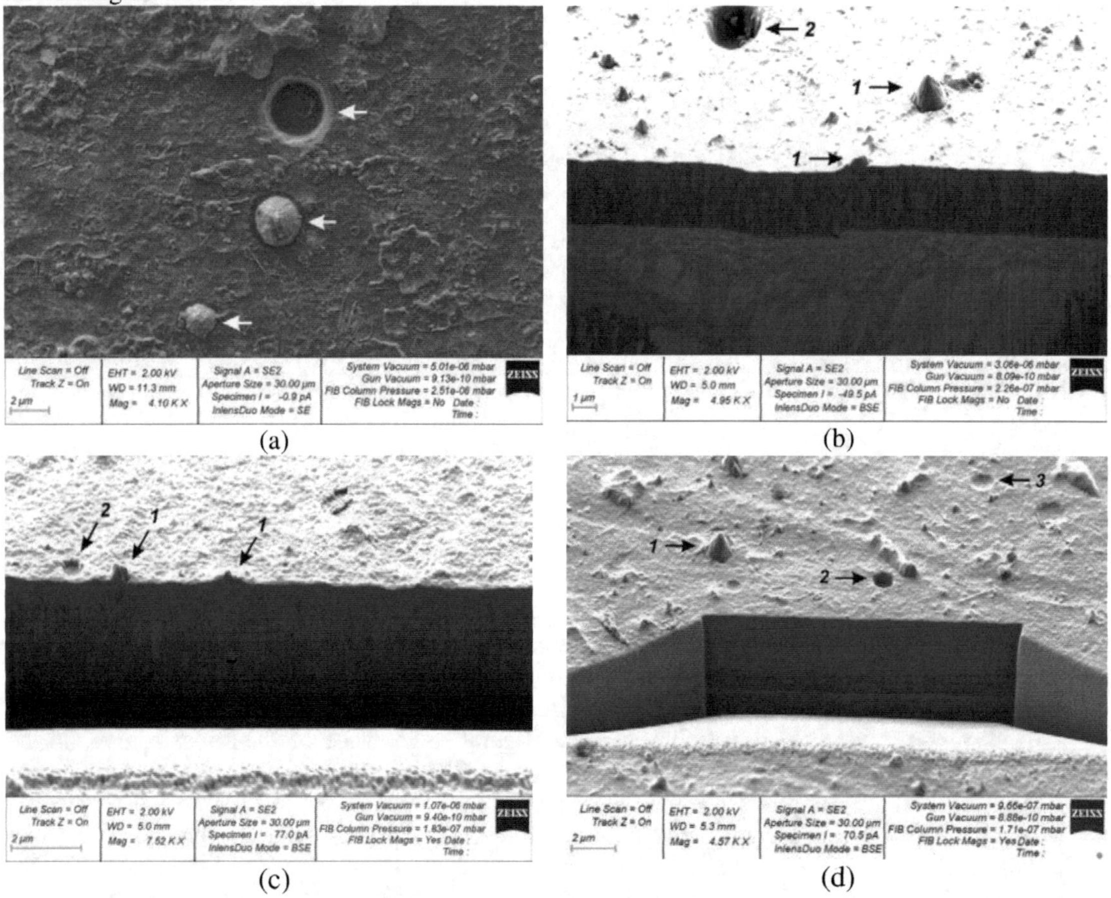

Figure 2. Substructural defects of ion-plasma TiAlN coatings, SEM: a - on the surface of the coating in normal projection (shown by arrows); b-d – on the surface of the coating and in section (cross-sections FIB, explanations in the text).

In the literature, both scientific and technological, concerning the process of formation of ion-plasma coatings, the considered substructural defects are often attributed to the droplet phase. It is believed that the drop formed in the chamber, getting on the coating surface, burns through it and "gets stuck" in the surface layer. When a drop crystallizes, its volume decreases and it "falls out", leaving a cylindrical niche. This mechanism of formation of the considered defects contradicts the growth kinetics described above and based on the above micrographs. The droplet version is also refuted by the experimental fact that defects of a similar geometry are observed not only during thermal evaporation, when the presence of the droplet phase cannot be avoided, but also during magnetron evaporation, the regime of which eliminates the formation of droplets in the chamber. Moreover, the morphology of the described defects in an arc and magnetron evaporator is identical. In addition, for coatings obtained by magnetron evaporation, these defects are also encountered in the case when all the coating components are refractory and their transition temperature to the liquid state cannot be achieved under the applied coating conditions [1, 2].

Based on the experimental data presented, a hypothesis is stated forward about the endogenous (internal with respect to the coating) origin of the considered defects. A further part of the work is devoted to the substantiation of the hypothesis stated, which is based on the following physical model.

As you know, there are no real defect-free materials (including coatings). During the formation of the crystal structure of the coating, its inherent defects of various geometries (point, linear, surface, bulk) are formed, even the most dense PVD coatings with an ordered structure (for example, monolayer thin films) contain various defects in the crystal structure [3-6]. The formation of dislocations during coating growth is natural like any crystallization process. Given the nature of the deposition of ion-plasma coatings, when the modes of shear stresses in the coating are practically absent, the formation of screw dislocations is most likely. If such a dislocation comes to the surface of the coating, it forms a helical step. During deposition, as a result of the addition of ions to the step, the latter will move along the surface, rotating around a fixed axis. This implements a well-known mechanism of spiral (helicoidal) crystal growth in the direction of the dislocation axis. The idea of this mechanism – crystal growth without two-dimensional nucleation, – belongs to Sir Frederick Charles Frank (1949), like many creative points of the theory of dislocations. In 1951, F. Frank, W. Barton, and N. Cabrera proved the dislocation growth mechanism experimentally [7]. They also managed to explain the high growth rates of the crystal faces, their relief, many morphological features in the volume of crystals, their defects, physical properties, etc. Later, the dislocational growth mechanism was used to explain the formation of "crystalline whiskers".

With a high density of screw dislocations on the surface, the distance between them is small and the atomic layers formed on adjacent steps merge. In this case, the surface growth occurs as a united front and gives a relatively flat plane of the coating [8]. Dislocations remote over long distances form single cones. This is due to the fact that the sections of the step closer to the dislocation axis rotate around it faster and require a smaller amount of deposited ions (per unit time) for growth than the remote sections. Thus, spiraling during helical growth occurs from the periphery to the axis of dislocation, which determines the conical shape of the protruding part ("head") of the growing crystal in figure 2. Schematically, the mechanism of spiral (helicoidal) growth with its main parameters is shown in figure 3. Growth spiral steps with sizes λ and h occurs in the direction of speed υ, and the growth of the helicoid (conical "head" of the defect) – in the direction ω. The shape of helicoid helices (i.e. step sizes) is determined by their growth rate, which, in turn, depends on the concentration of components in the growing crystallite. The shape of the areas of growing spirals at a fast growth rate (with large supersaturations) is close to circular, at a lower growth rate (with small supersaturations) polygonal. With an increase in the growth rate (with an increase in supersaturation), the angle of the growth cone becomes steeper; the height of the steps in such spirals is large [9-12].

The above features of helicoidal growth suggest that the considered defects of ion-plasma coatings, shown in Figure 2, are formed with a high growth rate. This is indicated by the cylindrical shape of their "base" buried in the coating and the sharp conical shape of the "head" protruding above the surface of the coating. These geometrical features also suggest the enrichment of conical crystallites with one of the coating components. Apparently, a higher growth rate of the spiral faces of the considered defect in comparison with the surrounding volume of the coating is the main reason for the isolation of the considered substructural defects in the coating and their subsequent extrusion.

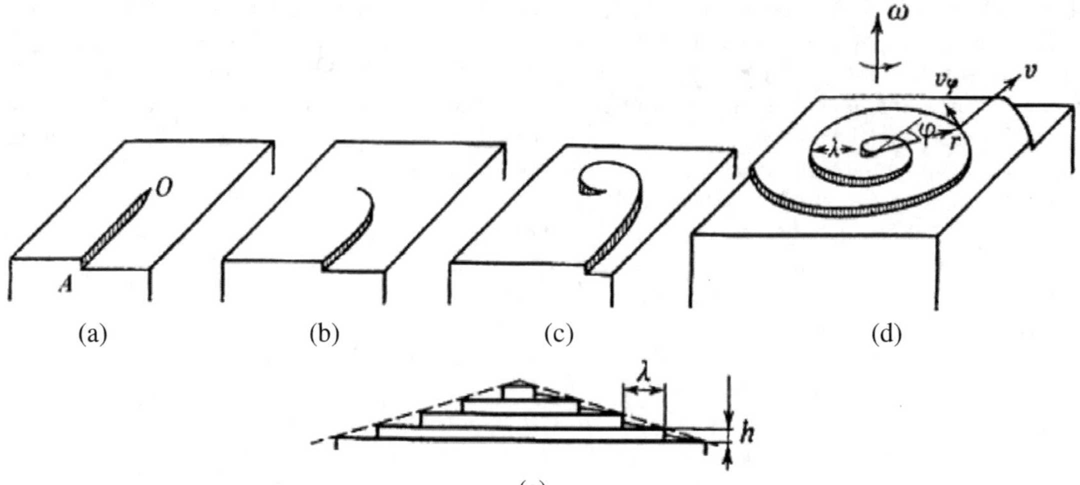

Figure 3. Successive stages of step formation around the point of exit of a screw dislocation to the surface.

AO - the initial step formed by the exit to the surface of the screw dislocation; the dislocation line and its Burgers vector are located at point O perpendicular to the surface; the definition of the parameters in figures **d** and **e** are given in the text [13].

Thus, from a terminological point of view, the defects of PVD coatings that we are considering are substructural in origin, since they arise at dislocations; at the place of localization they are stochastic, since the occurrence of dislocations in the coating is a random process; according to morphology, they are helicoid, since they form on screw dislocations and develop according to the spiral growth mechanism.

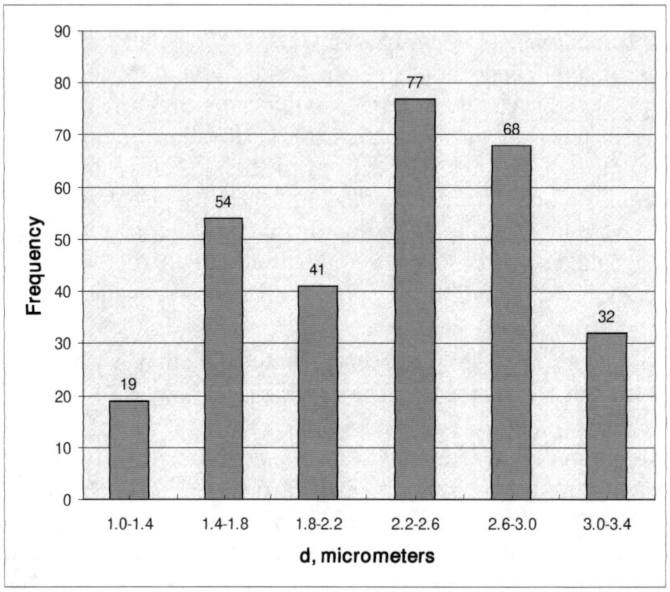

Figure 4. Frequency size distribution of substructural defects in Ti-Mo coatings.

Statistical data on the geometrical characteristics of substructural defects in the Ti-Mo coating are shown in figure 4. In the statistical analysis, the maximum diameter of the "head" measured on the surface of the coating was considered as the defect diameter *d*. The scatter of *d* values is quite significant 0.95...3.3 μm, the average value was $d_{med.}$=2,26 μm. For TiAlN coatings, similar statistical results: d=1,2…4,0 μm; $d_{med.}$=1,91 μm. The average statistical value d/h_0 (where h_0 is the height of the conical "head" of the defect projecting above the surface of the coating) for all investigated coating compositions was 1.48, which corresponds to an angle of 73 degrees at the apex of the cone "head".

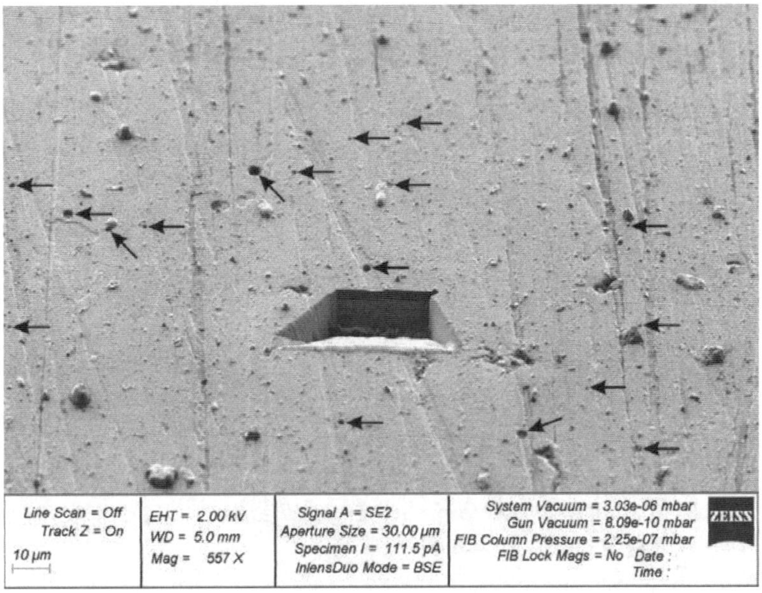

Figure 5. The surface of the TiAlN coating with substructural defects, SEM: arrows indicate extrusion defects; in the center – a rectangular cross-section FIB.

A similar statistical analysis based on the results of electron microscopic studies of the diameters of extruded defects (niches) d_e showed that the scatter of the d_e values is quite large to suggest that the extrusion of the defect occurs when a certain fixed (critical) value of d_e is reached. Figure 5 shows that extrusions of various diameters were found in the TiAlN coating, the range of values of which was d_e = 1.3...4.6 μm. This interval overlaps with the interval of values of d, and the average values are also close: $d_{med.}$ = 1.91 μm and $d_{e\ med.}$ = 2.17 μm. However, the available experimental data do not answer the question "At what geometric parameters of a growing substructural defect does it extrude from the coating?" But they form the basis for further theoretical studies of this aspect, for example, on the basis of classical theories of dislocation and nucleation.

4. Conclusion

According to our aims, the work gives certain intermediate results both on the classification features of defects in ion-plasma coatings and on the study of defects of substructural origin. However, the first of these areas involves addition and discussion, and the second can develop in two directions. First, the substructural defects that we examined are found in coatings of various compositions and structures. Therefore, they can, claim the title of "systemic technology error", despite their stochastic nature, which does not, however, have a fatal effect on the structure and properties of the coating due to its small size. We more or less fully studied this type of defects only in the TiAlN coatings; therefore, the first direction involves expanding the range of coatings. Secondly, the conducted studies leave open the question of the physical mechanism of extrusion of substructural defects. The high surface energy of the boundaries between the defect and the coating in the TiAlN system indicates their low strength, which facilitates the defect extrusion mechanism. But this does not reveal the physical nature of the mechanism itself. The issue of the mechanism requires a separate study, which is beyond the scope of this work because it lies more in the field of mesomechanics and computer modeling [14-17] than in the field of materials science.

Acknowledgments
The authors are deeply grateful to prof. L.M. Petrov (JSC "NIAT") for a fruitful discussion of the topic and valuable comments. The work was carried out with financial support from Russian Foundation for Basic Research (project code 18-08-00546).

References
[1] Varavka V N, Kudryakov O V and Ryzhenkov A V 2015 Multilayered Nanocomposite Coatings for Anti-Erosive Protection *Piezoelectrics and Nanomaterials: Fundamentals, Developments and Applications* ed I A Parinov (New York: Nova Science Publishers) chapter 5 pp 105-132
[2] Kudryakov O V, Varavka V N, Zabiyaka I Yu and Morozkin I S 2016 *Int. J. Res.* **46** (4) 117-20
[3] 2006 *Nanostructure Coatings* eds A Cavaleiro, J T de Hosson (Springer Science+Business Media, LLC) 752 p
[4] Freund L B and Suresh S 2009 *Thin film materials: stress, defect formation and surface evolution* (Boston: Cambridge University Press) 750 p
[5] Cleland A N 2003 *Foundations of Nanomechanics* (Berlin, Heidelberg, New York: Springer-Verlag) 436 p
[6] Ilyin A A, Plihunov V V, Petrov L M and Spector V S 2014 *Vacuum ion-plasma treatment* (Moscow: INFRA-M) 160 p [in Russian]
[7] Burton W K, Cabrera N and Frank F C 1951 *Phil. Trans. R. Soc. Lond.* A **243** 299
[8] Chernov A A 1980 *Modern crystallography* (*Crystallization processes* vol 3) ed B K Vainshtein (Moscow: Nauka) 408 p [in Russian]
[9] Rashkovich L N, Petrova E V, Shustin O A and Chernevich T 2003 *Sol. St. Phys.* **45** (2) 400-7
[10] Ester G R, Price R and Halfpenny P J 1999 *JOP: Applied Physics* **32** (10A) 128-33
[11] Ester G R and Halfpenny P J 1998 *J. Cryst. Growth* **187** (1) 111-7
[12] Murugakoothan P, Kumar R M, Ushasree P M, Jayavel R, Dhanasekaran R and Ramasamy P 1999 *J. Cryst. Growth* **207** (4) 325-9
[13] Available at: www.chem.msu.su/rus/teaching/materials/7cryst.pdf (accessed: 20.07.2019)
[14] Vasiliev A S, Sadyrin E V, Mitrin B I, Aizikovich S M and Nikolaev A L 2018 *Rus. Eng. Res.* **38** (9) 735-7
[15] Volkov S S, Vasiliev A S and Sadyrin E V 2018 *MATEC Web of Conf.* **226** 03018
[16] Argatov I and Mishuris G 2016 *Applied Mathematical Modelling* **40** (4) 2541-8
[17] Śliwa A, Mikuła J and Dobrzański L A 2010 *Journal of Achievements in Materials and Manufacturing Engineering* **41** (2) 164-71

IOP Conf. Series: Materials Science and Engineering **680** (2019) 012022 doi:10.1088/1757-899X/680/1/012022

Investigation of optical properties of QD-InAs/GaAs heterostructures obtained by ion-beam deposition

E E Blokhin and V A Irkha

Federal Research Center The Southern Scientific Centre of The Russian Academy of Sciences (SSC RAS), Chekhova str. 41, Rostov-on-Don 344006, Russian Federation

E-mail: holele@mail.ru

Abstract. In this paper, we investigate heterostructures with an array of InAs quantum dots encapsulated by GaAs barrier layers obtained by ion-beam epitaxy. The thickness of the layers was less than 30 nm. It is shown that this technique allows to obtain quantum dots with lateral dimensions up to 50 nm with a height of 10 nm. The density of the obtained array of quantum dots was 10^9 cm^{-2}. The studies performed using photoluminescence methods revealed the peaks of the main transitions for quantum dots at 1.1 eV (1150 nm) for samples with GaAs barrier, which corresponds to the near-infrared. The width of the main peak of the samples was about 0.2-0.25 eV, which is associated with the dispersion of quantum dots sizes. Dark current-voltage characteristics of the structures proved that the value of dark current density at 90 K is about 10^{-6} A/cm^2. The asymmetry of the dark current curves at positive and negative shifts is determined. The samples also showed that an increase of temperature leads to degradation of characteristics. When the operating temperature rises to 300 K, the density of the dark current changes in the range from 0.1 to 0.01 A/cm^2.

1. Introduction

Modern photodetectors are subject to a number of requirements, namely, speed, minimal internal noise, high operating temperatures, good detectability. At present, QDIP (with quantum dots) and QWIP (with quantum pits) photodetectors based on A3B5 multicomponent solid solutions are perspective and meet the requirements [1]. Localization of photogenerated electrons in a quantum dot in three directions contributes to the reduction of thermionic generation and dark current [1]. The creation of A3B5 nanoheterostructures with quantum dots gave rise to the development of high-efficiency injection lasers [2], solar cells [3], and photodiodes [4]. The growth of QDs occurs in the Stranski–Krastanov mode, when the substrate wetting layer is elastically stressed. An InAs/GaAs heteropair is ideal for such growth conditions. However, in terms of technology, it has limitations when used in photodetectors of infrared radiation. The most frequently used methods for growing photodetecting structures are molecular-beam epitaxy (MBE) and vapor-phase epitaxy (MOCVD) [2-8]. Disadvantages of these techniques are high cost and technological complexity of equipment. The subject of this paper is to obtain heterostructures with an array of InAs quantum dots by ion-beam deposition (IBD). This technique allows us to obtain epitaxial layers [9-10] and quantum dots [11-13] of complex semiconductors of high quality and density. The advantages of this technique include the hardware simplicity and the low-cost equipment compared to MBE and MOCVD. Additionally, the IBD technique allows ion-beam etching processes to be carried out before the crystallization of semiconductor heterostructures, which has a positive effect on their quality.

IOP Conf. Series: Materials Science and Engineering **680** (2019) 012022 doi:10.1088/1757-899X/680/1/012022

2. Experimental

Single crystal GaAs and InAs targets were used to obtain heterostructures with quantum dots. The targets of the bottom conductive material n+ contained an admixture of tellurium with a concentration of $3 \cdot 10^{17}$ cm^{-3}. For the top conductive layer, the targets were doped with zinc to a concentration of $3 \cdot 10^{17}$ cm^{-3}. Targets of closing layers and matrices for the quantum dots array did not contain doping impurities. The bottom conductive layer was 200 nm thick, the spacer layer was 180 nm, the top conductive layer was 200 nm, the barrier layer was 30 nm and 50 nm. The substrate temperature before cleaning was 450 °C. Annealing of the samples was carried out at the temperature of 550 °C. Growing heterointerfaces with quantum dots requires constant layer growth control, low growth rates and low sputtering ratio. The low sputtering ratio can be achieved by reducing the ion-beam energy. However, it is important to consider the change in the geometry of the ion-beam-target-substrate system. It was established that at the ion-beam energy E_{Ar} of 180 eV (sputtering ratio was 0.4 for GaAs and 0.5 for InAs), the ion-beam current I_{Ar} is 140 μA. At the given parameters of the ion-beam, the minimum loss factor was achieved at the incidence angle of 55obetween the ion-beam and the target. The voltage of the cathode control unit was 190 V, voltage of the ion beam was 110 V. Parameters of the discharge control unit: U = 60 V, I = 0.5 A. These conditions allowed to reach the lower limit of growth rate control of 0.7 nm/s.

The substrate temperature was 500°C when the conductive and buffer layers were applied. The temperature was reduced to 480°C before the InAs layer was applied. At the selected growth rate, the InAs wetting layer was formed in 40 seconds (by the Stranski–Krastanov mode). The wetting layer usually contains early nucleation centers. According to Stranski-Krastanov growth mode, further sputtering of InAs during 125 s allows to form islands of quantum dots of necessary density.

At the final stage of island formation, the resulting layer should be «covered» with a barrier layer to obtain a field of elastic deformations. Monocrystalline targets of unalloyed GaAs were used for barrier layer deposition. The barrier layer has been deposited at the following process parameters: ion beam energy E_{Ar} = 280 eV, ion beam current I_{Ar} = 500 μA, incidence angle between the ion-beam and the target - 65°, substrate temperature was 490 °C, growth rate was 2 nm/s.

Surface morphology and elemental composition of the grown heterostructures was investigated on a scanning high-vacuum atomic force microscope NT-MDT Solver HV (Russia) and a high-resolution two-beam scanning electron microscope with the Schottky cathode Zeiss CrossBeam 340 (Germany).

3. Results and Discussion

It was established that when InAs is deposited for 40 s (figure 1a), the first small nanoislands spread over the sample surface are observed. Further deposition of the active material (figure 1b) results in the formation of three-dimensional nanometer-sized islands (quantum dots). The formation of quantum dots occurs due to the appearance of the elastic deformation force field. The difference between the lattice constant of the active material and the lattice constant of the substrate material results in elastic stresses that increase the internal energy of the system. For achievement of equilibrium it is energetically advantageous to pass to the island growth mode with increase of wetting layer thickness.

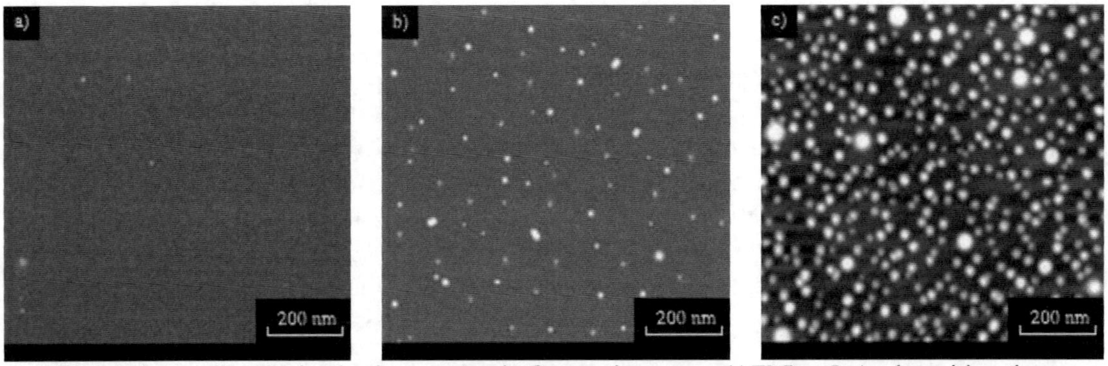

Figure 1. Results obtained using an atomic-force microscope (AFM) at InAs deposition time a) 40 sec., b) 70 sec., c) 125 sec.

When InAs is deposited for 70 s, the density of quantum dots remains insufficient for technological applications. InAs deposition time was increased to achieve higher density of quantum dots (figure 1c). Two-minute deposition leads to the development of the dot shape and formation of an array with a density of ~10^{10} cm^{-2}. Further deposition of the active material leads to the development of dot shape and high size dispersion. The studies conducted using AFM show that the best conditions for the growth of InAs quantum dots are: substrate temperature 480 °C, InAs deposition time 125 seconds.

Analysis of statistical data on the surface allowed us to determine that the majority of InAs nanoclusters are smaller than 50 nm, but there are also larger islands (more than 50 nm) on the surface, which indicates some heterogeneity. The graph of quantum dots density distribution by lateral sizes over the considered section of 10x10 microns is shown in figure 2. As shown in the histogram, the density of quantum dots within the study area was about 10^3 pcs/μm^2, which corresponds to a density of about 10^9 cm^{-2}.

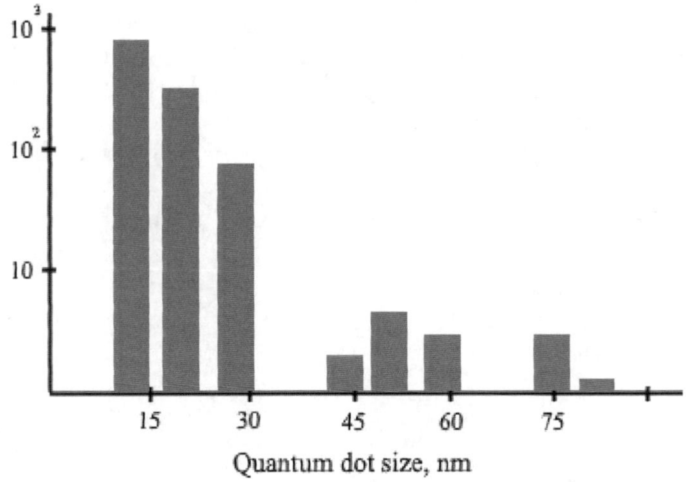

Figure 2. Size distributions of InAs quantum dots

The obtained results of the surface morphology of the samples coincide with the works of third-party studies on this subject [18-20].

Figure 3 shows the photoluminescence spectra obtained for samples with GaAs barrier layers. The peak of the main transitions in QD is at 1.12 eV, which corresponds to λ = 1.2 μm. The large width of the main peak is related to the dispersion of quantum dots sizes. With a barrier layer thickness of 30 nm, all quantum dots are in the electrostatic field of this barrier, which provides the emission of

IOP Conf. Series: Materials Science and Engineering **680** (2019) 012022 doi:10.1088/1757-899X/680/1/012022

excited electrons from QD levels into the heterostructure matrix by two main mechanisms: over-barrier thermal emission and tunneling. A decrease in the thickness of the barrier layer was accompanied by a red shift of the main peak. However, when applying a thin cover layer, the red shift disappears, which is probably due to the elastic deformation forces in the quantum dot array.

The secondary peak is shifted to the short-wave region of the spectrum (blue shift) of 1.3 eV (λ = 953 nm). This peak characterizes the emission transitions in the wetting layer. As the barrier layer is applied to an array of quantum dots, the wetting layer changes its composition and represents a $Ga_xIn_{1-x}As$ solution. This peak with a wavelength of 953 nm corresponds to 35% indium content in the solution.

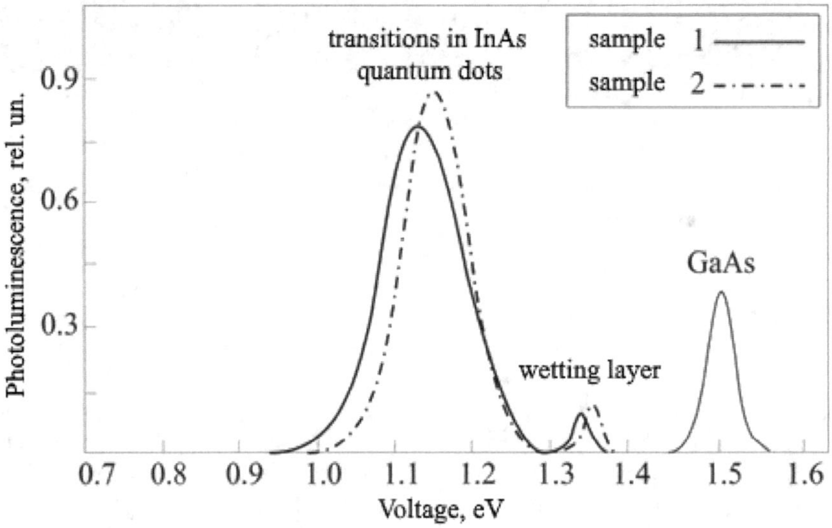

Figure 3. Photoluminescence of samples with InAs quantum dots and GaAs barrier layers in different technological modes

The last peak (1.5 eV, 890 nm) corresponds to the edge of self-absorption in the GaAs layer. This peak is a low-intensity peak because not much of the radiation reaches this area of the structure.

Figure 4 shows the dark current density dependencies for samples with GaAs barriers. Asymmetry of the dark current curves at positive and negative shift is typical. It is related to the asymmetry of quantum dots sizes.

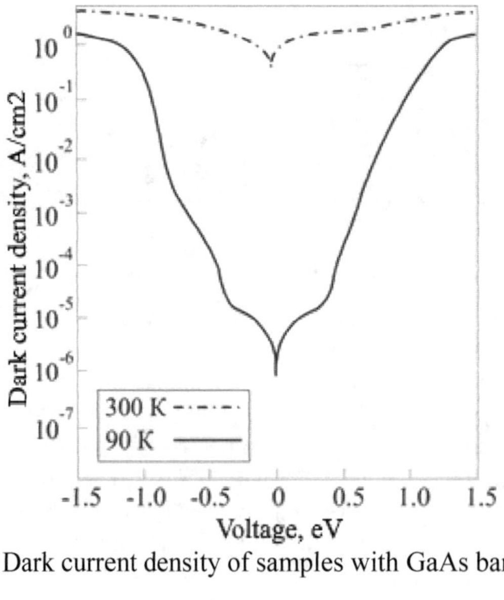

Figure 4. Dark current density of samples with GaAs barrier layer

The number of electron-filled states in quantum dots increases with a small shift. When most of the states are occupied, no increase in dark current is observed. A further shift increase leads to a reduction in the barrier and an exponential increase in dark current. At zero shift the density of dark current structure with GaAs barrier is 10^{-6} A/cm^2 (at 90 K). If the shift is increased to 0.8 V, the dark current increases by three orders of magnitude (10^{-3} A/cm^2). Further shift leads to an increase in dark current up to 100 A/cm^2, whereupon external voltage does not affect the characteristic. At 90 K, only the tunnel effect contributes to the dark current. As the temperature increases, the dark current increases due to the contribution of thermionic emission. The graph shows that the dark current value at room temperature (300 K) is about 10^{-1} A/cm^2.

4. Conclusion

The paper demonstrates the possibility of obtaining photodetecting InAs/GaAs heterostructures with quantum dots via ion-beam deposition. The investigations of the surface morphology of the samples performed using an atomic-force microscope (AFM) and a scanning electron microscope (SEM) showed the presence of InAs quantum dots with lateral sizes up to 50 nm, height of about 10 nm and density of 10^9 cm^{-2}.

According to the studies of of dark current-voltage characteristics, it is established that the dark current density in structures with GaAs barrier layer is 10^{-6} A/cm^2 at the temperature of 90 K. A sharp increase in the density of the dark current to 10^{-1} A/cm^2 at 300 K was observed. Photoluminescence peaks for the main transitions in the array of InAs quantum dot of 1.1 eV (for the GaAs barrier) were found. It was established that the usage of thinner barriers leads to a red-shift. The obtained results of photoluminescence and dark current density dependences demonstrate the possibility of applying the used heterostructures for near-infrared photodetectors.

It was revealed that it is necessary to increase the number of InAs QD layers and reduce the thickness of barrier and cover layers of the grown heterostructures in order to achieve dark current densities of less than 10^{-8} A/cm^2.

Acknowledgments

This work was carried out within the framework of the state assignment of the SSC RAS for 2019, (the project registration No. AAAA-A19-119040390081-2.

References

[1] Alfimova D L, Lunin L S, Lunina M L 2014 *Inorganic Materials* **50** 113-119
[2] Shih-Yen LIN, Yao-Jen 2001 *J. Appl. Phys.* **40** 1290–1292
[3] Kwang Moo KIM, Young Ju Park, Cheong Hyun ROH 2003 *J. Appl. Phys.* **42** 54–57
[4] Wei-Hsun Lin, Kuang-Ping Chao 2009 *J. Appl. Phys.* **106** 054512-1–054512-3
[5] David A. Ramirez, Jiayi Shao, Majeed M Hayat, and Sanjay Krishna 2010 *Appl. Phys. Lett.* **97** 212–215
[6] Rajeev V Shenoi, Jessie Rosenberg, Thomas E Vandervelde 2010 *Ieee Journal of Quantum Electronics* **46** 1051–1057
[7] Perera A G U, Aytac Y, Ariyawansa G, Matsik S G, Buchanan M 2011 *IEEE International Conf. on Nanotechnology* (Portland) pp 286–291
[8] Eui-Tae Kim, Zhonghui Chen and Anupam Madhukar 2001 *Applied physics letters.* **79** 3441–3443
[9] Lunin L S, Sysoev I A, Alfimova D L, Pashchenko A S 2011 *Journal of Surface Investigation. X-ray, Synchrotron and Neutron Techniques* **5** 559-562
[10] Pashchenko A S, Lunin L S, Irkha V A 2013 *Technical Physics Letters* **39** 726-729
[11] Lunin L S, Chebotarev S N, Pashchenko A S, Bolobanova L N 2012 *Inorganic Materials* **48** 439-444
[12] Pashchenko A S, Chebotarev S N, Lunin L S 2015 *Inorganic Materials* **51** 197-200
[13] Chebotarev S N, Pashchenko A S, Lunin L S, Williamson A, Irkha V A, Gamidov V A 2015

IOP Conf. Series: Materials Science and Engineering **680** (2019) 012022 doi:10.1088/1757-899X/680/1/012022

Technical Physics Letters **41** 661-664
[14] Joseph E Greene 2006 *Epitaxial crystal growth by sputter deposition: Applications to semiconductors.* Part I (Department of metallurgy coordinated science laboratory materials research laboratory) vol 11 pp 47–97
[15] Zhengmao Ye, Joe C 2002 *IEEE journal of quantum electronics* **38** 1234-1237
[16] Phillips J, Bhattacharya P, Kennerly S W, Beekman D W and Dutta M 1999 *IEEE J. quantum electron* **35** 936–943
[17] Lunin L S, Chebotarev S N, Pashchenko A S, Dudnikov S A 2013 *Journal of Surface Investigation. X-ray, Synchrotron and Neutron Techniques* **7** 36-40
[18] Amtout A, Raghavan S, Rotella P 2004 *J. Appl. Phys.* **96** 3781-3786
[19] Beattie N S 2014 *Solar energy materials & solar cells* **130** 241-245
[20] Wang S Y, Lin S D, Wu H W 2001 *Appl. Phys. Lett.* **78** 1023–1025

IOP Conf. Series: Materials Science and Engineering **680** (2019) 012023 doi:10.1088/1757-899X/680/1/012023

The use of the atomic force microscopy to investigate the structure of steel 14G2

V V Duka, V N Pustovoit, D A Ostapenko, L P Aref`eva and Yu M Dombrovskij

Department of Physical and applied material science, faculty «Machinery engineering and equipment» , Don state technical university, Gagarin's square 1, Rostov-on-Don 344000, Russia

E-mail: valentina.duka.92@mail.ru

Abstract. In this work, we investigated the structure of doeutectoid steel 14G2, which is mainly used for welded structures. Two methods were used to analyze the structure of this steel: optical microscopy (Metam RV-22 and Neophot - 21) and atomic force microscopy (PHYWE Compact AFM) using the semicontact mode. The investigation of the microstructure of steel was carried out for two samples obtained by different heat treatment. The first sample was quenched in water and subsequent low-temperature tempering and had a uniform troost-martensitic structure. The second sample was quenched from the intercritical interval with subsequent low-temperature tempering and had a stitch structure. The results of studying the microstructure of both samples of steel 14G2 are presented. The analysis of AFM images made it possible to establish the geometric parameters of the structural components, to estimate their shape, relief, roughness, and fractal dimension of surface.

1. Introduction

The most important problem of modern materials science is the creation of new and improving the quality of existing materials, the development of methods for analysis and diagnostics of materials. To obtain the necessary properties, the steel is subjected to various types of processing (thermal, thermomechanical, chemical-thermal, etc.), during which the phase composition and microstructure change. Traditionally, structural changes are studied using optical microscopy, which studies the substructure with sizes up to 1 μm. One of the actual directions for improving research methods is to increase the resolution of tools. As was shown in a works [1–9], scanning probe microscopy allows one to obtain information not only about the morphology and fractal properties of thin films and nonmetallic materials, but also about the micro- and nanostructure of steels. For example, in [6], the AFM method was used to investigation the structure of structural steels before and after prolonged exposure to hydrogen-containing media at elevated temperatures and pressures and showed a change in the micro- and nanostructures. In [9], AFM images of steel 45 samples were obtained in the delivery state and after long-term exploitation, and an analysis was made of the change typical of steel 45 in the delivery state of a ferrite-pearlite structure to a structure consisting of a pearlite component surrounded by a ferrite network from a disk-shaped polyhedral ferrite which the Widmannstätt ferrite plates extend.

The aim of this work is to analyze the use of the atomic force microscopy method to determine the state of steel according to microstructure data. To achieve this aim, the following problems were successively solved: 1) preparing the surface of the thin section, 2) obtaining AFM images, 3)

determining the geometric parameters and fractal dimension of the microstructure of the surface of the samples, 4) comparing the parameters of different samples.

2. The morphology and fractal analysis of steel samples

Metallographic researchs were carried out on longitudinal sections. The preparation was completed on sanding paper with final polishing on a felt wheel using GOI paste. Samples were etched with a 4% solution of nitric acid (HNO_3) in alcohol.

To analyze the state of the steel structure of two samples of structural steel 14G2 with the initial stitch ferrite-pearlite structure, which were processed at different temperature conditions, were investigated using atomic force microscopy. The first sample was quenched at 860°C in water and subsequent low-temperature tempering (200°C) and had a uniform troost-martensitic structure (figure 1a). The second sample was quenched from the intercritical interval at 730°C with subsequent low-temperature tempering (200°C) and had a stitch ferrito-martensitic structure, natural ferrite-martensitic composite. With this processing mode, it becomes possible to control the structure and properties of the composite. By changing the heating temperature for quenching from the intercritical temperature range, it is possible to change the volume ratio of the phases of the composite — ferrite and martensite. The carbon content in martensite and, consequently, its mechanical properties also change. A quantitative assessment of the volume fraction of the hardening phase in sample 2 was carried out using the KOI-1 program (figure 1b). Stitch structure on the one hand, is a defect in metallurgical production. On the other hand, samples with stitch structure can be considered as a natural composite, which has anisotropy of mechanical properties [10, 11] and can be used for highly loaded welded building structures, boiler plants, high pressure pipes, steel armor, etc.

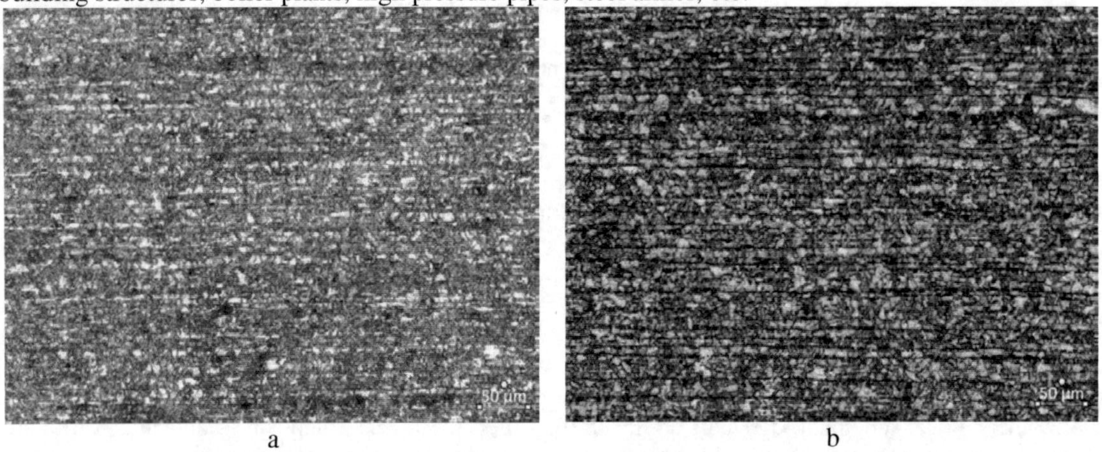

a b

Figure 1. The microstructures of the sample 1 (a) and sample 2 (b).

Microstructural analysis of metals and alloys is carried out on the basis of the analysis of the sizes of structural components and their volume fraction using Euclidean dimension. The method for describing microstructures based on defect densities has significant drawbacks - on the one hand, for many microstructures, the defect density does not determine the behavior of the material under load, and on the other hand, in the process of changing the system, the dimension and energy of defect change. Fractal geometry studies rough surfaces, which are difficult to describe using traditional methods. Fractal dimension allows to quantitatively describe the microstructures and their constituent elements, to determine the structural parameters associated with the properties of the material [12].

Figures 2 and 3 show three-dimensional AFM images of the surface of both samples. We found that at the nanoscale the structural surface formations of both samples are located fairly uniformly and are smaller than the nanotechnological boundary of 100 nm in all directions. The structural formations of the surface of sample 2 in all directions are smaller than those of sample 1.

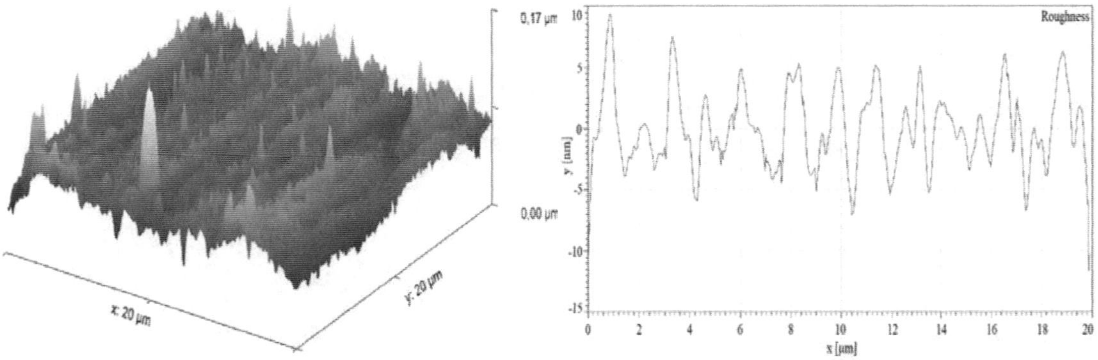

a) b)

Figure 2. Three-dimensional AFM image (a) and surface roughness (b) of the microstructure of steel 14G2 after quenching at 860°C and low tempering.

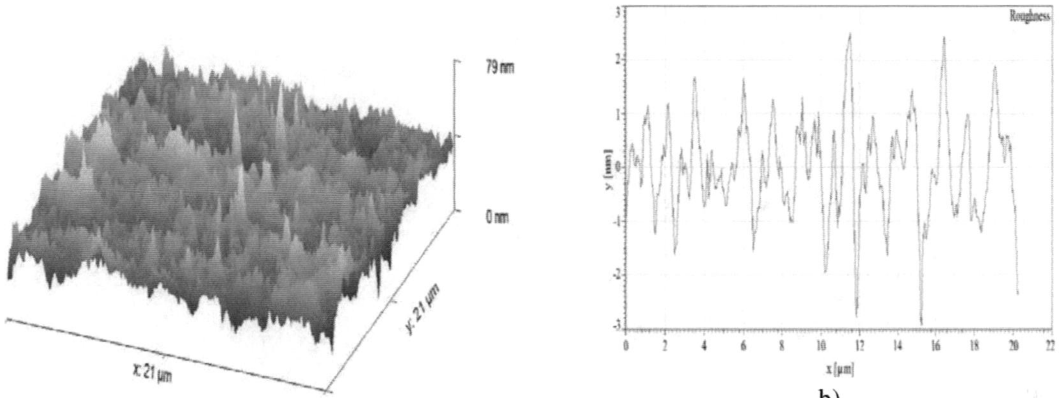

a) b)

Figure 3. Three-dimensional AFM image (a) and surface roughness (b) of the microstructure of steel 14G2 after quenching at 730°C and low tempering.

A comparative analysis of the roughness and waviness of the surface of the samples was held (table 1). The data show that sample 1 has separate small areas that differ in size, and also quite large areas of the "plateau" type (figure 2b). Sample 2 has an almost smooth surface.

Table 1. Roughness and waviness parameters of the studied samples.

Parameters (nm)	Sample 1	Sample 2
Average roughness	2,686	0,686
RMS Roughness	3,45	0,855
Maximum roughness height,	21,371	5,400
Maximum roughness depth	10,999	3,057
Maximum roughness peak height	10,373	2,881
Average maximum roughness height	14,244	3,743
Average distance between profile irregularities	957,857	878,820

Both samples are of interest from the point of view of detecting a fractal structure. The fractal dimension of the surface was determined by the method of counting cubes. Figure 4 shows graphs of fractal dimension on a logarithmic scale. As you can see, both graphs are linear, that is, the structure is fractal. The slope of the graph directly gives the fractal dimension D_f. For sample 1, $D_f = 2.30433$, for sample 2, $D_f = 2.31783$. According to [4, 12], values of fractal dimension tending to 2 are characteristic for a fairly smooth surface, values tending to 3 speak of a more complex surface structure. A comparison of the results shows that the fractal dimensions of the samples differ slightly and the samples have a fairly smooth surface. Since both samples were polished, this conclusion is valid.

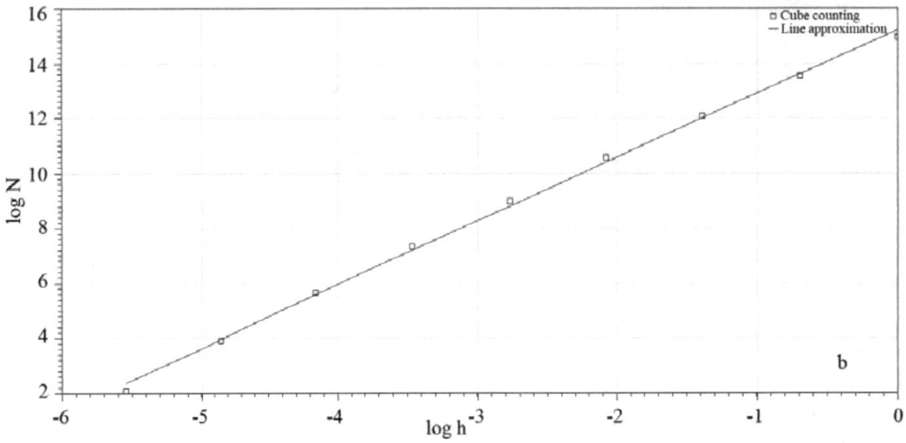

a)

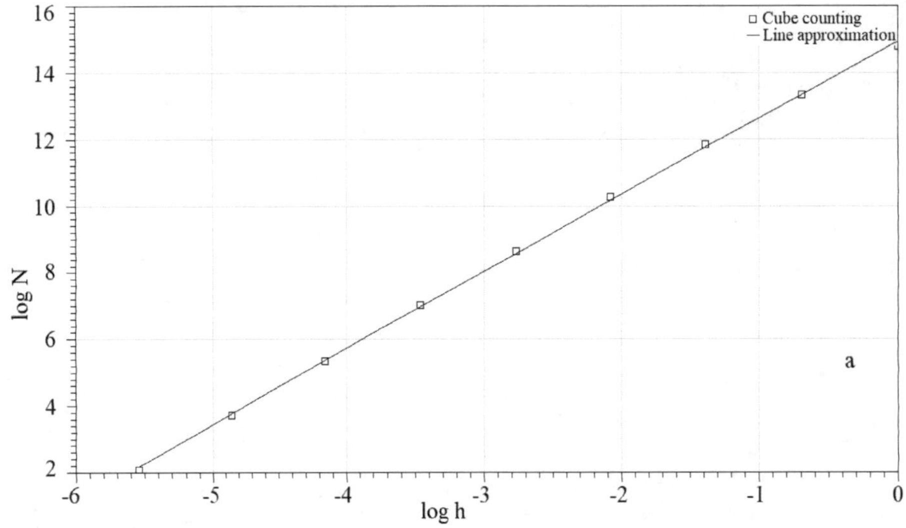

b)

Figure 4. The fractal dimension of sample 1 (a) and sample 2 (b).

3. Conclusions

Thus, researches of structural steel 14G2 with troostro-martensitic and line-wise ferritic-martensitic structures showed that atomic force microscopy is an informative method along with optical microscopy and less time consuming than electron microscopy. Atomic force microscopy gives a three-dimensional image of the structural elements of the surface of the material and allows us to evaluate their fractal dimension. At the same time, AFM does not require additional sample preparation. On the other hand, the combination of optical microscopy and AFM is optimal for achieving the main goals of metal science.

IOP Conf. Series: Materials Science and Engineering **680** (2019) 012023 doi:10.1088/1757-899X/680/1/012023

References

[1] Memnonov V P, Ul'yanov P G, 2011 *Technical Physics. The Russian Journal of Applied Physics*. **56** 12 1802
[2] Aref`eva L P, Blinov A V, Kravtsov A A, Shebzukhova I G, Serov A V *Matec Web of Conferences* 2018 **226** 03009
[3] Aref`eva L P, Shebzukhova I G 2018 *Physical and chemical aspects of the study of clusters, nanostructures and nanomaterials* **10** 27
[4] Sdobnyakov N Yu, Zykov T Yu, Bazulev A N, Antonov A S, 2009 *Vestnik Tverskogo gosudarstvennogo universiteta. Seriya «Fizika»* **6** 112
[5] Brylkin Yu V, Kusov A L 2013 *Physical and chemical aspects of the study of clusters, nanostructures and nanomaterials* **5** 33
[6] Ulyanov P.G., Usachov D. Yu., Senkovskiy B.V., Borygina K.I., Nikolaev Ph. A., Adamchuk V.K., Pushko S.V., Maltsev A.A., Balij K.S. 2012 *Vestnik of Saint Petersburg University. Physics and Chemistry* **4** 43
[7] Zuev L B, Shlyakhova G V 2014 *Materials Science* **7** 7
[8] Barannikova S A, Shlyakhova G V, Zuev L B, 2016 *Vestnik Tambovskogo universiteta. Seriya Yestestvennyye i tekhnicheskiye nauki. Prilozheniye k zhurnalu* **21** 3 882
[9] Shlyakhova G V, Zuev L B, Popova E A Vestnik 2018 *Tambovskogo universiteta. Seriya Yestestvennyye i tekhnicheskiye nauki. Prilozheniye k zhurnalu* **23** 123 581
[10] Pustovoit V N, Dolgachev Yu V, Aref`eva L P, Duka V V, Fedosov V V, Salynskih V M, 2018 *MATEC Web of Conferences* Vol. **226**. 03006.
[11] Pustovoit V N, Dolgachev Yu V, Duka V V, 2017 *Izvestiya Volgogradskogo gosudarstvennogo tekhnicheskogo universiteta* **10 (205)**. 118
[12] Ivanova V S, Balakin A S, Bunin I Zh, Oksogoev A A 1994 Sinergetika i fraktaly v materialovedenii (*Moskow: Nauka)* p 383

IOP Conf. Series: Materials Science and Engineering **680** (2019) 012024 doi:10.1088/1757-899X/680/1/012024

Development of Absorbing Coatings for Protection Against Electromagnetic Pollution of the Environment

Yu Bakhracheva

Department of Information Security, Priority Technologies Institute, Volgograd State University, Prosp. Universitetsky, 100, Volgograd, 400062, Russian Federation

E-mail: bakhracheva@volsu.ru

Abstract. Radioabsorbing materials are materials whose composition and structure provides effective absorption (with little reflection) of electromagnetic energy in a certain range of radio wavelengths.The paper investigates the weakening effects of screens of various kinds on the electromagnetic waves passing through them. With such solutions, it is possible to make estimates for the weakening effect of electromagnetic screens of arbitrary shape and for radiation with arbitrarily given characteristics.

1. Introduction

Radioabsorbing materials are materials whose composition and structure ensures effective absorption (with little reflection) of electromagnetic energy in a certain range of radio wavelengths. There is no universal classification of radioabsorbing materials. Conventionally, they can be classified by composition and principle of action.

In modern developments of radioabsorbing materials for absorption of energy of electromagnetic waves the traditional electroconducting disperse (soot, graphite, metal particles), fibrous (carbon, metal, metallized polymeric) and magnetic (sintered ferrite plates, powders of ferrites, carbonyl iron, etc.) fillers applied both separately, and together are used for absorption, forming difficult composite structures. Radioabsorbing materials made in the form of varnishes, paints, sealants, polymers, fabrics, tiles, foams, filled rubbers, building slabs, loose mixtures and other variants of various compositions are the main components in the creation of electromagnetic wave absorbers, which are used for the equipment of anechoic chambers.

In the simplest case, the screens are the windows and walls of the room where the device is working; in other cases, the role of the screen will perform specially created for this design.

In this regard, there is an urgent task to study the weakening effects of screens of various kinds on the electromagnetic waves passing through them. The solution of such problems depends on the characteristics spurious emissions, primarily from wave, its spectral composition and the radiation pattern and characteristics, namely its geometry and electromagnetic parameters of the material from which it is made.

Mathematically, problems of this kind are extremely complex and, in General, can only be solved approximately by numerical methods.

Accurate analytical solutions can be obtained only in the analysis of the simplest situations, such as the normal incidence of a plane harmonic linearly polarized wave on the screen in the form of a homogeneous plate.

IOP Conf. Series: Materials Science and Engineering **680** (2019) 012024 doi:10.1088/1757-899X/680/1/012024

2. Mathematical model

We formulate the initial relations of electrodynamics and the approximations that will be used in the calculations. All space we will divide into three areas - the material of the screen and the air environment on both sides of it.

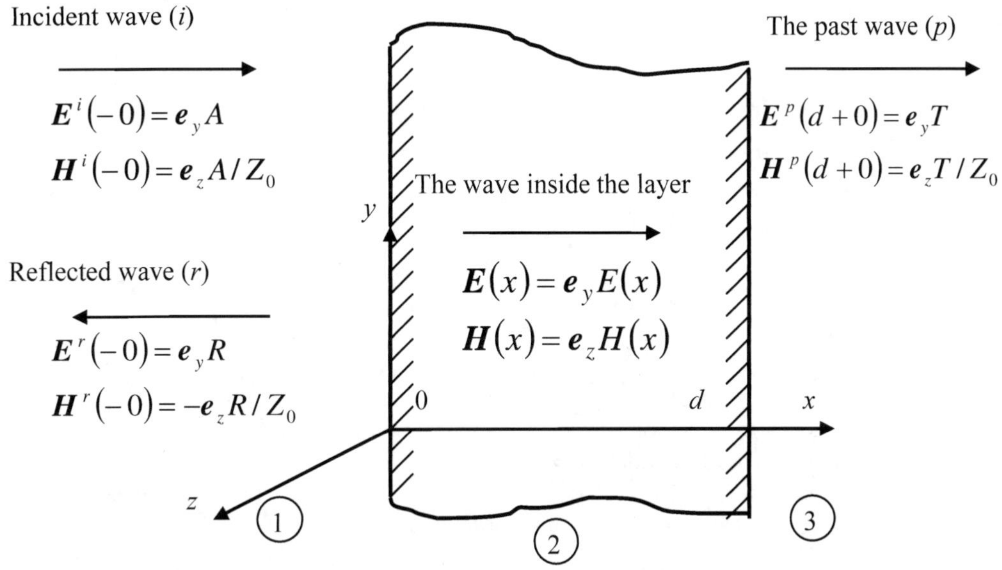

Figure 1. Passage of a plane harmonic electromagnetic wave through a homogeneous plate.

An important characteristic of radioabsorbing materials is the absorption coefficient (A), which is found by formula 1:

$$A = 1 - R - T; \tag{1}$$

R and T are the reflection and absorption coefficients, respectively, and are found by formulas 2 and 3.

$$R = \frac{U_R}{U_i} \tag{2}$$

$$T = \frac{U_T}{U_i} \tag{3}$$

U_i is the tension of the amplitude of the incident wave; U_R is the tension of the amplitude of the reflected wave; U_T is the tension of the amplitude of the transmitted wave

Relationship between electric $E(x)$ and magnetic $H(x)$ field functions at the outer boundaries of the plate:

$$E^i(-0)=A, \quad E^r(-0)=R, \quad E^p(d+0)=T \tag{4}$$

$$H^i(-0)=A/Z_0, \; H^r(-0)=-R/Z_0, \; H^p(d+0)=T/Z_0 \tag{5}$$

Z_0 is characteristic impedance of vacuum

$$Z_0 = \sqrt{\mu_0/\varepsilon_0} \tag{6}$$

ε_0 is dielectric capacitivity, μ_0 is magnetic permeability

Conjugation conditions for functions $E(x)$ and $H(x)$ at plate boundaries:

$$A+R=E(+0); \quad A-R=Z_0H(+0); \quad T=E(d-0); \quad T=Z_0H(d-0) \tag{7}$$

The amplitude reflection r_a and transmission t_a coefficients are calculated by the formulas:

$$r_a \equiv \frac{R}{A} = \frac{m_{11}Z_0 + m_{12} - m_{21} - m_{22}Z_0}{-m_{11}Z_0 + m_{12} + m_{21} - m_{22}Z_0} \quad (8)$$

$$t_a \equiv \frac{T}{A} = \frac{2Z_0}{m_{11}Z_0 - m_{12} - m_{21} + m_{22}Z_0} \quad (9)$$

Here for brevity symbols m_{ij} denote elements of the Cauchy matrix M:

$$m_{11} = \cos k_0 nd;\ m_{12} = -\frac{iZ_0}{n}\sin k_0 nd;\ m_{21} = -\frac{in}{Z_0}\sin k_0 nd;\ m_{22} = \cos k_0 nd. \quad (10)$$

The reflection r_e and transmittance t_e energy coefficients can be defined as:

$$r_e \equiv \frac{|S^r|}{|S^i|} = |r_a|^2;\quad t_e \equiv \frac{|S^p|}{|S^i|} = |t_a|^2 \quad (11)$$

S is Poynting vector for the corresponding wave

3. Discussion

An example of the reflection and transmission coefficients calculation is shown in figures 2 and 3. The graphs are constructed as a function of the ratio d/λ at a constant real part of the refractive index n for three values of the imaginary part of this index (n = 0; 3,5; 7).

Thus, the graphs give an idea of the behavior of the reflection and transmission coefficients depending on the wavelength of radiation and the absorption capacity of the material.

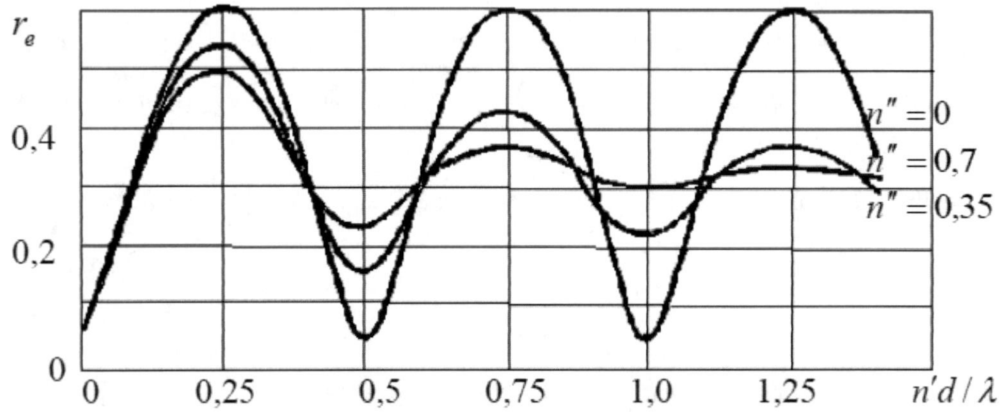

Figure 2. The calculation of the reflection coefficient.

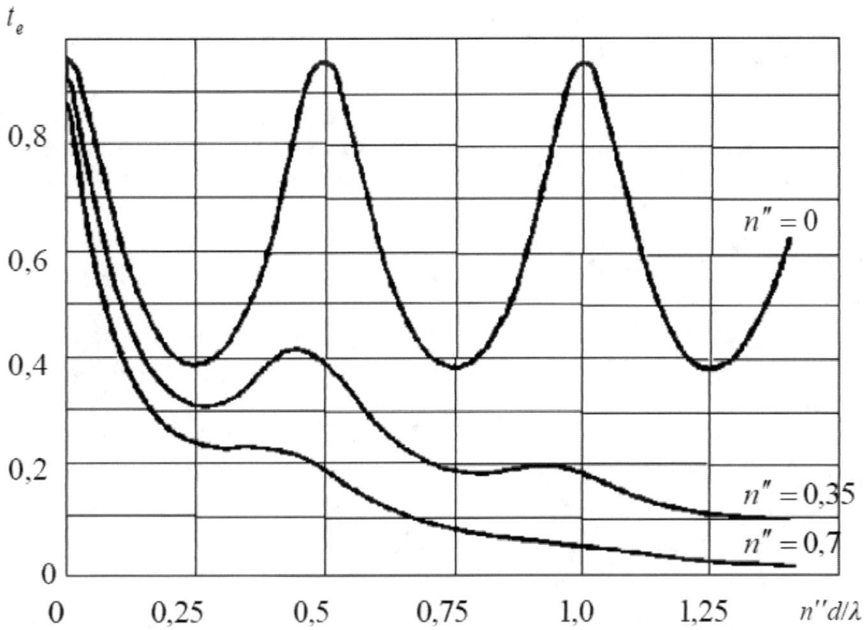

Figure 3. The calculation of the transmittance coefficient.

4. Conclusion

The proposed algorithm for calculating the reflection and transmission coefficients is easily generalized to the case where instead of a plate we have a layered medium. Thus, it is possible to solve the problem of calculating the coefficients of attenuation of electromagnetic radiation by dielectric screens of different designs.

Acknowledgments

The reported study was funded by RFBR and Volgograd region according to the research project № 19-43-340005.

5. References

[1] Kim J S 2011 Improved electrical. conductivity of very long multi-walled carbon nanotube bundle/poly(methyl methacrylate) composites *Carbon* **49** pp 2127-2133.

[2] Semaan Ch. 2013 Influence of wrapping on some properties of MWCNT-PMMA and MWCNT-PE composites *Polymer Bulletin*. **70** pp 1919-1936.

[3] Zaporotskova I V 2013 About adsorption of the polyethylene monomer unit on the single-walled corbon nanotubes surface, European Polymer Congress *Book of Abstracts Pisa* pp. 3-6.

[4] Elbakyan L S 2013 The polymers filled eith carbon nanotubes as new materials in stomatology *European Polymer Congress* pp 30-31

[5] Belonenko M B 2010 Vlijanie peremennogo.jelektricheskogo polja na. provodimost'. odnoslojnyh uglerodnyh nanotrubok poluprovodnikovogo tipa *Fizika i tehnika poluprovodnikov* **44** pp 1248-1253.

[6] Boroznin S V, Zaporockova I V, Boroznina E V 2013 Kapilljarnye jeffekty v borouglerodnyh nanotrubkah *Vestnik Volgogradskogo gosudarstvennogo universiteta. Seriya 10, Innovatsionnaya deyatelnost [Science Journal of Volgograd State University. Innovation Activity]* **8** pp 38-43.

[7] Semenova L M et al. 2013 Laws of formation of diffusion layers and solution of the diffusion problem in temperature-cycle carbonitriding of steel *Metal Science and Heat Treatment* 55 pp 34–37.

IOP Conf. Series: Materials Science and Engineering **680** (2019) 012024 doi:10.1088/1757-899X/680/1/012024

[8] Khokhlov V M 1996 Wear laws at elastic interaction *Russia Engineering Research.* **16** pp 11–12.
[9] Bakhracheva Ju S 2013 Fracture toughness prediction by means of indentation test *International Journal for Computational Civil and Structural Engineering* **9** pp 21–24.
[10] Baron A, Bakhracheva J 2007 A method for impact strength estimation *Mechanika* **66** pp 31–35
[11] Bakhracheva Yu S 2014 The Method for Lif etime Estimation Through the Mechanical Properties in Tension *Science Journal of Volgograd State University. Technology and Innovations* **11** pp 27-32. DOI: 10.15688/jvolsu10.2014.2.4
[12] Bakhracheva Y S, Vasilyev A V, Petikova T N 2016 Solution to the diffusion problem in the thermocyclic nitrocementation of steel *J. Materials science forum* **870** pp 545-549 DOI: 10.4028/www.scientific.net/MSF.870.545
[13] Vasilyev A V, et al. 2014 Valve Cam Design Using Numerical Step-by-Step Method. *Vestnik Volgogradskogo gosudarstvennogo universiteta. Seriya 10, Innovatsionnaya deyatelnost [Science Journal of Volgograd State University. Innovation Activity]* **10** pp 26-32.
[14] Vasilyev A V et al. 2016 Dynamics simulation model for the internal combustion engine valve gear *Procedia engineering* **150** pp 312-317. doi:10.1016/j.proeng.2016.07.017

Modernized RC4 encryption algorithm

V Semenov, S Makov, I Grinev

Don State Technical University, Gagarin square 1, Rostov-on-Don 344000, Russia

E-mail: vvsemenov@mail.ru

Abstract. A modernized RC4 encryption algorithm with implementation of the resulting algorithm in program for encrypting files is proposed. The program has simple and intuitive interface Testing showed that the modernized encryption algorithm is slower for 4 milliseconds than the standard RC4, which is explained by a more complex key expansion algorithm. Proposed algorithm can replace RC4 without substantial loss in file encryption.

1. Introduction

Advent of new information technologies and development of powerful computer systems for storing and processing information have increased the levels of information security and necessitated that effectiveness of information security grows along with complexity of data storage architecture. Protection of economic information becomes necessary: documents are developed to protect information; recommendations on information protection are formed, etc [1].

Nowadays, a wide range of information storage and processing systems is existed, where the information security factor is taken into account in the process of their design. Such information systems include, for example, banking or law systems for safe document management and other information systems for which the protection of information is very important [2].

Designing secure systems is impossible without use of various data encryption algorithms. Different algorithms approach different areas. There is no one universal encryption algorithm. The cryptographic strength of any algorithm is examined by the time and the number of attacks undertaken on it to crack it. More unsuccessful attempts mean more stability working of this algorithm [5-8].

Relevance of using modern encryption algorithms in information systems is that it increases security of information storage and transmission. The purpose of this work is to develop an encryption algorithm based on standard RC4 with the performance of a standard RC4 encryption algorithm and eliminating its vulnerabilities, also implementation of the modernized algorithm as a file encryption program.

In 2001, Fluhrer, Mantin and Shamir published a paper on vulnerability of RC4 key schedule algorithm [5]. They showed that among all possible keys, first few bytes of key stream are completely nonrandom. From these bytes, can be obtained information about the key used by the cipher, which substantial reduces the reliability of the algorithm. RC4 is a stream cipher and is widely used in various information protection systems in computer systems. Main features of the cipher are high speed and variable key size. A typical implementation performs 19 machine instructions per byte text. Like any stream cipher the RC4 algorithm is constructed on the basis of a uniformly distributed pseudo-random bit generator parameterized by the key.

This paper describes a possible method of algorithm modernization for eliminating the identified vulnerability.

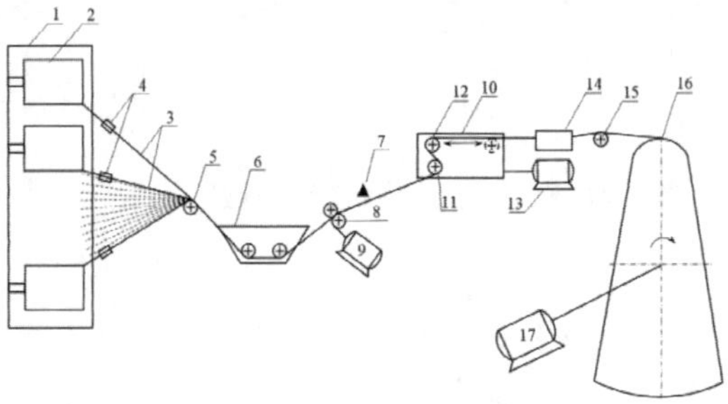

Figure 1. The "wet" winding method

The squeezing rollers 8 remove the excess of binder. The electric drive 9 changes the gap between the rollers. The tension device 10 consists of a fixed roller 11 and a roller 12 that moves along the tensioner using an electric drive 13. By this way, the "wet" tape coverage angle changes. The measurer 14 measures the tape tension. The roller 15 lays the tape on the product 16 along a special path. For this purpose, it is equipped with stacking mechanisms, which is not shown on the figure. The drive 17 rotates the product 16. Several interconnected systems control the winding.

The change of the movable roller 12 position maintains the winding tension at a predetermined level.

The improvement of the winding technology imposes stringent requirements for the drive speed and their adaptation to changing operation modes. This article is devoted to the development of such a system. The experience of using the tension control systems allowed us to determine the basic requirements for the developed systems. In particular, the transition time must not exceed the period of 0,1 second and the excessive correction must not be more than 10 %.

2. RC4 algorithm

Consider the operation of the RC4 algorithm. The RC4 algorithm consists of two parts [3-4]:

1. Key creation (also called key extension).
2. Encryption.

The key in RC4 is a sequence of bytes of arbitrary length, which is used to build the initial state of the cipher S - permutation of all 256 bytes. (Fig. 1, a).

Initially, S is filled with sequential values from 0 ... 255, and K is filled with a key (if necessary, the key is repeated to fill the entire array). Then each element S_c swaps with an element with i number, which is determined by the key element K, the element itself and the sum of the numbers of elements exchanged at previous iterations, i.e. $j = S_i + K_i + j$. The values of the counters i and j are initially equal to 0. (Fig. 1, b).

The following actions are performed for encryption. The next element of the pseudo-random permutation S_i of all bytes is swapped with the element S_j, where $i = i + 1 \bmod 256$, and $j = j + S_i \bmod 256$. As the next byte, the value of third element S is given, whose number is equal to the sum S_i and S_j. The value of counter i is initially 0, but it increases by 1 already before the first sample S_x. The value of j is also initially 0.

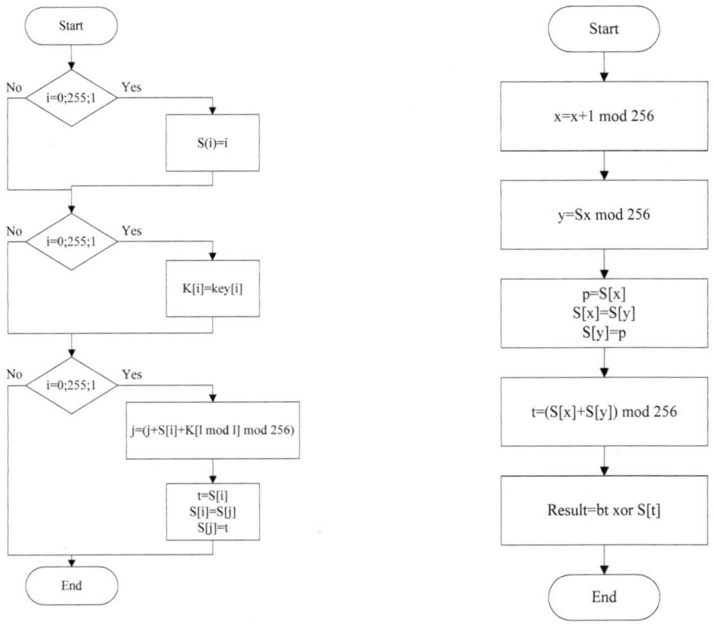

a) Flowchart of algorithm key extension b) Flowchart of algorithm byte encryption

Figure 2. RC4 encryption algorithm

3. Modernized RC4 algorithm

A lot of researches for ferrous metallurgy, pulp and paper, textile, chemical and electrical industries deal with the mathematical description of elastic tape. Differential equations obtained by a number of authors [2–8] describe the behavior of textile materials, metal tape and ropes when their winding or rewinding. There is a mathematical description of the "dry" and "wet" elastic tape [9–14].

The modernization of the encryption algorithm consists in changing the algorithm of key expansion and byte encryption.

Initially, the KArray is filled with sequentially entered key from 0 to 255. If the key is less than 256 characters, then it is repeated until the entire array is full. Next, KArray is split into two equal arrays K1Array and K2Array. Array K1Array is filled with elements using the formula $K1Array[k] = KArray[(i0 + iMixing) \bmod l]$, array K2Array is filled with elements using the formula $K2Array[r] = KArray[i0 \bmod 64]$. Next, K3Array and K4Array arrays are filled. Array K3Array is filled with elements using the formula:

$$K3Array = 2 \cdot K1Array[i] + K2Array[i] \bmod 256 .$$

Array K4Array is filled with elements using the formula:

$$K4Array = K1Array[i] + K2Array[i] \bmod 64 .$$

Then K3Array array and K4Array array are combined into a single SessionArray array. Elements from K4Array are put on odd positions and elements from K3Array are put on even position in the SessionArray array. Each element is superimposed with the serial number of this element in the SessionArray using bitwise logical operation XOR

Theoretical part

The following expression shows the winding tension [12, 13]:

IOP Conf. Series: Materials Science and Engineering **680** (2019) 012025 doi:10.1088/1757-899X/680/1/012025

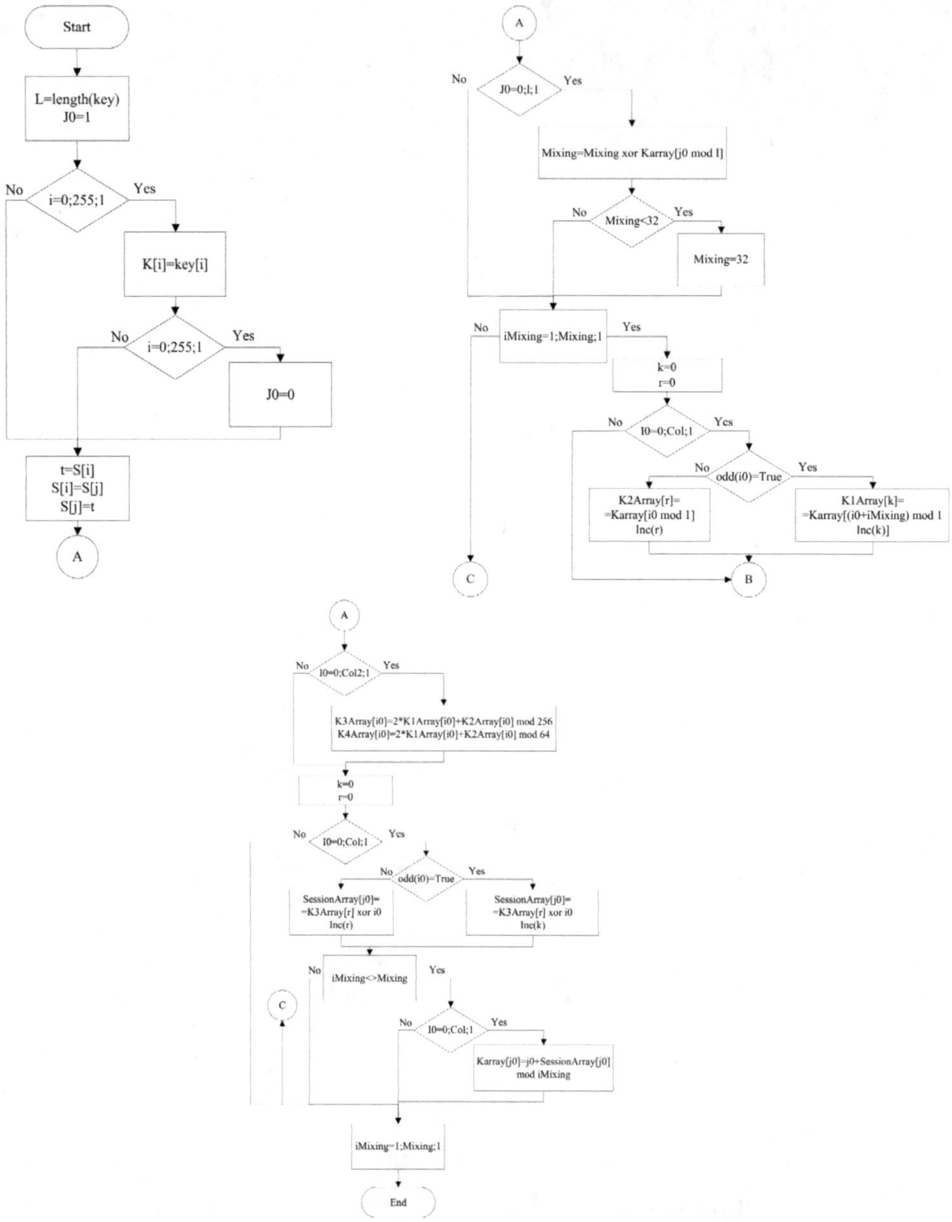

Figure 3. Flowchart modernized RC4 encryption algorithm

The actions described above are repeated Mixing times until the key is finally formed. If the Mixing value is not reached, then the values from SessionArray are transferred to KArray using the formula $KArray[j0] = j0 + SessionArray[j0] \bmod iMixing$. The value of Mixing is calculated using XOR operation of superimposing the elements of the user key. If received value is less than 32, then Mixing is equal to 32 (Fig. 3). The byte encryption process occurs as in the standard algorithm (Fig. 2, b). The initial value of the counters is calculated by the formulas: $i = Mixing \bmod l$ and $j = (Mixing + l) \bmod 64$.

As a result of modernized the algorithm, the previous vulnerability is fixed, while the speed of the algorithm does not change substantial. The algorithm can be used in commercial applications for data protection, since the software implementation of the proposed modified algorithm does not present substantial difficulties.

The appearance of the program is shown in Fig. 4

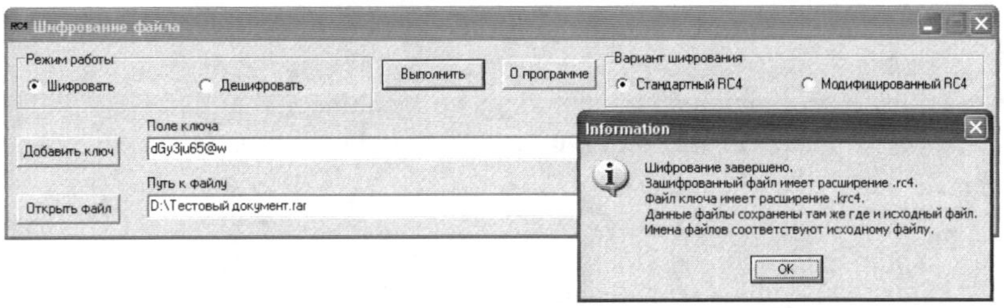

Figure 4. Appearance of data encryption program

Where, $\Delta M_д$ is the tensioner drive moment, Hm; $\Delta\Omega_д$ is the motor shaft rotation frequency, rad/s; $J_Д$ is the equivalent moment of inertia, kg/m2; k7 is the reduction gear ratio.

We use the following formulas to calculate the coefficients:

$$\left.k_1 = -\frac{\left(\frac{\partial F}{\partial \alpha}\right)^0}{\left(\frac{\partial F}{\partial S_1}\right)^0}; k_2 = -\frac{\left(\frac{\partial F}{\partial v_2}\right)^0}{\left(\frac{\partial F}{\partial S_1}\right)^0}; k_3 = -\frac{\left(\frac{\partial F}{\partial S_0}\right)^0}{\left(\frac{\partial F}{\partial S_1}\right)^0}; k_4 = -\frac{\left(\frac{\partial F}{\partial \iota}\right)^0}{\left(\frac{\partial F}{\partial S_1}\right)^0}; \iota_1 = \frac{1}{\left(\frac{\partial F}{\partial S_1}\right)^0}\right\} \quad (1)$$

The program [9] has a simple and intuitive interface, which greatly facilitates the work of unprepared user. Possible errors that user can make are processed by the program. Also, the program displays a message describing error.

One of the tasks in developing the program was to maintain speed of encryption and decryption of files, regardless of algorithm used. To examination the speed of the program was taken a 1 MB file which was encrypted and then decrypted by two encryption algorithms. The results are shown in table 1.

Table 1. Runtime in milliseconds

Algorithm	Encrypted	Decrypted
RC4	172	172
Modernized RC4	176	176

As can be seen from table 1, the modernized encryption algorithm is slower for 4 milliseconds than the standard RC4, which is explained by a more complex key expansion algorithm. Therefore, this algorithm can be used to encrypt a continuous data stream.

Security of the encryption algorithm used depends on the key length and the cryptographic strength of the algorithm itself. users exchanging ciphertexts between themselves must exchange keys so that a potential attacker could not intercept the key.

4. Conclusions

In this paper, we propose the modernization of the RC4 encryption algorithm, which allows us to fixed the existing vulnerability while maintaining the data encryption speed.

References

[1] Subashini S, Kavitha V, Journal of Network and Computer Applications, **34 (1)** (2011) 1-11
[2] Mister S, Tavares S, Proceedings of SAC 1998, LNCS, **1556** (1999)

IOP Conf. Series: Materials Science and Engineering **680** (2019) 012025 doi:10.1088/1757-899X/680/1/012025

[3] Prasithsangaree P, Krishnamurthy P, Telecommunications Program University of Pittsburgh, Global Telecommunications Conference, (2003) 1445-1449
[4] Mister S, Tavares S, Selected Areas in Cryptography (Kingston, ON, 1998), **1556** (1999) 121–143
[5] Wenbo Mao. *Modern Cryptography: Theory and Practice,* (Prentice Hall, 2003)
[6] Paul S, Preneel B, Fast Software Encryption 2004, **3017** (2004). 245–259
[7] Ferguson N, Schneier B, *Practical Cryptography: Designing and Implementing Secure Cryptographic Systems*. (Wiley, 2003)
[8] Smart N. *Cryptography: An Introduction*. (McGraw-Hill College, 2004)
[9] Semenov V V,. Kryuchkov I M, Modernized encryption algorithm RC4. Certificate of state registration of computer programs №2012616992 (03.08.2012).

IOP Conf. Series: Materials Science and Engineering **680** (2019) 012026 doi:10.1088/1757-899X/680/1/012026

Using Big data and artificial intelligence to solve the fire safety problem of potentially dangerous objects

M Shahramanyan[1], V Marchuk[2], I Shrayfel[2], D Chernyshov[2], B Meskhi[2] and O Ananova[2]

[1] Research institute "Aerocosmos", lane Gorohoviy 4, Moscow 105094, Russia
[2] Don State Technical University, Gagarin's square 1, Rostov-on-Don 344000, Russia

E-mail: 7283763@mail.ru

Abstract. In this paper, we consider a hardware-software system for evaluating the level of fire safety in real-time, which allows us to proactively monitor the dynamics of changes in the level of fire safety and provide the supervision authorities with remote access to the information about the level of fire safety

1. Introduction

Currently, the evaluation of the fire safety level is carried out based on a peri-odic inspection by the fire inspectors, issuing them instructions on the elimination of the revealed violations [1]. The inspections rate to evaluate the level of fire safety depends on the risk category of an object:

- For the high-risk category - once every 3 years;
- For the significant risk category - once every 4 years;
- For the medium risk category - no more than once every 5 years;
- For the moderate risk category - no more than once every 10 years;

The disadvantage of this approach is the significant time interval between inspections, during which a huge fire may occur causing the death of a significant number of people and great economic damage, which is unacceptable for the objects of the national economy.

2. Problem statement

To solve this problem, it is proposed to develop and implement a hardware-software system for evaluating the level of fire safety in real-time (HSS ELFS) at capital construction facilities, primarily with a massive presence of people, the block diagram of such system (HSS ELFS) is presented in Fig.1.

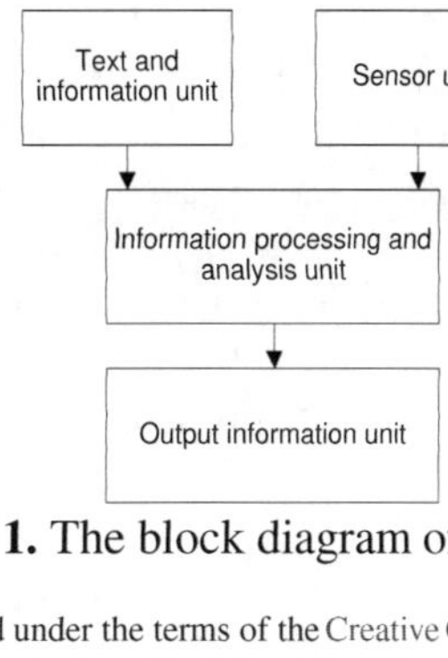

Figure 1. The block diagram of HSS ELFS.

Consider the main blocks in the presented block diagram (Fig. 1).

The sensor unit includes:

Video cameras monitoring the presence/absence of emergency exits and evacuation routes for people at the facility, the absence of obstacles on the access roads to the facility, recording the presence of people indoors, monitoring the presence and other factors affecting fire safety;

Sensors that detect the presence of locks on the exit doors;

Sensors that monitor the performance of alarm and fire extinguishing systems, the state of insulation of conductive networks, other parameters of various equipment of the facility that affect fire safety [2-3].

The text and other information unit includes instructions of the fire supervision authorities, measures aimed at eliminating the identified comments and other information related to the issue of ensuring fire safety. Essentially the data from the test and information unit together with the data from the output of the sensor unit can be considered as big data due to the large amount of data from video observation cameras in real-time and the need for high-speed processing, as well as data various types of information (continuous video data as binary codes, text and information). Processing of this information in real-time is carried out in the processing and analysis of information unit and based on the use of artificial intelligence systems [4-6].

The output information block contains an automated workstation of an operator of the facility operation service, the computer screen displays the output information of the processing and analysis unit, for example, in the form of simple semaphore signals: **green** – high level of fire safety, **yellow** – medium fire safety, **red** – low fire safety.

3. Solving the construction problem of HSS ELFS

Information from the sensor unit and the text and information unit is fed to the information processing and analysis unit, which is used as a computer program that implements artificial intelligence technology. The operation of the processing and analysis of information is as follows:

Three types of data are processed: video information from cameras, sensors data and the data from the text and information unit.

The first input of the artificial intelligence system receives video information from cameras. based on the analysis of video information it generates a signal whose amplitude is proportional to the number of people indoors, unfavorable factors that impede the evacuation of people through emergency exits, unfavorable factors that impede access of fire equipment to the facility and other adverse factors reducing the level of fire safety [5-7].

The second input receives information in the form of binary codes (0, 1) on the status of the output doors (locked/unlocked), on the status of individual elements of the alarm and fire extinguishing systems (operational/inoperative), on the insulation status of the electrical wiring (meets accepted standards/does not correspond to accepted standards) and other information affecting the level of fire safety.

The third input receives textual and information, including the instructions of the fire control authorities, information on measures taken to ensure fire safety adopted and scheduled by the administration of the facility.

For a predefined sliding time interval (ex. 30 days), the artificial intelligence system analyzes all gathered data and generates a signal depends on a combination of various factors affecting the level of fire safety. For example, during the predefined time interval there is a day/night accumulation of the large number of people indoors of the facility, the presence of packing emergency exits and evacuation routes at the facility, the presence of obstacles on the access roads to the facility, the presence of locks on the exit doors, the inoperative individual elements of alarm systems and fire. In this case, the system generates a signal which has a low level of fire safety at the facility. Other combinations of output data from the sensor unit are possible, on the basis of which all possible combinations are divided into three groups [6-8].

IOP Conf. Series: Materials Science and Engineering **680** (2019) 012026 doi:10.1088/1757-899X/680/1/012026

1. The combination of output data from the sensor unit for a specified time interval, recording a high level of fire safety.
2. The combination of output data from the sensor unit for a specified time interval, recording a medium level of fire safety.
3. The combination of output data from the sensor unit for a specified time interval, recording a low level of fire safety.

The output information unit contains a 3-D model of the object. In case of a medium and low level of fire safety, the 3-D model of the object displays conventional signs (ex. flashing) showing that there are factors that reduce the level of fire safety and these conventional signs will act as long as these factors are not eliminated.

Information about the level of fire safety of an object can be remotely represented in a single system of operational and Supervisory control and has a central authority - the unified duty dispatch service of the city . This will allow to rank all potentially flammable objects by the level of their fire safety and to take preventive measures to improve the level of fire safety at those facilities which revealed low to medium level of fire safety.

4. Conclusions

Thus, we can conclude that the HSS ELFS allows us to proactively monitor the dynamics of changes in the level of fire safety and provide the supervision authorities with remote access to the information about the level of fire safety.

References

[1] The federal law of the Russian Federation, July 22, 2008. №123-FL «Technical regulations on fire safety requirements».
[2] Brushlinsky N N, *Fire risks* (Moscow, 2005).
[3] Smetankina G I, *Mathematical modeling as a factor for improving the information support of the state fire service Russian Emergency Situations Ministry.* (Novosibirsk, 2011)
[4] Marchuk V I , Modern information technology. **19** (2014) 153-159
[5] Marchuk V I, Rumyantsev K E, Aerospace Instrument Engineering. **4**, (2004) 48-51
[6] Marchuk V I, Voronin V V, Franz V A, Scientific and technical statements of St. Petersburg State Polytechnic University. Informatics. Telecommunications. Control, **1 (115)**, (2011) 31-36
[7] Frantc V A, Voronin V V, Marchuk V I, Sherstobitov A I, Agaian S, Egiazarian K, Proceedings of SPIE - The International Society for Optical Engineering Cep. "Mobile Multimedia/Image Processing, Security, and Applications 2014", (2014) 91200
[8] Marchuk V I, Measurement Techniques. **12** (2003) 1121-1126

IOP Conf. Series: Materials Science and Engineering **680** (2019) 012027 doi:10.1088/1757-899X/680/1/012027

A mathematical model of switched-reluctance motor drive for electrical spindle in synthetic yarn machine

P Kolpakhchyan[1], A Kochin[1], B Lobov[2] and O Peregudov[2]

[1]Rostov State Transport University, 344038 Rostov-on-Don, Russia
[2]Platov South-Russian State Polytechnic University (NPI), str. Prosvesheniya 132, Novocherkassk 346428, Russia

E-mail: kolpahchyan@mail.ru

Abstract. The article deals with the modeling of switched-reluctance drive motor for electrical spindle in synthetic yarn machine. We propose to use the backward differentiation formula to solve a system of algebraic-differential equations that governs electromagnetic processes in switched-reluctance electrical machine. The considered approach allows us to avoid the use of numerical differentiation of current and turning angle dependence of the phase flux linkage. As a result, the mathematical model does not significantly depends on basic data and has good convergence. This article shows the switched-reluctance electrical machine modeling. The use of method proves the correctness of the chosen approach

1. Introduction

The production of composite materials of synthetic yarn is one of the most important areas of modern technologies. Synthetic yarns properties largely determine the characteristics of composite materials made of them. There for the improvement of synthetic yarn quality is the crucial task. One of the way to solve this problem is the spinning machines automation. The tension uniformity during winding has a significant effect on the yarn quality. We must use an adjustable electric drive to achieve the required quality of winding.

The switched-reluctance electrical machine is the most appropriate as an electrical spindle drive in synthetic yarn spinning machines. Its design simplicity and precise torque and speed control are its main advantages. The switched-reluctance electrical machine consists on electromechanical converter and power converter. Principles and methods of control have a decisive influence on the characteristics of such a machine. To realize all the benefits of switched-reluctance electrical machine, we need to optimize the active layer configuration of the electromechanical converter, laws, algorithms and parameters of control system. For this purpose, it is necessary to use the mathematical model of switched-reluctance electrical machine. This model must allow analysis of electromagnetic and electromechanical processes in an electromechanical converter and a power converter, determination of the parameters and characteristics of switched-reluctance electrical machine with various algorithms and control parameters.

The determination of current pulses parameters and control system testing at the stage of designing are the most difficult tasks of electric drives creation. This article describe the model that allows solving these tasks and takes into account the design and control system particularities of the switched-reluctance electrical machine.

2. Problem statement.

The use of models based on field theory allows us to receive the most adequate information about electromagnetic processes in the electric machine under consideration. The disadvantages of such models is a calculation complexity. Therefore, in most cases, the mathematical model uses the approaches based on the circuit theory [1-5]. The equation below describes the phase processes:

$$\frac{d\Psi_f}{dt} = U_f - i_f R_f, \quad (1)$$

where Ψ_f, i_f, U_f are flux linkage, phase current and applied voltage; R_f is phase resistance.

The flux linkage depends on phase current and the stator and rotor relative position:

$$\Psi_f = f(i_f, \gamma_r), \quad (2)$$

The joint solution of equations 1 and 2 models switched-reluctance electrical machine. The switched-reluctance electrical machine has a significant magnetic saturation [5]. The saturation is especially manifested when the rotor and phase are in a consistent position and when the stator and rotor teeth axis match. The characteristics phase flux linkage versus phase current is non-linear. Equations (1) and (2) are the system of nonlinear algebraic-differential equations, which is difficult to solve directly. Therefore, the flux linkage is the product of phase current and phase inductance [1, 5]. Then, the phase flux linkage derivative is:

$$\frac{d\Psi_f}{dt} = \frac{\partial\Psi(i_f,\gamma)}{\partial i} \cdot \frac{di_f}{dt} + i_f\omega \cdot \frac{\partial\Psi(i_f,\gamma)}{\partial\gamma}, \quad (3)$$

where $\omega = \frac{d\gamma}{dt}$ is rotor rotational speed.

This change allows us to go to the solution of the phase current differential equation.

The expression (3) consist of derivatives of phase flux linkage versus current and rotor rotation angle. Dependencies (2) are currently determined using field theory methods. The phase weber-ampere characteristics are calculated at different rotor positions. The interpolation methods is used to determine the flux linkage at intermediate values of current and rotor position. When differentiating the resulting dependencies, significant errors may appear.

Another factor cumbering the switched-reluctance electrical machine modeling using expression (3) is the fact that a voltage in the form of rectangular pulses with a steep edge is applied to the phase. The edge has breaks of the first kind. It renders difficult to apply multistep methods for solving differential equations.

For the indicated reasons, the modeling time significantly increases when using such a software as MatLAB Simulink and in some cases, the solving becomes impossible due to problems of numerical methods convergence [1]. Therefore, when switched-reluctance electrical machine modeling, we must not use the numerical determination of phase linkage derivatives versus current and relative position of the stator and rotor.

3. Switched-reluctance electrical machine mathematical model.

A switched-reluctance electrical machine consist on the mechanical converter which is synchronous reluctance machine with concentrated tooth windings on the stator and tooth rotor without winding. Unipolar current pulses provide power supply of the switched-reluctance electrical machine. Phases are not galvanically connected to each other. The current pulses are formed by individual, autonomous half-bridge voltage inverters, that are connected to the common DC link.

Each phase module of a power converter consists of two power keys (S1 and S2, S3 and S4, S5 and S6) and two diodes (D1 and D2, D3 and D4, D5 and D6). figure. 1 shows the structural-functional diagram of the switched-reluctance electrical drive for electric spindle.

IOP Conf. Series: Materials Science and Engineering **680** (2019) 012027 doi:10.1088/1757-899X/680/1/012027

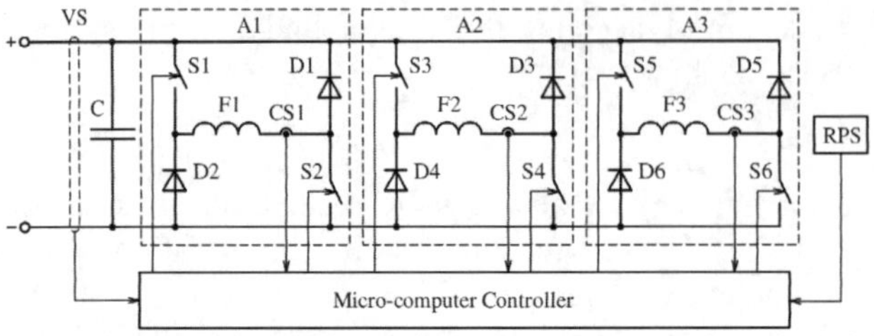

Figure 1. Structural-functional diagram of the switched-reluctance electrical drive for electric spindle/

When mathematical modeling of the switched-reluctance electrical drive we accept the following assumptions:

- phases in switched-reluctance electrical machine are not magnetically coupled, we consider independently all processes in each of them;
- power semiconductor devices in power converter present key elements with great resistance in off state, and with low resistance in on-state; state transition is prompt;
- constant voltage source power a switched-reluctance electrical machine, in the DC link we take into account the E.M.F., active resistance, inductance and capacitance;
- design model topology does not change during calculations.

Figure 2 shows the design model of the switched-reluctance electrical drive for electric spindle.

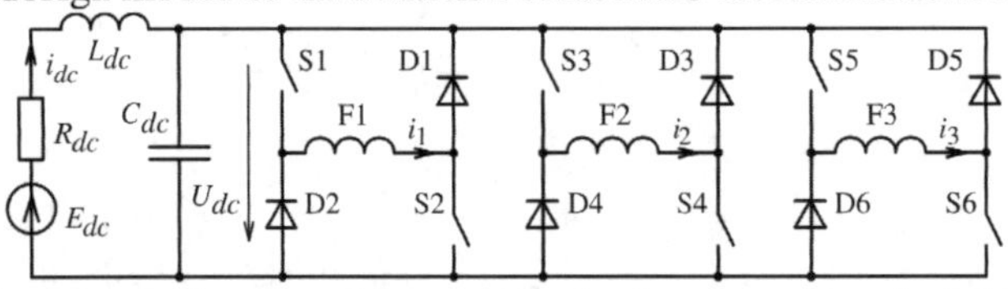

Figure 2. Structural-functional diagram of the switched-reluctance electrical drive for electric spindle

The equation system (1) and (2), that describe electromagnetic processes in phase of a switched-reluctance electrical machine, can be solved by Backward Differentiation Formula (BDF). In this case, the solution of the algebraic-differential equations system comes down to the consistent solution of an algebraic equations system. When using the first order BDF, phase flux linkage is as follows:

$$\frac{d\Psi_f}{dt} = \frac{\Psi_{fk+1} - \Psi_{fk}}{\Delta t}, \tag{4}$$

where Ψ_{fk+1}, Ψ_{fk} is the stator phase flux linkage at time moment t_{k+1} and t_k; Δt is interval of time.

Then the equations (1) and (2) can be as follows:

$$\begin{cases} \frac{1}{\Delta t}\Psi_{fk+1} = U_{fk+1} - R_f i_{fk+1} + \frac{1}{\Delta t}\Psi_{fk}; \\ \Psi_{fk+1} = L_f i_{fk+1}, \end{cases}, \tag{5}$$

where U_{fk+1}, i_{fk+1} are the applied voltage and phase current at time moment t_{k+1}; R_f is the phase active resistance;

$L_f = \frac{\Psi_f(i_f,\gamma)}{i_f}$ is the phase static inductance at current rotor position.

Provided that the switched-reluctance electrical machine magnetic system is saturated, the equation system (5) is non-linear, since the value of static inductance depends on the phase current. We used the iterative method with the forecast – correction method to solve the equation system in each modeling stage. At the beginning of the iterative process, we use the static inductance value that was received in previous modeling stage. Then goes the forecast that means the equation system (5) solving. We use the obtained current value to specify the inductance value. Then we use the specified inductance value to correct and resolve the equation system (5). The correction stage can be performed several times to achieve the required accuracy.

The power semiconductor devices of power converter define the switched-reluctance electrical machine phase voltage. It can be defined as follows:

$$U_f = c_f U_{dc},$$

where c_f is phase switching function and U_{dc} is capacitor voltage in the DC link.

The control of switched-reluctance electrical machine is carried out by the formation in phases of unipolar current pulses. After the set current becomes nonzero, both phase power keys are in on-state, the positive voltage is applied to the phase and the switching function takes on value "1". After the measured phase current reaches the set value, one of the power keys is at the off state and the phase is shorted through the other power key in on-stage and a diode, phase current decreases. The switching function takes the value "0". After the current reducing to the lower permissible value, both power keys are in on-stage and the cycle repeats. When the current becomes zero, the power keys are in on-stage, and the phase gives the accumulated energy through the diodes to the capacitance in the DC link. The switching function has a value of "0".

The equations (5) can be presented in the form of matrices

$$\begin{bmatrix} \frac{1}{\Delta t} & R_f \\ 1 & -L \end{bmatrix} \cdot \begin{bmatrix} \Psi_{fk+1} \\ i_{fk+1} \end{bmatrix} = \begin{bmatrix} \frac{1}{\Delta t} & 0 \\ 0 & 0 \end{bmatrix} \cdot \begin{bmatrix} \Psi_f \\ i_f \end{bmatrix} + \begin{bmatrix} c_f U_{dck+1} \\ 0 \end{bmatrix}, \quad (6)$$

Equations (6) are the algebraic-linear equations system.

The following equations describe processes in a DC link:

$$\begin{cases} C_{dc} \frac{dU_{dc}}{dt} = i_{dc} - \sum_{m=1}^{3} c_{f\,m} i_{fm}; \\ L_{dc} \frac{dI_{dc}}{dt} = E_{dc} - R_{dc} i_{dc} - U_{dc}. \end{cases}, \quad (7)$$

After using the BDF, formulas and transformations to the matrix form of the equation (7) are as follows:

$$\begin{bmatrix} \frac{C_{dc}}{\Delta t} & -1 \\ 1 & R_{dc} + \frac{L_{dc}}{\Delta t} \end{bmatrix} \cdot \begin{bmatrix} U_{dck+1} \\ i_{dck+1} \end{bmatrix} = \begin{bmatrix} \frac{C_{dc}}{\Delta t} & 0 \\ 0 & \frac{L_{dc}}{\Delta t} \end{bmatrix} \cdot \begin{bmatrix} U_{dck} \\ i_{dck} \end{bmatrix} + \begin{bmatrix} -\sum_{m=1}^{3} c_{f\,m} i_{fmk} \\ E_{dc} \end{bmatrix}, \quad (8)$$

Equations (6), completed with the equations (8) describe the electromagnetic processes in the switched-reluctance electrical machine

4. Electric spindle motor drive model

We designed the one of the variant of the spinning machine electric spindle motor. Figure 3 shows the magnetic system drawing.

There is a three-phase switched-reluctance electrical machine with two pairs of poles. In has 12 teeth on the stator and 8 teeth on the rotor. The stator outer diameter is 100 mm, the rotor inner diameter is 30 mm, the clearance diameter is 60 mm, the air gap is 0.2 mm, the active core length is 100 mm. Each winding has 20 turns.

We determined the phase flux linkage versus the current at different rotor position angles. For this purpose, we calculated the distribution of the magnetic field in the magnetic system using the FEMM software package [6]. Figure 4 shows the magnetic field distribution in the computational space. Figure 5 shows the phase flux linkage and electromagnetic moment versus the stator phase current at different rotor position angles.

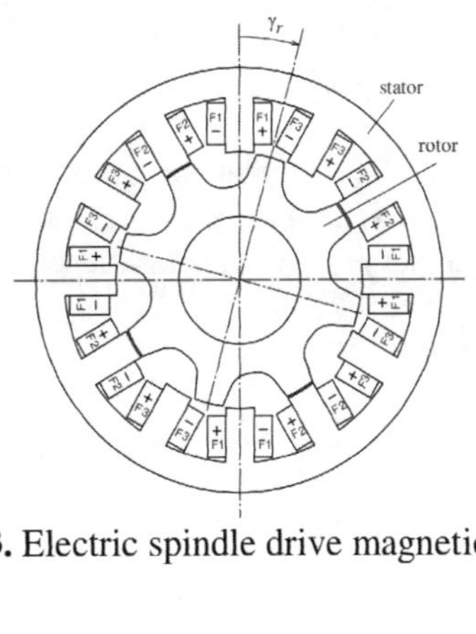

Figure 3. Electric spindle drive magnetic system.

IOP Conf. Series: Materials Science and Engineering **680** (2019) 012027 doi:10.1088/1757-899X/680/1/012027

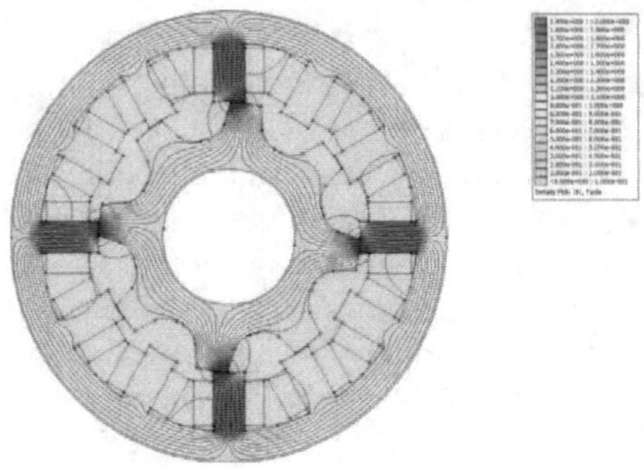

Figure 4. The magnetic field distribution in the computational space of the switched-reluctance electrical machine.

We simulated processes in the switched-reluctance electrical machine. Fig. 6 and 7 shows the results of simulation, the dependence of the phase current, current in the DC link, the switched-reluctance electrical machine electromagnetic moment created by one phase and the total electromagnetic moment.

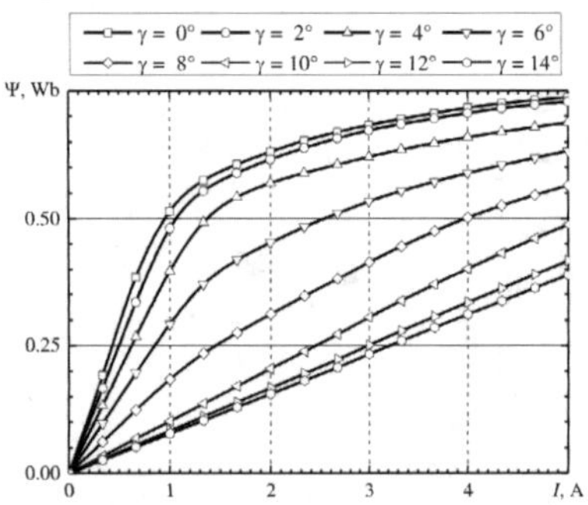

a)

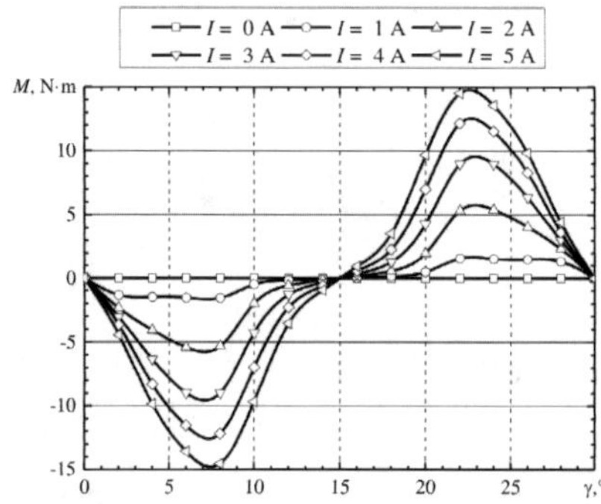

b)

Figure 5. The phase flux linkage (a) and electromagnetic moment (b) versus current at different rotor position angles.

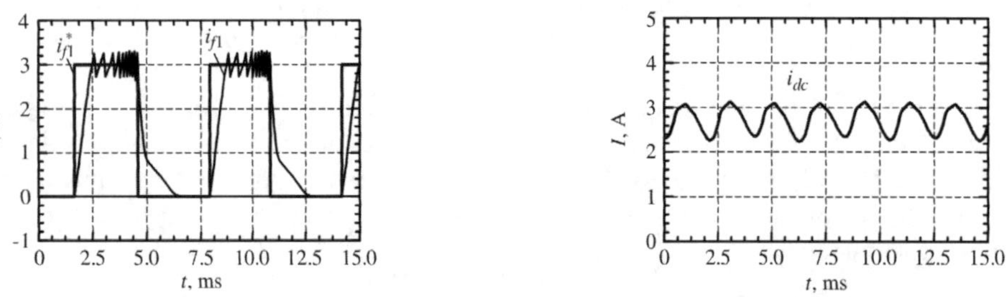

Figure 6. The simulation results: phase current and current in the DC link

The analysis of the obtained results shows that the proposed modeling approach provide a good convergence and stability of the model. With a selected simulation step of 100 μs we needed only 2 or 3 μs in each stage of modeling to achieve the required accuracy. The described model can be used in such modeling system as MatLAB Simulink.

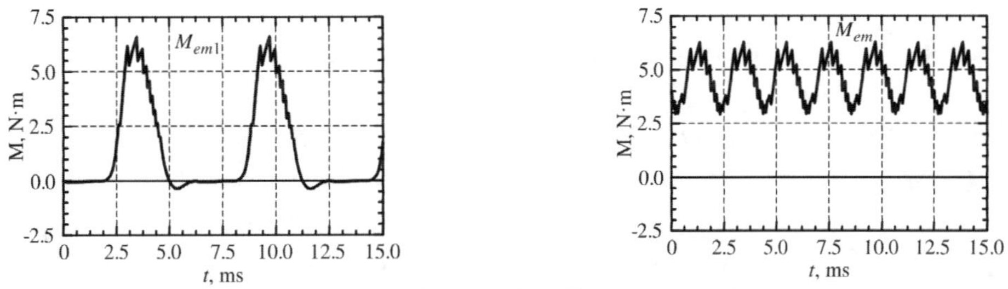

Figure 7. The simulation results: Electromagnetic moment

5. Conclusions.

The spinning machines automation is one of main ways of improving the synthetic yarn quality. The application of controlled individual electric drive of spinning machine spindles regulates winding. For this purpose, the use of switched-reluctance electrical machine is efficient. At the design stage, the use of mathematical model can help to optimize the design and determine the control system settings. The use of BDF allows us to reduce the impact of errors when determining the phase flux linkage versus phase current and rotor position. We defined that the use of first order BDF formulas is the most appropriate for this task. The model received by using such an approach has a good convergence and a high operation speed.

References

[1] Lyubarsky B G, Electrical engineering and electromechanics, **5**, (2009) 67-72
[2] Do V L, Minh Cao Ta, Proceedings of “Energy Conversion Congress and Exposition, 2009. ECCE 2009. IEEE”, (2009) 1-6
[3] Chi H P. Lin R L, Chen J F, IEEE Proc. - Electr. Power Appl, **vol. 152, no. 3**, (2005) 577
[4] Fahimi B, Suresh G, Mahdavi J, and Ehsami M, 29th Annual IEEE Power Electronics Specialists Conference, Fukuoka, **2**, (1998) 2097–2102
[5] Krassovsky A B, Bulletin of MSTU. N.E. Bauman. Series "Natural Sciences", **3**, (2005) 74-91
[6] Meeker D. Finite Element Method Magnetics. Version 4.2. User’s Manual (2018).

IOP Conf. Series: Materials Science and Engineering **680** (2019) 012028 doi:10.1088/1757-899X/680/1/012028

Tension control adaptive system synthesis used in winding of products made of “dry” composite material

A Mikitinsky

Platov South-Russian State Polytechnic University (NPI), str. Prosvesheniya 132, Novocherkassk 346428, Russia

E-mail: kolpahchyan@mail.ru, mialexp@mail.ru

Abstract. Products made by winding are widely used in many industries. The winding control systems requires synthesis of tension control regulators taking into account the specific features of this process. They determine the requirements for the synthesis of control regulators made of modern materials. Synthesis methods do not take into account the peculiarities of the “wet” composite tape. This article deals with the tension control adaptive system used in winding.

1. Introduction

The protective cover of the contact electrical bollards (CEB) is not hermetical and that is why the moisture, dust, gaseous substances and other deleterious matter can get in it. The CEB’s consists of a multicomponent mixture consists of a multicomponent mixture circulating in a turbulent air stream. Destructive factors play a substantial role in the electric contacts’ wear, fundamentally change the idea of contact wear processes and make the results make unfair the results based on the deterministic models [1-3,6]. The studies of the electrical contacts wear in seaports confirm this conclusion.

Gradual failure of high-current electrical contacts under the influence of high-entropy destabilizing factors consist of two processes. The first one characterizes the impact of internal destabilizing factors (mechanical abrasion, temperature). The second one is connected with external high-entropy destabilizing factors described above. The first process needs the use of deterministic model; the second one needs the probabilistic approach.

Products made of composite materials by the winding have found application due to its unique chemical and mechanical characteristics in many industries such as chemical industry, spacecraft and rocket production [1]. There are “dry” and “wet” methods of winding, both methods are widely used.

Figure 1 shows the scheme of the “wet” winding.

Bobbins with material 2 are installed on the creel 1. The number of bobbins depends on the technological process and can change from 6 to 1200. Dry material 3 passes through tensioners 4 and the string (ровница) is formed on the special comb 5. The string passes through the bath 6 with the binder. A special sensor 7 controls the amount of binder.

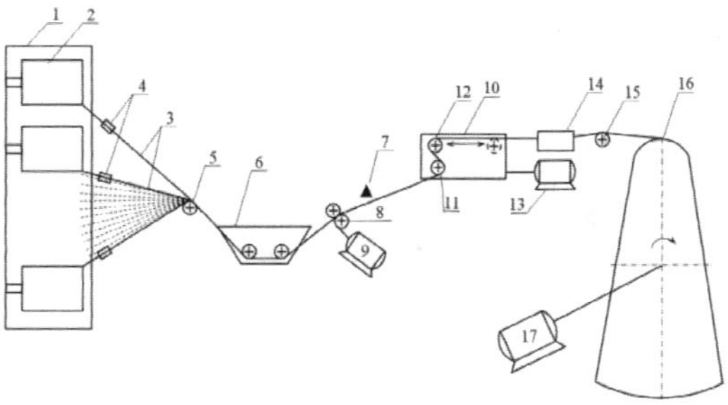

Figure 1. The "wet" winding method

The squeezing rollers 8 remove the excess of binder. The electric drive 9 changes the gap between the rollers. The tension device 10 consists of a fixed roller 11 and a roller 12 that moves along the tensioner using an electric drive 13. By this way, the "wet" tape coverage angle changes. The measurer 14 measures the tape tension. The roller 15 lays the tape on the product 16 along a special path. For this purpose, it is equipped with stacking mechanisms, which is not shown on the figure. The drive 17 rotates the product 16. Several interconnected systems control the winding.

The change of the movable roller 12 position maintains the winding tension at a predetermined level.

The improvement of the winding technology imposes stringent requirements for the drive speed and their adaptation to changing operation modes. This article is devoted to the development of such a system. The experience of using the tension control systems allowed us to determine the basic requirements for the developed systems. In particular, the transition time must not exceed the period of 0,1 second and the excessive correction must not be more than 10 %.

2. Terms of reference

It is necessary to develop a modern control system for the high-quality composite production using the "wet" winding method. The analysis showed that such a system should have three control loops (the inner loop of drive moment, the central loop of tensioner roller position, the external loop of tape tension). To create and study such a system we need a mathematical description of tape path taking into account the main elastic wound material features. This description must allow the control system synthesis and the analysis of static, dynamic and energy performance of system.

3. Literary sources overview

A lot of researches for ferrous metallurgy, pulp and paper, textile, chemical and electrical industries deal with the mathematical description of elastic tape. Differential equations obtained by a number of authors [2–8] describe the behavior of textile materials, metal tape and ropes when their winding or rewinding. There is a mathematical description of the "dry" and "wet" elastic tape [9–14].

4. Theoretical part

The following expression shows the winding tension [12, 13]:

$$\frac{dS_1}{dt}=\frac{1}{l_1(t)}\cdot\frac{dl_1(t)}{dt}\cdot S_1-\frac{\left(E\cdot F_S-S_0\right)^2}{l_1(t)\cdot E\cdot F_S}\cdot v_1+\frac{E\cdot F_S-S_0}{l_1(t)}\cdot v_2-\frac{2\cdot\left(E\cdot F_S-S_0\right)}{l_1(t)\cdot E\cdot F_S}\cdot v_1\cdot S_1+$$
$$+\frac{1}{l_1(t)}\cdot v_2\cdot S_1-\frac{1}{l_1(t)\cdot E\cdot F_S}\cdot v_1\cdot S_1^2+\frac{E\cdot F_S-S_0}{l_1(t)}\cdot\frac{dl_1(t)}{dt}+\frac{dS_0}{dt}, \tag{1}$$

where, S_1 is the winding tape tension; $l_1(t)$ is the length of the material warp zone at the moment of time t; v_1, is the linear velocity of the material entrance in the winding zone; v_2 is the winding linear velocity; S_0 is the tape tension in the previous section; E is the elastic modulus; FS is the tape cross-section area.

This expression is non-linear and its use for the regulatory systems analysis and synthesis is difficult. Therefore, the expressions are linearized [12, 13], and we can see that the linearized mathematical model accurately describes the elastic tape transition processes.

There is the linearized equitation below:

$$\Delta\dot{S}(t)+\left(\frac{\partial F}{\partial S_1}\right)^0\cdot\Delta S_1(t)+\left(\frac{\partial F}{\partial v_2}\right)^0\cdot\Delta v_2(t)+\left(\frac{\partial F}{\partial \alpha}\right)^0\cdot\Delta\alpha(t)+ \\ +\left(\frac{\partial F}{\partial S_0}\right)^0\cdot\Delta S_0(t)+\left(\frac{\partial F}{\partial \dot{\alpha}}\right)^0\cdot\Delta\dot{\alpha}(t)-\Delta\dot{S}_0=0. \tag{2}$$

The dependence of "wet" tape coverage angle of the tension roller α on the rolling roller moving along the x-axis is also non-linear function [12, 13]. After its linearizing, we receives the following expression:

$$\Delta\alpha(t)=k_5\cdot\Delta x(t), \tag{3}$$

where, $k_5=\left(\frac{\partial\alpha}{\partial x^0}\right)^0$, x^0 is the rolling roller position.

Figure 2 shows the block diagram of the linearized generalized mathematical model of the tensioner mechanical part.

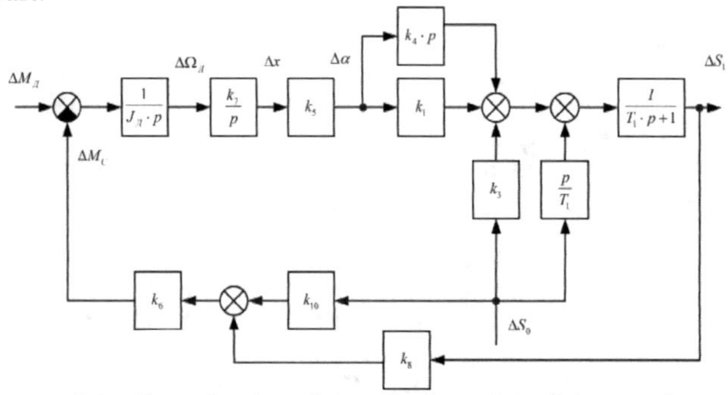

Figure 2. Block diagram of the linearized mathematical model of the tensioner mechanical part with the "wet" tape

Where, $\Delta M_{Д}$ is the tensioner drive moment, Hm; $\Delta\Omega_{Д}$ is the motor shaft rotation frequency, rad/s; $J_{Д}$ is the equivalent moment of inertia, kg/m2; k7 is the reduction gear ratio.

We use the following formulas to calculate the coefficients:

$$\left.k_1=-\frac{\left(\frac{\partial F}{\partial \alpha}\right)^0}{\left(\frac{\partial F}{\partial S_1}\right)^0};k_2=-\frac{\left(\frac{\partial F}{\partial v_2}\right)^0}{\left(\frac{\partial F}{\partial S_1}\right)^0};k_3=-\frac{\left(\frac{\partial F}{\partial S_0}\right)^0}{\left(\frac{\partial F}{\partial S_1}\right)^0};k_4=-\frac{\left(\frac{\partial F}{\partial \dot{\alpha}}\right)^0}{\left(\frac{\partial F}{\partial S_1}\right)^0};T_1=\frac{1}{\left(\frac{\partial F}{\partial S_1}\right)^0}\right\} \tag{4}$$

The carried out studies justify that the linearization coefficients depend on the winding speed, the tape tension during the winding and the tape tension in the previous section, the rolling roller position (tension roller coverage angle):

$$T_1=f(v_2);k_1=f(\alpha,S_0);k_2=1;k_3=f(\alpha,S_0);k_4=f(v_2). \tag{5}$$

We think that the use of brushless DC motor [14-16] or synchronous electric motor with permanent magnets [17] is the most appropriate as an electromechanical chopper in tension device electric motor.

We have enhanced the block diagram of the tensioning device mechanical part with the description of the electric part of electric drive and we've got a functional diagram of the tension control system (figure 3). There is a three-loop control system and it includes the internal current loop, the rotor speed control loop and the external loop of tension winding control.

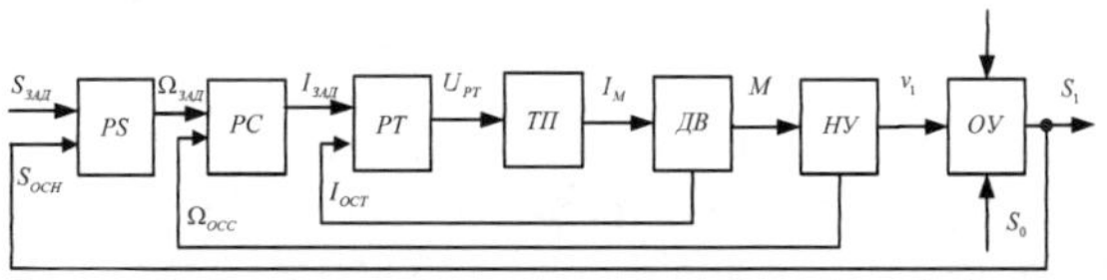

Figure 3. Tension control system functional diagram

The calculations of current control and subordinate control systems speed are well described [16,17] and therefore, we did not consider them in this paper. It should be noted that first-order aperiodic link [17,18] with transfer function $W(p)=\frac{1/k_{OCC}}{4T_{\Pi}p+1}$. (where, k_{OCC} is the feedback sensor gain; T_{Π} is the converter time constant) can replace two internal loops.

There are legends on figure 3: *PS* is the tension controller; *PC* is the engine rotation speed controller; *PT* is the motor current controller; T_{Π} is the transistor transducer; НУ is the tensioner; *OУ* is the controlled object (elastic tape).

Figure 4 shows the simplified structural diagram of a tension control system.

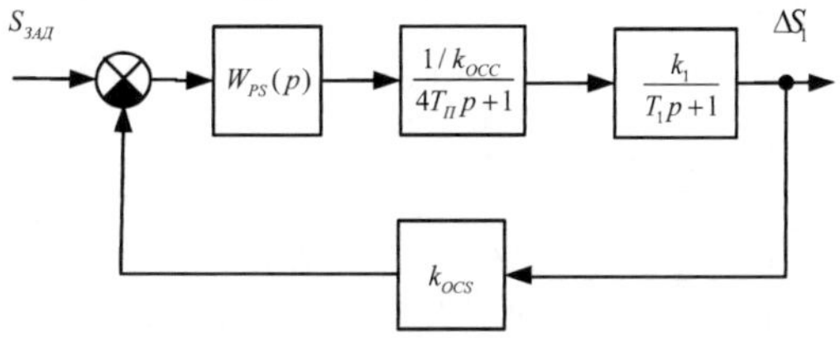

Figure 4. Tension control system structural diagram

The transfer function of the tension regulator is as follows:

$$W_{PS}=\frac{T_1p+1}{k_1\cdot(1/k_{OCC})\cdot k_{OCS}\cdot 8\cdot T_{\Pi}\cdot p}=\frac{T_1p+1}{k_1\cdot k_{CYM}\cdot p}, \tag{6}$$

were, is the coefficient of tension control feedback sensor gain loop; $k_{CYM}=8\frac{k_{OCS}}{k_{OCC}}T_{\Pi}$.

In the resulting function, the values of T1 and k1 change in wide range. It negatively affects the tension control quality during the winding process. The study was realized using the MatLab Simulink software. Tension regulator does not provide satisfactory quality of tension control when changing the parameters of the controlled object. In previously developed tension control systems, the regulator was set up to the most optimal parameters combination of the controlled object [10, 11, 22, 23]. But the tension control quality worsened when the controlled object parameters was changed.

To reduce the influence of parameters on the regulatory system it is advisable to introduce a tension regulator correction.

Obviously, we can use expression (3) to do it. However, they are quite complex and the calculations require a lot of time.

We simplify these expressions. For this purpose, we used the approaches that are described in the theory of experiment planning [19-21].

In order to build a standard matrix of the experiment we converted the factor level values to natural numbers. Table 2 shows calculation data.

Table 1. Variable parameters

	from	to	
α	*2*	*6*	*rad*
S_0	*100*	*500*	*H*
v_2	*0,1*	*0,3*	*m/s*

First, we considered the varied parameters that change during the winding. They are variable parameters. Table 1 shows variable parameters.

The regression coefficients values allow us to estimate factors and their interactions influence on the model parameter. The coefficient value corresponds to this factor influence on the model parameter value when the factor value moves from zero to upper or lower level.

We use the first-order model to assess the influence of these factors and the mathematical description of the process:

$$\tilde{y} = b_0 + b_1 X_1 + b_2 X_2 + b_3 X_3 + b_4 X_4 + b_{12} X_1 X_2 + b_{13} X_1 X_3 + b_{14} X_1 X_4 + b_{23} X_2 X_3 + \\ + b_{24} X_2 X_4 + b_{34} X_3 X_4 + b_{123} X_1 X_2 X_3 + b_{234} X_2 X_3 X_4 + b_{124} X_1 X_2 X_4 + b_{134} X_1 X_3 X_4, \quad (7)$$

where, $b_u = \dfrac{\sum_{i=1}^{N} X_{ui} y_i}{N}$; $b_{uj} = \dfrac{\sum_{i,j=1}^{N} X_{ui} X_{uj} y_i}{N}$; $b_{uj} = \dfrac{\sum_{i,j,m=1}^{N} X_{uim} X_{ujm} y_i}{N}$.

We have the following results for k1 coefficient:

$b_0 = 71{,}89$; $b_1 = -52{,}79$; $b_2 = 43{,}3$; $b_3 = -0{,}01$; $b_{12} = -34{,}6$; b23 = 0,012; b13 = 0,012; b123 = –0,012;

We have the following results for k3coefficient:

$b_0 = -525{,}89$; $b_1 = -126{,}46$; $b_2 = -306{,}29$; $b_3 = 0$; $b_{12} = -82{,}31$; $b_{23} = 0$; b13 = 0; $b_{123} = 0$;

We have the following results for k4 coefficient:

$b_0 = 22378{,}78$; $b_1 = 2671{,}0$; $b_2 = 2702{,}11$; $b_3 = 3227{,}88$; $b_{12} = -3007{,}88$; $b_{13} = -2339{,}88$; $b_{23} = -2339{,}88$;

$b_{123} = 3227{,}88$;

Table 2. Experiment planning matrix (calculations) and calculation results

1	*2*	*3*	*k1*	*k3*	*k4*	*T1*
1	*1*	*1*	*45,99*	*-133,30*	*25400*	*10,05*
	1	*1*	*5,01*	*-232,64*	*25330*	*10,15*
1		*1*	*221,2*	*-637,27*	*25400*	*10,05*
		1	*24,58*	*-1107,00*	*22644*	*10,15*
1	*1*		*45,88*	*-133,30*	*25080*	*3,38*
	1		*5,01*	*-232,64*	*25330*	*3,48*
1			*221,2*	*-637,27*	*25400*	*3,48*
			24,58	*-1107,00*	*26640*	*3,48*
			53,59	*-512,59*	*25560*	*5,1*

We have the following results for T1 time constant:

$b_0 = 6{,}59$; $b_1 = 0{,}033$; $b_2 = 0{,}011$; $b_3 = -2{,}95$; $b_{12} = -0{,}011$; $b_{13} = -0{,}011$; $b_{23} = 0{,}011$; $b_{123} = -0{,}011$.

These results are obtained for coded variables, that are varying from -1 to +1.

Obviously, the lower the coefficient before the coded variable the less influence this variable has on the overall result. We think that the test significance is 5 % of b0 coefficient.

We have the following result for k1 coefficient:

$$k_1 = b_0 + b_1X_1 + b_2X_2 + b_{12}X_1X_2 = 71,89 - 52,79X_1 + 43,3X_2 - 34,6X_1X_2;$$

We have the following result for k3 coefficient:

$$k_3 = b_0 + b_1X_1 + b_2X_2 + b_{12}X_1X_2 = -525,89 - 126,46X_1 - 306,29X_2 - 82,31X_1X_2;$$

We have the following result for k4 coefficient:

$$k_4 = b_0 + b_1X_1 + b_2X_2 + b_3X_3 + b_{12}X_1X_2 + b_{13}X_1X_3 + b_{23}X_2X_3 + b_{123}X_1X_2X_3 =$$
$$= 22378,78 + 2671X_1 + 2702,11X_2 + 3227,88X_3 - 3007,88X_1X_2 - 2339,88X_1X_3 -$$
$$-2339,88X_2X_3 + 3227,88X_1X_2X_3;$$

We have the following result for T1coefficient:

$$T_1 = b_0 + b_3X_3 = 6,59 - 2,95X_3.$$

The obtained results show that the parameters vary in a wide range:

k_1 from 5,01 to 221,2, i.e. 44 times;

k_3 from -133,3 to -1107, i.e. 8,3 times;

k_4 from 25330 to 26640, i.e. 1,05 times;

T_1 from 3,38 to 10,15 i.e. 3 times.

Figure 4 shows the block diagram of tension control system with adaptive controller.

Block 2 converts the dimension value into dimensionless value. Block 1 counts the values of k_1 and T_1 coefficients.

Figure 6 shows the simulation results of the tension control system when the controlled object parameters are changed to a significant degree.

Adaptive system works well with wide changes of the controlled object parameters, the excessive correction in the system does not exceed 9%, and the transient time is 1.12 sec.

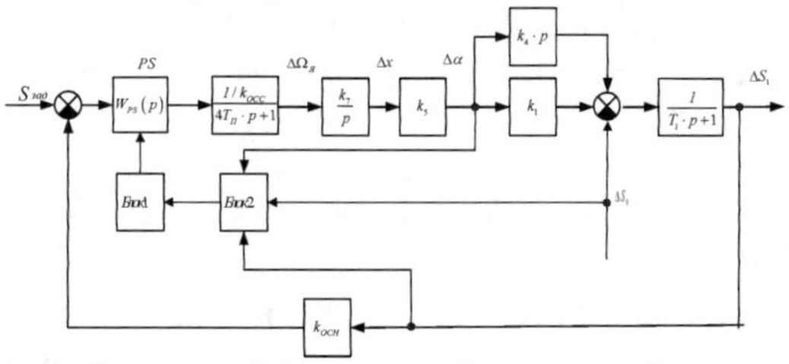

Figure 5. Block diagram of the composite tape tension control system

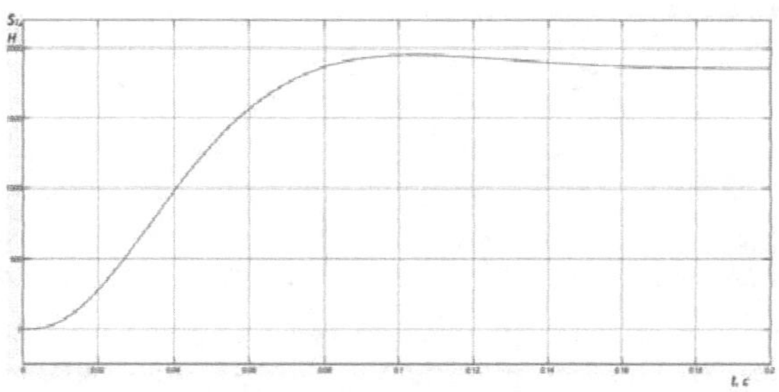

Figure 6. Simulation results of the tension control system

5. Conclusions

1. The block diagram and the tension control system of the products winding by "wet" composite tape with adaptive controller are developed.
2. Analytical expressions to calculate the tension control parameters are obtained. These expressions take into account the "wet" composite tape properties.
3. The synthesized adaptive controller takes into account the elastic properties of the tape.

IOP Conf. Series: Materials Science and Engineering **680** (2019) 012028 doi:10.1088/1757-899X/680/1/012028

References

[1] Rosato D V, *Glass fabric winding.* (Mechanical Engineering, Moscow, 1969)
[2] Ivanov G M, *Automated multimotor dc electric drive* (Power, Moscow, 1978)
[3] Druzhynin N N, *Continuous rolling mill as an object of automation.* (Metallurgy, Moscow, 1975)
[4] Basharin A V, Novikov V A, Sokolovsky G G, *Control of electric drives.* (. Energoizdat, 1982)
[5] Krysin N N, Krysin M V, *Technological processes of structures forming, winding and gluing.* (Mechanical Engineering, Moscow, 1989)
[6] Bondarev N I, *Electromechanical systems of automatic monitoring and control tape materials tension.* (Power, Moscow, 1980)
[7] Ilina S T, *Development and research of the automatic control system of tension basis on looms.* (Moscow, 1973).
[8] Fainberg U M, *Automatic control in cold rolling.* (Kharkov, 1960)
[9] Kholodenko N O, Economy, science and education in the 21 century. III regional theoretical and practical conference of scientists, students and postgraduates. (24-25 March 2011 Shakhty), (2011) 376-380
[10] Mikitinskiy A P,VII International conference on automatic electric drive-2014, , (2014) 516-520
[11] Kravchenko O A, Theory and practical work integration as the mechanism of home technology and production development: IV All-Russian theor.prac. conf.; Kamensk-Shakhtinsky, (2015) 130-132
[12] Mikitinskiy A.P. Izv..Electromechanics. **1**, (2016) 62-66
[13] Mikitinskiy A P, Transactions IX International (XX All-Russian) conference on automatic electric drive -2016, (2016) 447-450
[14] Kolpakhchyan P. Izv. higher educational. Electromechanics. **2**, (2017) 38-43
[15] Lobov B, 2016 2nd International Conference on Industrial Engineering, Applications and Manufacturing (ICIEAM), http://ieeexplore.ieee.org/document/7911499/
[16] Kolpahchyan Pm International research journal. **No. 6-2 (48),** (2016). 74-77
[17] Vinogradov A B, *Vector control of ac electric drives.* («Ivanovo State Power University », Ivanovo, 2008)
[18] Kolpakhchyan P, Rail Transport-Systems Approach. Studies in Systems, Decision and Control, **87**, (2017)
[19] Augenbaum M, Ivanov A, Terekhov Yu, *Fundamentals of planning a research experiment* (Tashkent, 2004)
[20] Adler Yu, *Experiment Planning in search of optimal conditions* (Nauka, Moscow 1976).
[21] Sydney N, *The theory of planning of experiment and analysis of statistical data. The tutorial and workshop for undergraduate and graduate programs.* (Yurayt, Moscow, 2019)
[22] Mikitinskiy A, Proceedings of X International conference on automated electric drive of AEP-2018. (2018) 81- 84
[23] Mikitinskiy A. Mechanical industry: network electronic scientific journal. **T60.No. 2**, (2018) 3–8

Probabilistic-entropy approach to ensure the operational reliability of the seaport berth elec-trical contact bollards

B Lobov[1], P Kolpakhchyan[2], S Litskevich[3]

[2]Platov South-Russian State Polytechnic University (NPI), str. Prosvesheniya 132, Novocherkassk 346428, Russia
[2]Department of Higher Mathematics, Rostov State Transport University,2 Rostovskogo Strelkovogo Polka Narodnogo Opolchenia Sq., Rostov –on- Don, 344038, Russia
[3]State Maritime University named after Admiral F. F. Ushakov ,Novocherkassk 353918, Russia

E-mail: kolpahchyan@mail.ru, blobov@yandex.ru

Abstract. Reliability analysis of the gradual failure of high current contacts connecting the con-veying units with the seaport power supply electrical network is largely based on deterministic ideas about their degradation. The use the seaport contact electrical equipment indicates the limitations and inaccuracies of such an approach. The long-term studies show that the high marine humidity, turbulent air moisture with aggressive chemical compounds play a substantial role in the wear and tear of electric contacts of seaport berth electrical bollards. The impact of these factors leads to the occasional wear of electric contacts. Thus, we need adequate mathematical apparatus of probability theory and random processes for the reliable results.

1. Introduction

The protective cover of the contact electrical bollards (CEB) is not hermetical and that is why the moisture, dust, gaseous substances and other deleterious matter can get in it. The CEB's consists of a multicomponent mixture consists of a multicomponent mixture circulating in a turbulent air stream. Destructive factors play a substantial role in the electric contacts' wear, fundamentally change the idea of contact wear processes and make the results make unfair the results based on the deterministic models [1-3,6]. The studies of the electrical contacts wear in seaports confirm this conclusion.

Gradual failure of high-current electrical contacts under the influence of high-entropy destabilizing factors consist of two processes. The first one characterizes the impact of internal destabilizing factors (mechanical abrasion, temperature). The second one is connected with external high-entropy destabilizing factors described above. The first process needs the use of deterministic model; the second one needs the probabilistic approach.

The probabilistic approach takes into account the time variations, the contact resistanceR_k. When this parameter reach the limit R_{kmax}there is a breakdown after a random timespan [5, 8]. In the initial period of the electrical contact's operation there is a dispersion of the parameter R_k regarding its mathematical expectationR_{k0}.

As multiple measurements show, there is the noticeable change of contact resistance after a certain period of time that is random value and is connected with accumulation of damage from external highly entropic destabilizing factors [7].

The probability that the parameter R_k exceeds the limitR_{kmax}, thus the breakdown probability is$F(t) = 1 - P(t)$, can be determined with the model which is calculated using the law of distribution $f(R_k, t)$ [9, 10].

The random rate of change $\gamma_{R_k}$is the element of the model of probability. As the rate of change $\gamma_{R_k}$of the parameter $R_k$depends on a large number of random factors, in view of central limit theorem of probability theory, the distribution law has the following form [7, 9]:

$$f(\gamma_{R_k}) = \frac{1}{\sigma_{\gamma_{R_k}} \cdot \sqrt{2 \cdot \pi}} \cdot exp\left[-\frac{\left(\gamma_{R_k} - \gamma_{R_{ks}}\right)^2}{\sigma_{\gamma_{R_k}}{}^2}\right]. \quad (1)$$

Ehe following expression can approximately determine the rate of change of the parameter:

$$\gamma_{R_k} = \frac{R_{k2} - R_{k1}}{t_2 - t_1},$$

where, R_{k1} is the resistance at the moment t_1; R_{k2} is the resistance at the moment t_2.

The average rate of change of the parameter during the whole operating time t_{max} до R_{kmax} is determined by the following expression:

$$\gamma_{R_{ksr}} = \frac{Rk0_{kmax}}{t_{max}}.$$

$$R_k(t) = R_{k0} + \gamma_{Rk} \cdot t, \quad (2)$$

where, R_{k0} is the initial value of contact resistance; γ_{R_k} is the contact resistance change rate.

Assuming that the resistance parameters of heavy-current contacts have the normal distribution, the expression for time $t = T$, determines the average contact resistance:

$$R_{ksr} = R_{k0} + \gamma_{Rksr} \cdot T, \quad (3)$$

and the standard deviation of the contact resistance is [9,10]:

$$\sigma_{R_k} = \sqrt{\sigma_{R_{k0}}{}^2 + \sigma_{\gamma_{Rk}}{}^2 \cdot T^2}, \quad (4)$$

where, $\sigma_{Rk}$is the standard deviation ofR_k; $\sigma_{Rk0}$is standard deviation of initial value ofR_{k0}; $\sigma_{\gamma_{Rk}}$is the standard deviation of average rate of change of R_k.

Under these conditions, the odds that $R_k$will not exceed the value R_{kmax} during the period of time $t = T$ will be represented as $P(T) = \mathrm{BEP}(R_k \leq R_{kmax}$ [9,10]. When external and internal destabilizing factors act together, the degradation becomes nonlinear (the contact heats up at high current, and the air moisture further contributes to the formation of the nonconducting film on the contacts). The rate of contact resistance change $\gamma_{Rk}(t)$ and the standard deviation $\sigma_{\gamma_{Rk}}(t)$ of the rate of change $R_k$become a function of time. In this case, we obtain a generalized function of the probability of no-failure operation [9, 10]:

$$P(t) = 0.5 + F\left[\frac{Rk0\gamma_{Rk_{kmax}}}{\sqrt{\sigma_{Rko}{}^2 + \sigma_{\gamma_{Rk}}(t)^2 \cdot t^2}} []\right] \quad (5)$$

where $F(x) = \frac{1}{\sqrt{2 \cdot \pi}} \int_0^x exp\left[-\frac{u^2}{2}\right] \cdot du$is the normalized Laplace function.

Figure 1 shows the contact resistance change rates versus time curve based on the experimental data$\gamma_{R_k}(t)$.

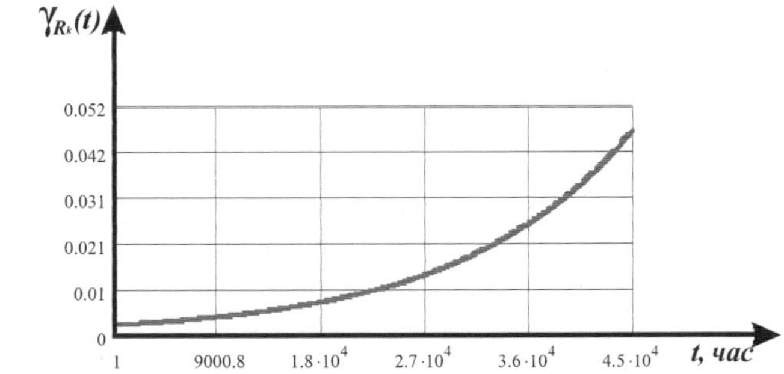

Figure1. Contact resistance change rates-vs-time curve

The approximation function for the standard deviation of the contact resistance change rate over time is $\sigma_{\gamma_{R_k}} = A \cdot [1 - exp(-B \cdot t^C)]$. We use the least-squares procedure and receive the expressions for parameters$A = 0.037, B = -2.393 \cdot 10^{-3}, C = 0.52$.

The approximated function for or the standard deviation of the contact resistance change rate over time is:

$$\sigma_{\gamma_{Rk}}(t) = 0.037 \cdot [1 - exp(-2.393 \cdot 10^{-3} \cdot t^{0.52})]. \quad (6)$$

The probability of bollard's one contact no-failure operation versus time curve is based on the statistical material, which is assembled on the large pier of the Novorossiysk seaport. Figure 2 shows the probability of bollard's one contact no-failure operation dependence on time $P(t)$, taking into account received approximations $\gamma_{R_k}(t)$, $\sigma_{\gamma_{R_k}}$ and the formula above.

$$P(t)_{reult} = \left\{0.5 + F\left[\frac{400-20-5.183\cdot10^{-3}\cdot exp(6.802\cdot10^{-5}\cdot t)\cdot t}{\sqrt{0.037\cdot[1-exp(-2.393\cdot10^{-3}\cdot t^{0.52})]^2\cdot t^2+0.02^2}}\right]\right\}^3$$

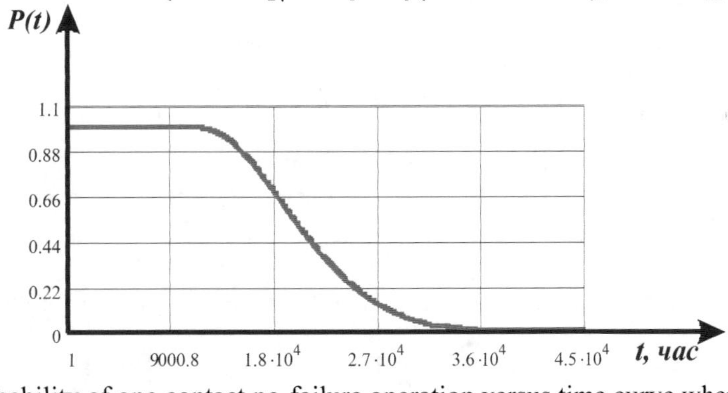

Figure 2. The probability of one contact no-failure operation versus time curve when the resistance limit is $R_k = 400$ μOhm

The expression for the probability of three-phase contact no-failure operation versus time is $P(t) \cdot P(t) \cdot P(t) = P(t)^3$, (since the operation of the three-phase disconnecting device is ensured only when three contacts are closed), and taking into account the obtained approximations $\gamma_{R_k}(t)$ and $\sigma_{\gamma_{R_k}}(t)$, is:

$$P(t)_{result} = \left\{0.5 + F\left[\frac{Rk0\gamma_{Rk_{kmax}}}{\sqrt{\sigma_{Rko}{}^2+\sigma_{\gamma_{Rk}}(t)^2\cdot t^2}}[]\right]\{\}^3\right\}, \quad (7)$$

Substituting into (7) the approximated values obtained above we get the expression for the probability of non-failure operation

$$P(t)_{reult} = \left\{0.5 + F\left[\frac{400-20-5.183\cdot 10^{-3}\cdot exp(6.802\cdot 10^{-5}\cdot t)\cdot t}{\sqrt{0.037\cdot[1-exp(-2.393\cdot 10^{-3}\cdot t^{0.52})]^2\cdot t^2+0.02^2}}\right]\right\}^3.$$

Figure 3 shows the $P(t)_{result}$ versus time curve:

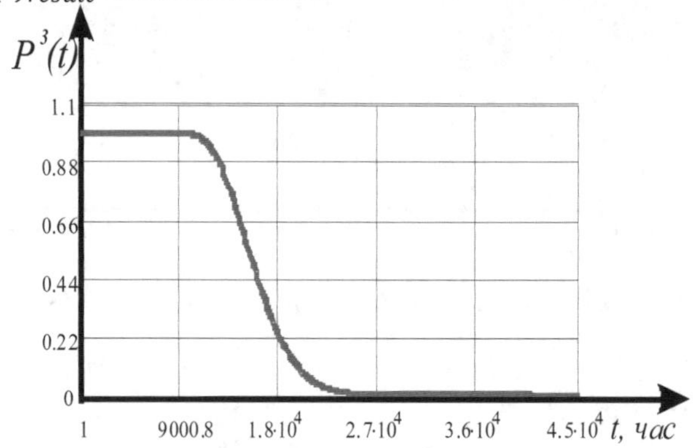

Figure 3. Probability of non-failure operation-vs-time curve of three-phase disconnecting device for three contacts of the circuit.

The initial time of service is the time when the entropy is maximum (figure 4):

$$H(t) = -\int_0^{t\int}\left\{0.5 + F\left[\frac{Rk0\gamma_{Rk}{}_{kmax}}{\sqrt{\sigma_{Rko}{}^2+\sigma_{\gamma_{Rk}}(u)^2\cdot u^2}}[]\right]\{\}^3\right\}\times\cdot\frac{log\left[\left\{0.5+F\left[\frac{Rko\gamma_{Rk}{}_{kmax}}{\sqrt{\sigma_{Rko}{}^2+\sigma_{\gamma_{Rk}}(u)^2\cdot u^2}}[]\right]\{\}^3\right\}[]\right]}{log(2)}. \quad (8)$$

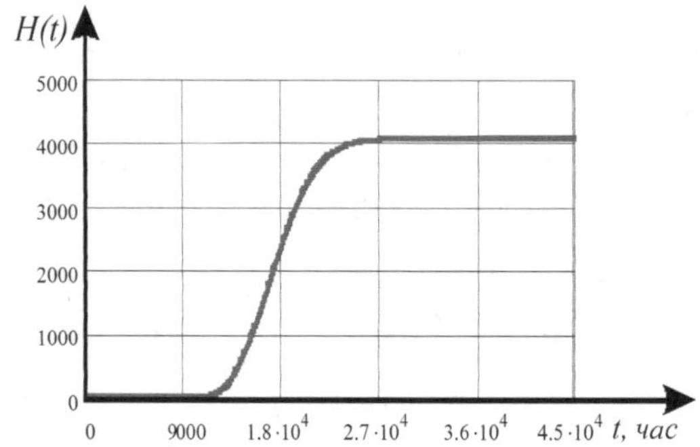

Figure 4. Entropy change of non-failure operation probability

The figure 4 shows that the entropy function is maximum when the operational time is $t =$ 24930hours. Figure 5 shows that the maximum entropy growth rate ($\frac{dH}{dt} = \frac{H(t+h)-H(t)}{h}$, $h = 0.01$ hour) is observed at an instant of $t = 16830$ hour.

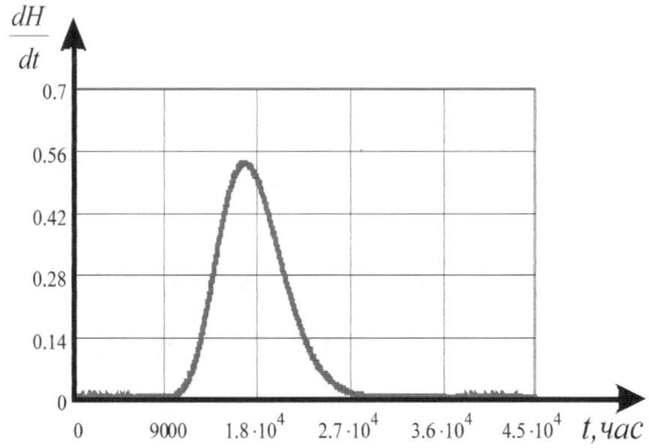

Figure 5. Entropy change rate of non-failure operation

Thus, the entropy characteristics values of probability model $max\, H\,(t)$ and $\max\frac{dH}{dt}$ can be used as criteria of critical operation of electrical contact bollards and determine the timescales of equipment maintenance or replacement.

Conclusion.

Expression that was suggested for the non-failure operation probability allow us to calculate the reliability of certain contacts, mechanisms, bollards and electricity distribution systems in its entirety for any point in time.

References

[1] Bredikhin A N, Khomyakov M V, *Electrical Contact Connections*, (Energy, Moscow, 1980)
[2] Kim E I, Omelchenko B T, Kharin C H, *Mathematical models of thermal processes in electrical contacts*, (Nauka, Alma-Ata, 1977)
[3] Bron O B, Friedman B E, Evseev M E, Electrical Engineering. **2**, (1978) 5-7
[4] Holm R, *Electrical Contacts*, (Publishing House of Foreign Literature, 1961)
[5] Dzektser N N, Vislenev Yu S, *Multi-ampere contact connections*. (Leningrad, 1987)
[6] Takano E, Mano K, IEEE Trans. Parts, Materials and Packaging. **1**, (1968) 51-55
[7] Wentzel E S, *Probability Theory*. (Moscow, 1964)
[8] Myshkin N K, Konchits V V, Braunovich M, *Electrical contacts*. (Intellect Publishing House, 2008)
[9] Litskevich S A, Demyanov V V, "Transport case of Russia", **6 part 2**, (2013) 86-91
[10] Litskevich S A, Lobov B N, Izv. institutions of higher education. Electromechanics. **1**, (2016) 42-47

IOP Conf. Series: Materials Science and Engineering **680** (2019) 012030 doi:10.1088/1757-899X/680/1/012030

Model and numerical experiment for calculating the theoretical current-voltage characteristic in electro-membrane systems

M Kh Urtenov[1], A V Kovalenko[1], A I Sukhinov[2], N O Chubyr[3], and V A Gudza[1]

[1]Kuban State University, 149 Stavropolskaya St, Krasnodar 350040, Russia
[2]Don State Technical University, Gagarin sq. 1, Rostov-on-Don 344000, Russia
[3]Kuban State Technological University, 2 Moskovskaya St, Krasnodar 350042, Russia

Abstract. This work is devoted to problems of numerical calculation of the theoretical current-voltage characteristics (CVC) and galvanography of electromembrane systems in the potentiometric mode. The purpose of this work is the derivation of the formula for calculating the CVC and galvanography of electromembrane systems in the potentiometric mode, which allows calculating stable relative to random errors and rounding errors in over-limiting current densities. To use the formulas, it is necessary to create a mathematical model for calculating the local current density. In this paper, as this model model is used transport of ions in binary salt in the desalination channel, taking into account electroconvection. The electromembrane systems have been used for water medium preparing in mobile mini systems for generating heat and electrical energy as well as vapour producing. The quality of clean water influences on reliability and life time of these systems, including vapour generator and dam-vapour turbine. The obtained formula in the form of a double integral of the local current density is derived, which allows calculating them steadily with respect to random errors and rounding errors. The formulas allowing considering CVC stably with respect to random errors and rounding errors are obtained, and the physical meaning of these formulas is clarified. The calculation of the theoretical CVC with the use of a mathematical model of ion transfer binary salts given electroconvection and shows that it qualitatively coincides with the experimental current-voltage curves. The quantitative difference can be explained by the fact that the mathematical model does not take into account the reaction of water dissociation / recombination, gravitational convection and other transport mechanisms.

1. Introduction

The current-voltage characteristic (CVC) is a common, integral and most important characteristic of heat and mass transfer processes in electromembrane systems. The CVC is used to evaluate and select the optimal design and effective operating modes. In theoretical studies, using the CVC, the fundamental concept of the limiting current is introduced; various modes of functioning of the electro-membrane systems are introduced and analyzed. Qualitative and quantitative correspondence of theoretical CVC calculated using mathematical models and experimental CVC is the most important argument in favor of the adequacy of the corresponding models. However, to date, the only way to study the CVC is to analyze experimental data. The study of the experimental CVC in the over-limiting regime, carried out in [1], by the methods of Fourier analysis, wavelet analyzes [2], and dynamic chaos methods, shows the complex, unsteady, and unstable behavior of the CVC. For a theoretical study of this behavior of the CVC, special mathematical methods and models are required. The absence of such models does not allow us to identify the relationship between the behavior of the CVC and the properties of the salt ion transfer process in electromembrane systems, for example, to adequately interpret the growth and vibrations, the instability of the CVC, with the size and velocity of the vortex structures, their interaction and bifurcations, etc. As a result of this, it is impossible to carry out engineering calculations to optimize the designs of electro-membrane systems, including optimizing their geometric and technological characteristics in order to intensify transfer processes in electro-membrane systems.

Here, for the first time, a formula is proposed that allows one to calculate the theoretical CVC based on some mathematical model. As an example, we use a 2D model of salt ion transport in the desalination channel of an electrodialysis apparatus (EDA) with homogeneous, perfectly selective ion-exchange membranes in an over-limiting current mode taking into account electroconvection, in the form of a boundary value problem for the system of Nernst-Planck-Poisson and Navier-Stokes equations [3].

Let the flow channel of desalination in the electrical circuit is indicated in the form:

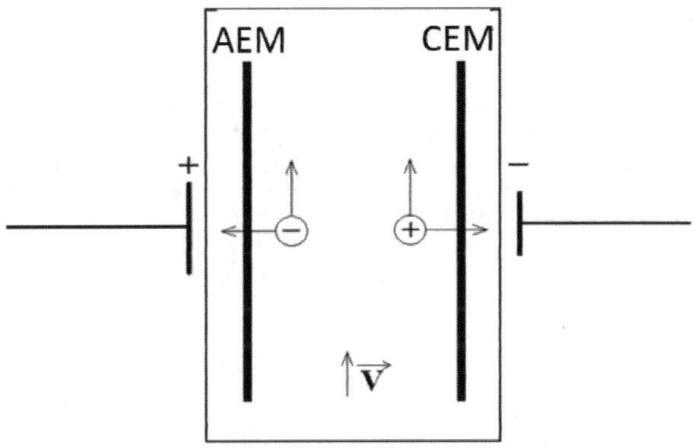

Figure 1. 2D scheme - images of the desalination channel in the electrical circuit.

Here AEM is an anion-exchange membrane, CEM is a cation-exchange membrane, the sign $\uparrow \vec{V} \rightarrow$ shows the direction of forced convection, the sign shows the directions of motion of anions, shows the directions of motion of cations. Suppose that the OY axis passes through the AEM, and the OX axis perpendicular to it is directed toward the CEM. Let the channel width be equal H and the length L. The current density at an arbitrary point in the channel at some point in time is denoted $\vec{I}(t,x,y)$. The rectangle Ω is used below when applying the Gauss theorem.

Electromembrane processes can operate in the potentiodynamic mode, when the potential jump is set and the current density is calculated, or in the galvanodynamic mode, when the current density passing through the electromembrane system is set and the corresponding potential jump is calculated. We will consider the potentiodynamic mode in this paper. For figure 1 shows a diagram of the flow channel desalination. With some accuracy, it can be assumed that the surfaces of ion-exchange membranes are equipotential, and then it is possible to combine the longitudinal walls of the rectangle with the surface of the membranes. In comparison with the experiment, it is necessary in this case to take into account the jumps of potential through the membranes. For the theoretical calculation of CVC potentiometric mode, assume that a known current density $\vec{I}(t,x,y)$ at each point (x,y) region at time t, given a jump in potential. We will denote

$$i_A = \frac{1}{L}\int_0^L I_x(t,0,y)dy, \qquad i_K = \frac{1}{L}\int_0^L I_x(t,H,y)dy,$$

In the stationary case, i_A and i_K are averaged densities of currents passing through AEM and CEM, so when the potential jump changes, they represent the current – voltage characteristic of AEM and CEM, respectively. In the unsteady case, at a fixed potential jump, they are averaged galvanic grams of AEM and CEM. The arithmetic mean of these values can be considered in many cases as a characteristic of the entire channel. It is i_A and i_K that are the required values. At first glance, i_A and i_K are easy to calculate using these formulas if $\vec{I}(t,x,y)$ is known. However, these formulas use the ordinate values of the current density at the boundary, which, as will be shown below, in turn depend on the concentration derivatives and the potential jump over the variable x at the boundary. Small errors in the computation of these derivatives lead to large errors in the computation of $I_x(t,x,y)$ at $x = 0$ and $x = H$. It would seem that integrals will correct these errors, but this does not happen since integrals are taken by the variable y. Therefore, it is necessary to derive a formula for calculating i_A and i_K, stable with respect to rounding errors.

2. Formula for calculating CVC

The formula for calculating CVC is based on Gauss' law[4-5], the formulation of which depends on whether the current density is a solenoid field or not.

2.1. *Gauss' Law*

In the two-dimensional case, the Gauss' law is written as [4]:

$$\oint\limits_{\partial\Omega} \vec{I}\cdot d\vec{l} = \iint\limits_{\Omega} div\vec{I}\, dS$$

Here $d\vec{l} = \vec{n}dl$ is the element of the oriented boundary, i.e. $\vec{n}$ – normal to the boundary, dl is an infinitesimal element of the boundary, and dS infinitesimal element of the area.

If we consider separately the potential jumps through the membranes and consider them constant as in [3,5,6], we can assume $\Omega = \{(x,y):\ 0 \le x \le H, 0 \le y \le L\}$, then:

$$\oint\limits_{\partial\Omega} \vec{I}\cdot d\vec{l} = \int_0^L I_x(t,H,y)dy - \int_0^L I_x(t,0,y)dy + \int_0^H I_y(t,x,L)dx - \int_0^H I_y(t,x,0)dx$$

and the integral Gauss' law is written as:

$$\int_0^L I_x(t,H,y)dy - \int_0^L I_x(t,0,y)dy +$$
$$+ \int_0^H I_y(t,x,L)dx - \int_0^H I_y(t,x,0)dy = \iint_\Omega div\vec{I}(t,x,y)dxdy$$

2.2. *Derivation of the formula for calculating the CVC in the electrical circuit for the solenoid case*

For solenoidal current density $div\ \vec{I} = 0$ and the integral Gauss' law, as follows from the previous formula becomes: $\oint_{\partial\Omega} \vec{I}\cdot d\vec{l} = 0$ or

$$\int_0^L \mathrm{I_x(t,H,y)dy} - \int_0^L \mathrm{I_x(t,0,y)dy} + \int_0^H \mathrm{I_y(t,x,L)dx} - \int_0^H \mathrm{I_y(t,x,0)dy} = 0$$

Let us take an arbitrary point $s \in (0,H)$ and consider the section $S = \{(x,y): x = s, 0 < y < L\}$ and region $\Omega_1 = \{(x,y): 0 \le x \le s, 0 \le y \le L\}$ and $\Omega_2 = \{(x,y):\ s \le x \le H, 0 \le y \le L\}$. It is obvious that $\Omega = \Omega_1 \cap \Omega_2$.

Since the field current density is solenoidal in a region Ω, then $div\vec{I} = 0$ in each of the areas Ω_1 and Ω_2, and therefore: $\iint_{\Omega_i} div\vec{I}dxdy = 0,\ \ i = 1,2$.

From the Gauss' law applied to each of the domains Ω_1 and Ω_2, it follows:

$$\int_0^s I_y(t,x,0)dx + \int_0^L I_x(t,0,y)dy - \int_0^s I_y(t,x,L)dx - \int_0^L I_x(t,s,y)dy = 0$$

$$\int_s^H I_y(t,x,0)dx - \int_0^L I_x(t,H,y)dy - \int_s^H I_y(t,x,L)dx + \int_0^L I_x(t,s,y)dy = 0$$

We multiply the first equation by -1 and add it to the second, then we get the equality valid for any $s \in [0,H]$:

$$\int_s^H I_y(t,x,0)dx - \int_0^L I_x(t,H,y)dy - \int_s^H I_y(t,x,L)dx + 2\int_0^L I_x(t,s,y)dy -$$

$$-\int_0^s I_y(t,x,0)dx - \int_0^L I_x(t,0,y)dy + \int_0^s I_y(t,x,L)dx = 0$$

We integrate this identity in s from 0 to H, and then we get:

$$2\int_0^H\int_0^L I_x(t,s,y)dy\,ds = -\int_0^H\int_s^H I_y(t,x,0)dx\,ds + H\int_0^L I_x(t,H,y)dy +$$

$$+\int_0^H\int_s^H I_y(t,x,L)dx\,ds + \int_0^H\int_0^s I_y(t,x,0)dx\,ds + H\int_0^L I_x(t,0,y)dy - \int_0^H\int_0^s I_y(t,x,L)dx\,ds$$

Changing the order of integration in integrals:

$$\int_0^H\int_s^H I_y(t,x,0)dx\,ds, \quad \int_0^H\int_s^H I_y(t,x,L)dx\,ds,$$

$$\int_0^H\int_0^s I_y(t,x,0)dx\,ds, \quad \int_0^H\int_0^s I_y(t,x,L)dx\,ds,$$

We got:

$$\int_0^H\int_s^H I_y(t,x,0)dx\,ds = \int_0^H\int_x^H I_y(t,x,0)ds\,dx = \int_0^H xI_y(t,x,0)dx,$$

$$\int_0^H\int_s^H I_y(t,x,L)dx\,ds = \int_0^H\int_x^H I_y(t,x,L)ds\,dx = \int_0^H xI_y(t,x,L)dx,$$

$$\int_0^H\int_0^s I_y(t,x,0)dx\,ds = \int_0^H\int_x^H I_y(t,x,0)ds\,dx = \int_0^H (H-x)I_y(t,x,0)dx,$$

$$\int_0^H\int_0^s I_y(t,x,L)dx\,ds = \int_0^H\int_x^H I_y(t,x,L)ds\,dx = \int_0^H (H-x)I_y(t,x,L)dx.$$

Therefore:

$$2\int_0^H\int_0^L I_x(t,s,y)dy\,ds = -\int_0^H I_y(t,x,0)x\,dx + H\int_0^L I_x(t,H,y)dy +$$

$$+\int_0^H I_y(t,x,L)x\,dx + \int_0^H (H-x)I_y(t,x,0)dx + H\int_0^L I_x(t,0,y)dy - \int_0^H (H-x)I_y(t,x,L)dx$$

Further:

$$2\int_0^H\int_0^L I_x(t,s,y)dy\,ds = H\int_0^L I_x(t,H,y)dy + H\int_0^L I_x(t,0,y)dy +$$
$$+\int_0^H (H-2x)I_y(t,x,0)dx - \int_0^H (H-2x)I_y(t,x,L)dx$$

Thus:

$$\frac{1}{2L}\left(\int_0^L I_x(t,H,y)dy + \int_0^L I_x(t,0,y)dy\right) = \frac{1}{HL}\int_0^H\int_0^L I_x(t,s,y)dy\,ds + \\ +\frac{1}{2HL}\left(\int_0^H (H-2x)I_y(t,x,L)dx - \int_0^H (H-2x)I_y(t,x,0)dx\right) \quad (1)$$

Denote by $i_{av} = \frac{1}{2}(i_A + i_K)$ the average current. Then from (1) follows:

$$i_{av} = \frac{1}{HL}\int_0^H\int_0^L I_x(t,s,y)dy\,ds + \frac{1}{2L}\left(\frac{1}{H}\int_0^H (H-2x)I_y(t,x,L)dx - \frac{1}{H}\int_0^H (H-2x)I_y(t,x,0)dx\right)$$

We denote $i_{in} = \frac{1}{H}\int_0^H I_y(t,x,0)dy$, $i_{out} = \frac{1}{H}\int_0^H I_y(t,x,L)dx$, which are averaged incoming and outgoing current densities.

Simple estimates show that

$$\left|\frac{1}{H}\int_0^H (H-2x)I_y(t,x,L)dx - \frac{1}{H}\int_0^H (H-2x)I_y(t,x,0)dx\right| \le |i_{in}| + |i_{out}|,$$

Because $\frac{1}{2L}\left(\left|\frac{1}{H}\int_0^H (H-2x)I_y(t,x,L)dx - \frac{1}{H}\int_0^H (H-2x)I_y(t,x,0)dx\right|\right) \le \frac{1}{L}\left(\frac{|i_{in}|+|i_{out}|}{2}\right)$.

If the input and output currents are small, or (and) the channel is long enough (in practice, the channel length is usually tens or hundreds of times longer than the width), then

$$i_{av} \approx \frac{1}{HL}\int_0^H\int_0^L I_x(t,s,y)dy\,ds \quad (2)$$

Unlike integrals (1) in the double integral (2), errors in the calculation of the derivative with respect to the variable x are compensated by integration with respect to the same variable.

Thus, the double integral (2) makes sense of the averaged current density (for the solenoid case) and allows calculating CVC accurately and adequately.

Formula (2) is quite accurate for long channels, but for short channels gives the value of the current with a significant error.

2.3. *Derivation of the formula for calculating the CVC in the electrical circuit for the non-solenoid case*

In general, for non-solenoid current density output formula for calculating the CVC is similar to the previous one and it has the form:

$$i_{av} \approx \frac{1}{HL}\int_0^H\int_0^L I_x(t,s,y)dy\,ds - \frac{1}{2HL}\int_0^L\int_0^H (H-2x)div\vec{I}dxdy \quad (3)$$

where the averaged current density $i_{acd} = \frac{1}{2HL}\int_0^L\int_0^H (H-2x)div\vec{I}dxdy$ associated with non-solenoidal current density.

Thus, from the formula (3) implies that the total average current density i_{av} is sum of the solenoidal part of the average current density $i_{sd} = \frac{1}{HL}\int_0^H\int_0^L I_x(t,s,y)dy\,ds$ and non-solenoidal part of the average current density i_{acd}:

$$i_{av} = i_{sd} + i_{acd}$$

It is easy to show that

$$|i_{acd}| \le \frac{1}{2HL}\int_0^L\int_0^H \left|(H-2x)div\vec{I}\right|dxdy \le \frac{1}{HL}\int_0^L\int_0^H \left|div\vec{I}\right|dxdy,$$

therefore, if $\max_{(x,y)\in\Omega}|div\vec{I}|$ is small, i.e. the current density is almost solenoidal, then again, get $i_{av} \approx \frac{1}{HL}\int_0^H \int_0^L I_x(t,s,y)dy\,ds$.

3. Mathematical model

For the theoretical calculation of CVC in the potentiodynamic mode, it is necessary to create a mathematical model to calculate the current density $\vec{I}(t,x,y)$ at each point (x,y) of the Ω region at any time t, for a given potential jump. The following secondary or conjugate phenomena of concentration polarization occur in over-limiting current regimes in membrane systems:

- the spatial electric charge occupies a macroscopic region, smaller, but comparable to the thickness of the diffusion layer [11],
- intensive generation of hydrogen and hydroxyl ions begins [12-17],
- microconvective flows, gravitational and electroconvection intensifying mass transfer appear in the system [18-22]

To assess the impact of each of these phenomena on the transport of salt ions, it is necessary to conduct a theoretical study using CVC, and to calculate the CVC it is necessary to use mathematical models of salt ion transport.

Currently, it is considered that electroconvection is the main mechanism of over-limiting mass transfer, so in this article we will limit the study to the calculation of CVC using the binary salt ion transport model taking into account electroconvection [3]. This model consisting of a system of partial differential equations is given below.

3.1. *System of equations*

The Nernst-Planck-Poisson and Navier-Stokes equations are used to describe the transfer process [4]:

$$\vec{j_i} = \frac{F}{RT} z_i D_i C_i \vec{E} - D_i \nabla\ C_i + C_i \vec{V}, \qquad i = 1,2$$

$$\frac{\partial\ C_i}{\partial\ t} = -div\,\vec{j_i}, \qquad i = 1,2$$

$$\varepsilon\Delta\varphi = -F(z_1 C_1 + z_2 C_2) \tag{4}$$

$$\vec{I} = F(z_1\vec{j_1} + z_2\vec{j_2})$$

$$\frac{\partial\vec{V}}{\partial t} + (\vec{V}\nabla)\vec{V} = -\frac{1}{\rho_0}\nabla P + \nu\Delta\vec{V} + \frac{1}{\rho_0}\vec{f},$$

$$div(\vec{V}) = 0.$$

where C_1, C_2 – concentrations of cations and anions in solution, respectively, z_1, z_2 – charge numbers of cations and anions, D_1, D_2 – diffusion coefficients of cations and anions, respectively, φ – electric field potential, ε_0 – dielectric permittivity of the electrolyte, F – Faraday constant, R – universal gas constant, T – absolute temperature. In this case $P, \vec{V}, \varphi, C_1, C_2$ – is the unknown function depending on time t and coordinates x, y, f – is the density of electric force.

The density of the electric field has the form [6]:

$$\vec{f} = \rho\vec{E}, \tag{5}$$

Where $\rho = F(z_1 C_1 + z_2 C_2)$ the density distribution of space charge, and $\vec{E} = -\nabla\varphi$ is the electric field strength. Formula (5) can be written in another way, using the Poisson equation (3):

$$\vec{f} = \rho\vec{E} = -\varepsilon\Delta\varphi\vec{E} = \varepsilon\Delta\varphi\nabla\varphi = \varepsilon\vec{E}\,div\vec{E}.$$

3.2. *Current density divergence:*
Multiply by z_i, and fold, then $\frac{\partial\,(z_1C_1+z_2C_2)}{\partial\,t} = -div\,(z_1\vec{j}_1 + z_2\vec{j}_2)$, or $\frac{\partial\,\rho}{\partial\,t} = -Fdiv\,(z_1\vec{j}_1 + z_2\vec{j}_2)$, where $\rho = F(z_1C_1 + z_2C_2)$ is the density of the charge distribution. Given (4), we obtain Gauss's law in differential form:

$$\frac{\partial\,\rho}{\partial\,t} = -div\,\vec{I} \qquad (6)$$

Therefore, $div\,\vec{I} = 0$ only if the condition of local electroneutrality $\rho = F(z_1C_1 + z_2C_2) = 0$, or is considered a stationary process $\rho = F(z_1C_1 + z_2C_2) = \text{constant}$.

Thus, the current density is a solenoid field only for the stationary problem and (or) in those areas where the local electroneutrality condition is satisfied.

3.3. *Formula for calculating the current density*
The current density in the desalination channel must be considered taking into account the cation and anion fluxes, according to the equation:

$$\vec{I} = F(z_1\vec{j}_1 + z_2\vec{j}_2)$$

Hence, taking into account the formula (6):

$$\vec{j}_i = \frac{F}{RT}z_iD_iC_i\vec{E} - D_i\nabla\,C_i + C_i\vec{V}, \qquad i = 1,2$$

We have

$$\vec{I} = \frac{F^2}{RT}(z_1^2D_1C_1 + z_2^2D_2C_2)\vec{E} - F(z_1D_1\nabla\,C_1 + z_2D_2\nabla\,C_2) + F(z_1C_1 + z_2C_2)\vec{V}$$

or

$$\vec{I} = -\frac{F^2}{RT}(z_1^2D_1C_1 + z_2^2D_2C_2)\nabla\varphi - F(z_1D_1\nabla\,C_1 + z_2D_2\nabla\,C_2) + F(z_1C_1 + z_2C_2)\vec{V}$$

3.4. *Results of numerical experiment*
In the paper the analysis of theoretical CVC and its comparison with experimental current-voltage curves (figure 2). The qualitative correspondence of theoretical and experimental CVC is seen. The numerical difference is explained by the idealized nature of the calculated CVC, since in the mathematical model membranes are considered ideal selective planes that have no thickness. In addition, the mathematical model does not take into account the gravitational convection and the effect of dissociation/recombination of water.

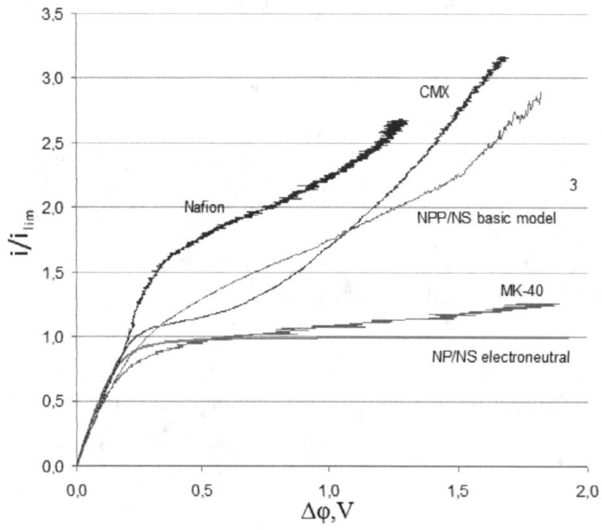

Figure 2 – Comparison of theoretical CVC (NPP/NS) and experimental CVC (other curves [23])

4. Discussion and conclusion

Formulas are obtained that allow us to count CVC stably with respect to random errors and rounding errors and the physical meaning of these formulas is elucidated. The calculation of the theoretical CVC using a mathematical model of ion transport of a binary salt taking into account electroconvection is given and it is shown that it qualitatively coincides with the experimental CVC. The quantitative difference can be explained by the fact that the mathematical model does not take into account the dissociation/recombination reaction of water, gravitational convection and other transport mechanisms. In this regard, it is necessary to formulate the appropriate models, to conduct their numerical study, to calculate the CVC for them and to conduct their comparative analysis.

References

[1] Budnikov E Yu 2000 Analysis of fluctuation phenomena in the field of transcendental currents in an electro-membrane system *Diss. Cand. Phys. Math.* p. 115

[2] Budnikov Y Y, Kukaev I F and Maximov A B 1999 *Measurement techniques* **11** pp.40-4

[3] Urtenov M K, Uzdenova A M, Nikonenko V V, Pismenskaya N D, Kovalenko A V, Vasileva V I, Sistat P and Pourcelly G 2013 *Journal of Membrane Science* **447** pp. 190-202

[4] Landau L D and Lifshitz E M 1982 *M.: Science*, 621

[5] Nikonenko V V, Mareev S A, Pismenskaya N D, Uzdenova A M, Kovalenko A V, Urtenov M H and Purcell 2017 *J Russ Electrochemistry* **53** No.10 pp. 1266-89

[6] Nikonenko V, Kovalenko A, Urtenov M, Pismenskaya N, Han J, Sistat P and Pourcelly G 2014 *Desalination. Elsevier* pp. 85–106

[7] Rubinstein I and Zaltzman B 2007 J. *Fluid Mech.* **579** pp. 173-226

[8] Babeshko V A, Zabolotsky V I, Korzhenko (Seidov) N M, Seyidov P P and Urtenov M H 1997 *J Russ Electrochemistry* **33** No.8 pp. 912-21

[9] Zaltzman B, Rubinstein I, Deng D S, Mani A and Bazant M Z 2011 *Phys. Rev. Let.* **107** p.

[10] Mani A and Bazant M Z 2011 *Physical Review E.* **84** p. 61504

[11] Rubinstein I and Maletzki F 1991 *J. Chem. Soc., Faraday Trans. II.* **87** No.13 pp. 2079-87

[12] Simons R 1979 *Nature, Land* **280** p. 41

[13] Rubinstein I and Zaltzman B 2017 *Physical Review Fluids* **2** No.9 p. 093702

[14] Chubyr N O 2009 *Ion transport in organic and inorganic membranes [Book of abstracts.]* (Krasnodar) pp. 38-40

[15] Kovalenko A V, Urtenov M H, Chubar N O, Uzdenova A M and Gudza V A 2018 *Ecological Bulletin of the scientific centers of the black sea economic cooperation* **3** pp. 78-86

[16] Kovalenko A V, Urtenov M H, Chubar N O, Uzdenova A M and Gudza V A 2018 *Ecological Bulletin of the scientific centers of the black sea economic cooperation* **4** pp. 67-84

[17] Urtenov M H, Pismenskij AV, Nikonenko V V and Kovalenko AV 2018 *Membranes and membrane technologies* **8** No.1 pp. 24-33

[18] Chubyr N O and Urtenov M A 2010 *Ion transport in organic and inorganic membranes Conference Proceedings [Book of abstracts.]* (Krasnodar) pp. 38-40

[19] Chubyr N O, Kovalenko A V and Urtenov M A H 2018 *Numerical and asymptotic methods for the analysis of 1:1 electrolyte transfer in membrane systems* p. 106

[20] Uzdenova A M, Kovalenko A V, Urtenov M K and Nikonenko V V 2015 *Electrochemistry Communications* **51** pp. 1-5

[21] Uzdenova A M, Kovalenko A V, Nikonenko V V and Urtenov M H 2011 *Condensed matter and phase interface* **13** No.4 pp. 317-24

[22] Kovalenko A V, Evdochenko E N and Urtenov M H 2015 *Scientific journal KubSAU* **05** No.109 pp. 1-13 http://ej.kubagro.ru/2015/05/pdf/66.pdf

[23] Kovalenko A V, Urtenov M H, Pismenskaya V A, Nikonenko V V, Sisto F and Pismenskaya N D 2012 *Electrochemistry* **48** No.7 pp. 830-42

IOP Conf. Series: Materials Science and Engineering **680** (2019) 012031 doi:10.1088/1757-899X/680/1/012031

Substantiation of parameters of the heat-shielding product for disabled people by using the system analysis and alternative synthesis

L G Babenko, M D Molev, A B Mikhailov, N Yu Savelyeva

Institute of service and business (branch) of Don state technical university in town Shakhty, 147 str. Shevchenko, Shakhty 346500, Russian Federation

E-mail: liana-babenko@mail.ru

Abstract. The article deals with topical issues of the development theory of heat-protective clothing for people with limited moving capabilities. In particular, the results of the study based on of the interaction of elements in the dynamic system «man – environment – heat protection product» are presented. The aim of the work was the scientific substantiation of heat-protective clothing parameters by means of using the author's model of the thermal state of the human lower limbs. The relevance of research work is due to the increased need for special heat-shielding products and insufficient knowledge of thermal processes in space and time. At the same time, the main difference of the proposed methodology is the solution of the problem within the framework of the analysis of the interaction of the human lower limbs and the product. In the process of solving the problem, the analysis of scientific publications and experimental results of the research conducted by the authors was carried out. At the same time to obtain complete and objective information, modern methods of mathematics and prognostics, including system analysis, alternatives synthesis, modeling and mathematical statistics are used. As part of the research work, an algorithm for evaluating the interaction of the elements of the «human – environment – heat-resistant product» system was substantiated and a model for predicting the thermal state of the human lower limbs was developed. As a result, temperature criteria were determined for the package of materials of an adaptation bag for people with limited moving capabilities. Scientific development has got experimental verification, which showed the effectiveness of the idea underlying the design. The obtained results represent a scientifically based and technically feasible method of selecting parameters and materials of a heat-shielding product. The implementation of the outlined development will contribute to improving the life quality of people with disabilities.

1. Introduction

The global integration of citizens with disabilities in all spheres of life of the modern Russian society is one of the strategic objectives of the federal and municipal authorities of the Russian Federation. In order to implement the strategy of socialization of people with disabilities, several important documents have been taken that regulate practical steps to improve the quality of life of this part of the population [1-3]. In particular, a special program was developed, according to which people with limited moving capabilities should be provided with modern rehabilitation products with relevant characteristics such as functionality, safety, operating reliability and high quality [3].

According to the Federal State Information System «Federal Register of Disabled Persons», more than 350 thousand Russian citizens have disabilities due to limited moving capabilities [4]. The specified group of the population moves with the help of technical means and it can be characterized with the concept «people with limited moving capabilities» (PLMC). It is important to note that PLMC represent a group of citizens with an increased risk of being in high and low ambient temperatures. This negative situation is associated with reduced thermoregulation and the lack of natural reaction of the dynamic system of lower limbs among people with limited moving capabilities. The described risk factors must be taken into account when designing heat-shielding products for PLMC.

Issues of designing special clothes for people with disabilities are pointed out in the scientific publications of Agafonova L. P., Bekmurzaev L. A., Brink I. Yu., Karabanova N. Yu., Savelieva N. Yu., Chashchina O. V., Cherunova I. V. and other scientists who specialize in this branch of science. A significant part of the works is devoted to the design of special clothes for people with limited moving capabilities, as well as to study the properties of materials used in the design of rehabilitation products [5-11].

At the same time, as the detailed consideration of the issue shows, the problem of designing special thermal protective clothing for PLMC with a thermal management device, remains insufficiently studied. So, one of the main reasons, according to the authors' opinion, is the incomplete account of the factors shaping the microclimate in the so-called «under clothing space». In particular, it can be stated that the incorrect assessment of the thermo-physical situation is associated with the problem of «scientific uncertainty» known to modern scientists. Thus, the established multi-factorial nature and specificity of risks caused by the influence of temperature and other physical fields from clothing and the external environment require specialists to develop scientifically based design solutions.

Summarizing the results of the analysis of the development of high-quality and functional products for people with diseases of lower limbs, it can be noted that there are no publications that consider scientific and methodological approaches to the design of thermal protective clothing at the current system level, taking into account the full range of negative factors and their interdependent influence on human organism. In connection with the current situation, the authors consider it possible to present in this article the results of their experimental and theoretical studies for achieving the goal aimed at the scientific substantiation of the parameters of heat-protective clothing, taking into account the thermal state of the human lower limbs.

2. Technique, methods and research materials of the interaction of human organs, clothes and environment

In the process of carrying out scientific research on the problem of assessing the impact of heat-protective clothing and external factors on people with moving disabilities, a detailed analysis of scientific works, regulatory and methodological documents published in Russian editions was carried out. The subject of close examination was articles on the assessment of thermal effects and their sources listed in the introduction.

A detailed analysis of publications convinced the authors that the implementation of research work (RW) should be based on the following assumptions:
a) all measurement and analytical procedures are carried out on the basis of a systematic approach; b) the object of research is a complex dynamic system «man – environment – heat-shielding product»; c) the subject of research is the impact of factors on the human body and their sources.

In the context of the presentation of the question under consideration, it is advisable to give some explanations. As the main method of research, the authors of the article selected integrated system analysis (ISA). This is done because, according to the fundamental principles of mathematics and prognostics, the level of the method must correspond to the complexity of the object (process) under study. Due to the fact that the object of study is a complex system, ISA is reasonably used [12-13].

Integrated system analysis, which currently represents, according to experts, the highest level of confidence, allows doing the most accurate and objective assessment of all physical and geometric parameters of the system, as well as the relationship between its elements. At the same time, using

ISA, it is possible to study the properties of the complex object under consideration in detail, due to its emergence and openness.

Based on the analysis of available materials, the authors came to the fundamental conclusion that the human body and specifically the lower limbs of people with disabilities, the environment and the heat-shielding article (clothing, bag) form a complex dynamic system (CDS). Within the framework of the CDS, it is possible, in our opinion, to consider all logical connections and processes, including those of a thermal nature. This concept is graphically shown in Figure.1

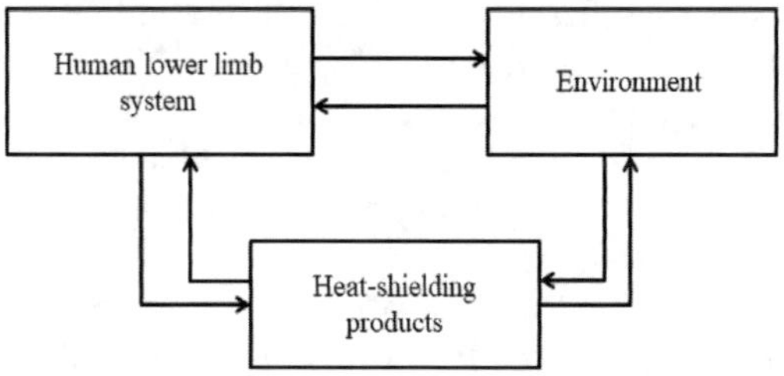

Figure 1. The block diagram of the relationship between the elements of the system «man – environment – heat protection product».

It can be reasonably stated that the proposed concept is in satisfactory agreement with the laws of physics and the theory of heat transfer.

Synthesis of alternatives when used together with ISA and the methods of prognostic theory allows you to choose the optimal solution to the problem, bearing in mind the following basic elements:

– the most probable forecast of the development of thermo-physical processes;
– material of heat-protective clothing;
– design of an adaptable heat-shielding product

The optimal solution chosen on the basis of consideration of a number of alternative proposals, as shown by many years of practice, ensures the quality of information by about 10-15% higher than using other methods [14].

In the process of carrying out the research work, methods of mathematical statistics and modeling were also used. Statistical methods were used by the authors to process the results of the experiment, the purpose of which was to determine the degree of influence of materials on the skin temperature of the hypo-dynamic system of the human lower limbs.

Large-scale experimental studies were conducted by the authors in real microclimatic conditions (at a temperature of 0°C, minus 15°C) with the involvement of people with disabilities as testers. During the tests, local and average temperatures of the skin surface of the lower limbs were determined using heat-proofing product layouts with various combinations of insulation materials. At the same time, besides quantitative parameters, the qualitative indicator of the comfortable well-being of PLMC, the so-called «heat sensation», was also assessed. The experiments were carried out in the amount of thirty-three tests, which, according to the theory of mathematical statistics, is quite enough for the sound formulation of conclusions based on the results of the analysis [15].

3. The results of the evaluation of thermophysical parameters

As mentioned above, the main purpose of the research was the scientific substantiation of the parameters of thermal protective clothing. In order to achieve this goal, an algorithm was developed for selecting the parameters of a heat-shielding product based on the analysis of theoretical materials and experimental results in Figure.2.

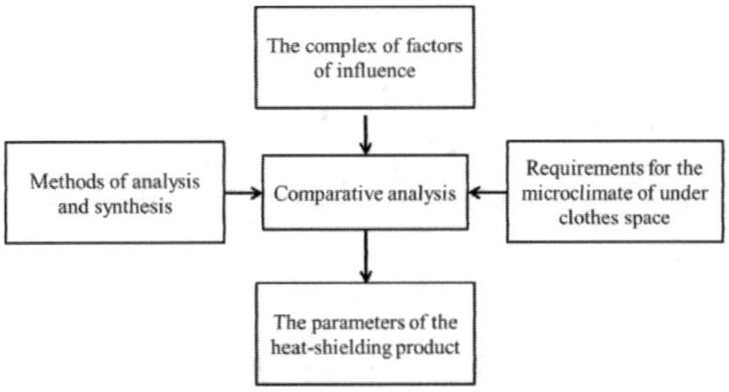

Figure 2. The algorithm for selecting the parameters of the heat-shielding product.

In the process of forming a set of clothing parameters, a comparative analysis of impact indicators was made, in particular temperature values, on under clothes space with the corresponding criteria defined by regulatory documents. As a result of a comprehensive assessment, a package of materials was selected that showed the best heat-shielding properties. When carrying out the experiments, this package was constructed from a membrane insulation of the «Tinsuleyt» brand and fleece lining.

For a more substantiated confirmation of the decisions made, the authors carried out mathematical modeling of the thermal state of the lower limbs of people with disabilities in the developed structure. The corresponding mathematical model for predicting the state of the system of the lower limbs of PLMC in a heat-shielding article is described by the formula:

$$\mathrm{T}_i(r_i,t)=\Theta_i(r_i,t)+\mathrm{T}_C=\sum_{k=1}^{\infty}\left(\int_0^t Q_k(t)e^{\mu_k^2 t}dt+L_k\right)e^{-\mu_k^2 t}\mathrm{M}_{ik}\left(J_0\left(\frac{\mu_k r_i}{\sqrt{a_i}}\right)+\mathrm{A}_{ik}Y_0\left(\frac{\mu_k r_i}{\sqrt{a_i}}\right)\right)+\mathrm{T}_C. \quad (1)$$

where t – is time (minutes) T_i – is i-th layer temperature, °C; T_C – is ambient temperature, °C; a_i – is thermal diffusivity of the i-th layer, m2 / h; J_0, Y_0 – are Bessel functions of zero order of the first and second kind. The above formula was used to determine the dependence of the surface temperature of the skin of the lower limbs in heat-protective clothing and the ambient temperature for a specific period of time (2 minutes). In the described way, the so-called thermal comfort of a person with limited moving capabilities is reasonably and objectively predicted as temperature criteria. The conditions, procedure, and results of this work are described in details in the previously published article by the authors [16]. In the framework of the presentation of the question we give Figure 3a, 3b and 3c, which show the predicted curves of dependence of the skin temperature of the tested people on the time of using the product at an ambient temperature of minus 15°C.

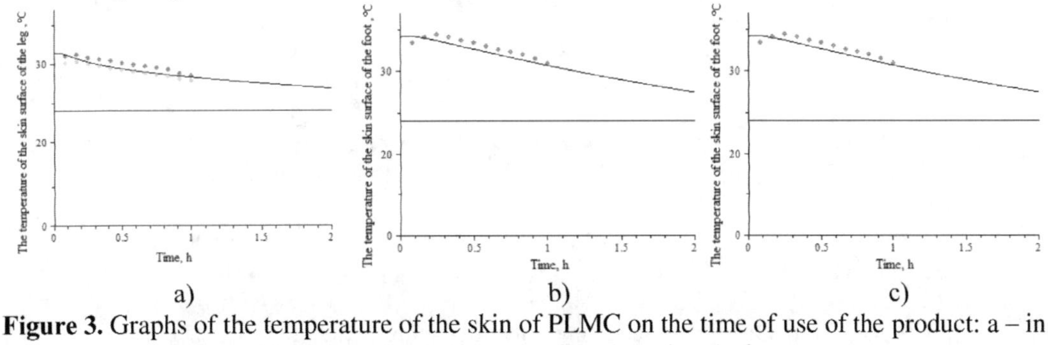

a) b) c)

Figure 3. Graphs of the temperature of the skin of PLMC on the time of use of the product: a – in the area of the knees; b – in the area of the legs; in – in foot areas.

The predicted curves made it possible to reliably find the freezing point of the lower limbs of PLMC in the coordinate system «temperature – time». The prediction results are confirmed by practical data with satisfactory convergence.

4. Conclusion

The use of a systematic approach and other modern methods of mathematics and prognostics to the study of processes taking place in the dynamic system «man – environment – heat protection product» allowed:

– to determine the nature of the impact of heat-protective clothing and the environment on people with limited moving capabilities;

– to develop an algorithm for selecting parameters of a heat-shielding product for people with disabilities;

– to establish the dependence of the temperature of the surface of the skin of the lower limbs in thermal protective clothing and the ambient using mathematical modeling,

– to justify the prediction of thermal comfort for people with limited moving capabilities (predictive assessment of temperature criteria).

It is important to point out that the socio-economic efficiency of this development is confirmed by the practice of using heat-shielding products in real conditions [16].

In the framework of the further research development of the problem considered in this article, it is planned to improve the methodological apparatus of the analysis of thermal processes and the selection of the structural elements of thermal protective clothing.

References

[1] The All-Russian Popular Front will report to the President of the Russian Federation on the problems of disabled people. [Electronic resource] Official website of the Government of Chelyabinsk region. Access mode:http://pravmin74.ru/novosti/obshcherossiyskiy-narodnyy-front-predstavit prezidentu-rossiyskoy-federacii-doklad-o-25647 (the date of the application 10.11.2018).

[2] Import Substitution. Text of the Message of the President of the Russian Federation V.V. Putin to the Federal Assembly of 2016 [Electronic resource]. Access mode http://importozamechenie.ru/poslanie-prezidenta-federalnomu-sobraniyu-polnyj-tekst/ (the date of the application 15.11. 2018).

[3] The order of the Government of the Russian Federation of November 22, 2017 N 2599-p «Strategy for the development of industrial production of oriented rehabilitation until 2025» [Electronic resource]. – Access mode: https://www.garant.ru/products/ipo/prime/doc/71719020/ (the date of the application 17.02. 2019).

[4] Federal Register of Disabled Persons [Electronic resource]. – Access mode: https://sfri.ru/ (date of the application 17.02. 2019 г.).

[5] Kozlova, Ye. V. Forming requirements for designing clothes for people with limited moving capabilities / Ye.V. Kozlova, O. M. Pleshkova // Clothing industry. - 2007. – № 2. – p. 42-44.

[6] Kharlova, O. N. Functions and requirements for hospital clothing / O. N. Kharlova, E. G. Andreeva, L. A. Shpagina, and T. V. Klimchuk // Clothing industry. – 2009. – № 2. – p. 42-45.

[7] Korobtseva, H. A. Special fashion, special design: state of the issue, problems, concept, perspectives / H. A. Korobtseva // Collection. – Tyumen: Publishing House of VOI LLP, 2008. – p. 3646.

[8] Savelieva, N. Yu. On the issue of creating adaptable, aesthetically harmonized clothing for various purposes for people with limited moving abilities / N.Yu. Savelyev // Clothing industry – 2011. – №5. – p. 35-36

[9] Brink, I. Yu. Study of the effect of wind on heat-protective clothing packages / I.Yu. Brink, E.O. Lebedeva // Clothing industry. – 2005. – №3. – p. 34-36.

[10] Bikbulatova, A. A. Fashion without borders. Unique practices of socio-cultural adaptation of

people with disabilities / A.A. Bikbulatova // Inclusive vocational education. Materials of the International scientific-practical conference, a collection of scientific articles: electronic resource. – 2015. – p. 99-109.

[11] Budeeva, O. N. Recommendations on the selection of materials in the design of garments for people with disabilities / O.N. Budeeva, Z.R. Grigorieva, T.S. Solodushenkova // Electronic scientific journal. – 2017. – № 1-3 (18). – p. 59-62

[12] Molev, M. D, Maslennikov, S. A. Technogenic risks of the population of large cities: a monograph. // MD Molev, S.A. Maslennikov. – Shakhty: Institute of service and business (branch) of DGTU, 2016. – 118 p.

[13] Lagosha B.A., Emelyanov A.A. Basics of system analysis // B.A. Lagosha, A.A. Yemelyanov. – M.: Publishing house MESI, 1998. – 106 p.

[14] Molev, M. D. Stradanchenko, S. G., Maslennikov, S.A. Theoretical and experimental substantiation of construction regional security monitoring systems technospheric / M. D. Molev, S. G. Stradanchenko, S.A. Maslennikov // ARPN Journal of Engineering and Applied Sciences. – 2015. – vol. 16. – P. 6787–6792.

[15] Rebrova, I. A. Planning of the experiment / I.A. Rebrova. – Omsk: SibADI, 2010. – 105 p.

[16] Babenko, L. G. Calculation of the thermal state of a person with limited moving abilities in an adaptation heat bag for legs / L. G. Babenko, A. B. Mikhailov, N. Yu. Savelieva, A. A. Kuchenova, I. D. Mikhailova // Design. Materials Technology. – 2019. – № 53. – p. 42-45

IOP Conf. Series: Materials Science and Engineering **680** (2019) 012032 doi:10.1088/1757-899X/680/1/012032

Development of a mathematical model for determining the EMF of the secondary winding of a measuring current transformer based on the Rogowski coil

R Oganyan, N Gorbatenko and M Lankin

Platov South-Russian State Polytechnic University (NPI), Department of Information and measuring systems and technologies, str. Prosvesheniya 132, Novocherkassk 346428, Russia

E-mail: roman.work18@gmail.com

Abstract. This article is devoted to the development of a mathematical model for determining the electromotive force (EMF) of the secondary winding of a measuring current transformer based on a Rogowski coil. The study focuses on determining the EMF in the secondary winding of the current transformer at different positions of the primary bus, its forms, as well as the overall dimensions of the whole measuring equipment. The study was conducted on the basis of finite element modeling and elements of the experimental design techniques. Finite-element models have been developed for various primary bus shapes in the Femm program. Regression models were obtained for determining the magnetic flux passing through the Rogowski coil in order to determine the EMF in the secondary winding using the Statistica program. The research results showed the importance of taking into account the position of the primary bus relative to the center of the Rogowski coil, as well as the influence of the shape of the primary bus on the measurement result of EMF in the secondary winding of the current transformer.

1. Introduction

Currently, the constitution of an intelligent energy system with an active-adaptive network is actively developing in the power industry. The key component of this network is the “digital” substation. Currently, active work is underway to study this technology [1-4], as well as its implementation [5-7]. Such substations are equipped with modern control systems, digital devices, including measuring devices, and technologies of conversion of information. However, these technologies are relatively new and do not fully meet the requirements for reliability and efficiency of work. This statement is also true for technologies for diagnosing the state of digital substations and its elements. According to [8], on the example of a current transformer, the most common innovative measuring equipment at the substations of the new generation is the current transformer based on Rogowski coil. This is due to a number of its advantages, for example, lack of saturation of core, low weight and size, meeting the requirements in a wide range of flowing currents, etc. Rogowski coils are becoming increasingly used as primary converters for relay protection [9-11]. Practical interest for the implementation of measuring current transformers based on Rogowski coil are open coils. This raises the question of taking into account the dimensions of the Rogowski coil and the shape of the primary winding of the current transformer for the optimal ratio of dimensions in order to achieve optimal measurement accuracy. Thus, this article proposes the development of a mathematical model of a current transformer based on a Rogowski coil, which allows to determine the EMF in the secondary winding, taking into account the shape of the primary winding, as well as the dimensions of the measuring equipment.

2. Object of research

In a current transformer, the Rogowski coil is used as a current sensor, which is a hollow coil of special design (Figure 1). The Rogowski coil is wound uniformly around the annular cross section of the nonmagnetic frame [12].

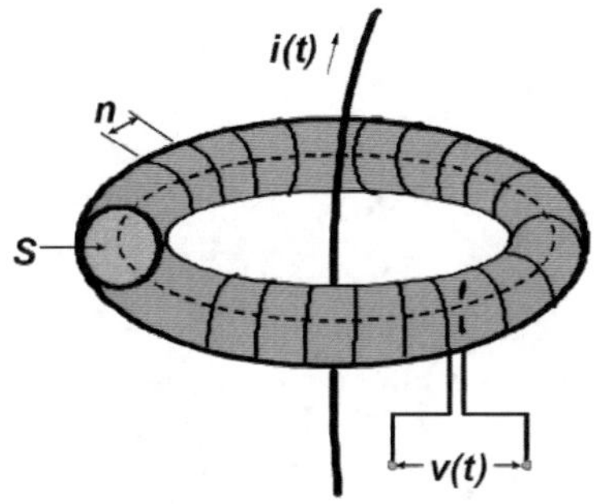

Figure 1. The principle of operation of the Rogowski coil

The output voltage induced in the coil is determined by the equation:

$$v(t) = -\frac{d\psi}{dt} \quad (1)$$

where ψ - total magnetic flux in the coil.

The total magnetic flux ψ in the calculation of the Rogowski coil is determined by the equation:

$$\Psi = \mu_0 \mu_r n \oint dl \int \vec{H} d\vec{S} \quad (2)$$

where

l - core length;

n - winding density (windings per unit length);

S - core cross-section area;

H - magnetic field intensity;

μ_0 - permeability of air;

μ_r - relative permeability of the core material.

3. Materials and methods

As can be seen from formulas (1.2), the determination of the output voltage in the secondary winding is made difficult by the complexity of the calculation formulas. To simplify the procedure for determining the EMF in the secondary winding and at the same time obtaining mathematical model, it is proposed to use physical modeling using elements of the experimental design techniques.

To develop a mathematical model of a current transformer based on a Rogowski coil, preliminary studies were conducted to identify the influence of the shape of the primary bus and its centering on the accuracy of measuring the magnetic flux in the secondary winding of the current transformer. According to [12], the Rogowski coil can be made round, oval, rectangular, teardrop-shaped, and also in the form of an open-ended fork, which must be considered when designing, but the most common are round or rectangular shape, due to the fact that other cross-sectional shapes are less practical. According to [13], the shape of the primary winding can also be different - round, rectangular, box-shaped and four-strip. Preliminary studies include an experiment with a round primary bus, in which the bus diameter, the distance between the bus and the Rogowski coil — R, the X-axis and Y-axis offset, and the diameter of the turn (Figure 2, a) were considered as factors. To solve the problem of determining the magnetic flux passing through the Rogowski coils, the numerical method is used - the finite element method [14]. For the finite element simulation, the Femm program was used. Based on the fact that the core of the secondary winding of the Rogowski coil is built on a non-magnetic material, air is chosen as the core material.

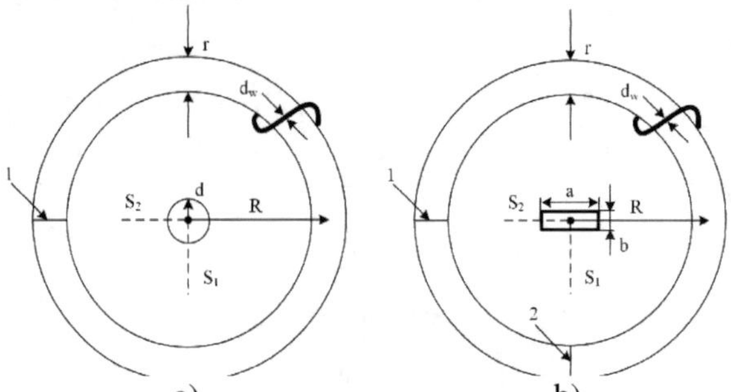

a) b)

Figure 2. Rogowski Coil:

a) Rogowski coil with a round primary winding; b) Rogowski coil with a rectangular primary winding

Specifying the parameters d = 51.687 mm, R = 78.329 mm, r = 11.867 mm and zero displacement of the primary winding relative to the center, finite-element modeling was performed to determine the magnetic flux in the Femm program:

$$\Phi = 3.771\ 10^{-6}\ \text{Wb}.$$

It was found that a statistically significant factor affecting the accuracy of determining the magnetic flux is the displacement of the primary winding relative to the center. The relative error at offset S1 = 20 mm, S2 = 20 mm is 2.56%, other factors are statistically insignificant, which indicate the importance of centering the primary winding.

To identify the influence of the shape of the primary bus on the result of determining the magnetic flux passing through the Rogowski coil, an additional experiment was conducted with a rectangular bus with dimensions a = 98.527 mm; b = 8.319 mm; r = 11,867 mm; R = 78.329 mm (Figure 2, b). The following result was obtained:

$$\Phi_{\Sigma} = \frac{\sum_{n=1}^{2} \Phi_n}{2} = 4.015\ 10^{-6}\ \text{Wb}.$$

Preliminary studies indicate the importance of taking into account the shape of the primary winding (for example, round and rectangular), since the difference between the measurement results is 6.5%. Thus, based on the obtained simulation results, it is required to develop a universal model of current transformer based on the Rogowski coil, which will take into account, in addition to the dimensions of the Rogowski coil, the shape of the primary winding of the current transformer.

With this consideration in mind, the EMF can be determined from the following system of equations:

$$\begin{cases} e = f(\omega, n, t, \psi) \\ \psi = f(i, G) \end{cases}, \quad (3)$$

where n - number of turns of the secondary winding of the Rogowski coil, ψ - magnetic flux, i - current passing through the primary winding, G - size of the entire measuring equipment, t - time, ω - cyclic frequency.

Rewrite the first equation using the basic formula of the transformer EMF:

$$\begin{cases} e_a = \omega \cdot n \cdot \psi_a \\ \psi_a = f(i, G) \end{cases}, \quad (4)$$

According to (4), to determine the potential, it is necessary to determine the magnetic flux. The method of evaluating the magnetic flux depends on the shape of the primary bus. With the rectangular shape of the primary bus, the measurement is carried out at positions 1 and 2, after which it is summed up and subsequently averaged; with a round form, measurement is carried out at position 1 (Figure 2, a, b).

4. Development of mathematical models of measuring current transformer based on Rogowski coil

4.1. Mathematical model of measuring current transformer on the basis of a Rogowski coil with a round bus

To obtain a regression model by applying the experimental design techniques, the following intervals of variation of the factors R, r, I with d = 10 mm were chosen: R = 101.5 ± 20 mm. r = 10 ± 3.5 mm, I = 100 ± 10 A. The planning matrix of experiment is presented in table 1.

Table 1. Planning matrix 2^3.

№	X1	R, mm	X2	r, mm	X3	I, A	Φ, 10^{-6} Wb
1	+	121.5	+	13.5	+	110	2.801
2	–	81.5	+	13.5	+	110	4.129
3	+	121.5	–	10.5	+	110	2.248
4	–	81.5	–	10.5	+	110	3.311
5	+	121.5	+	13.5	–	90	2.292
6	–	81.5	+	13.5	–	90	3.379
7	+	121.5	–	10.5	–	90	1.839
8	–	81.5	–	10.5	–	90	2.709

The results of the experiment were processed using the Statistica program (Figure 3):

Effect Estimates; Var.:Φ; R-sqr=,9559; Adj:,94929 (Rogovs 3f Rri 5% nov inter2)
2**(3-0) design; MS Pure Error=,0213437
DV: Φ

Factor	Effect	Std.Err. Pure Err	t(16)	p	-95,% Cnf.Limt	+95,% Cnf.Limt	Coeff.	Std.Err. Coeff.	-95,% Cnf.Limt	+95,% Cnf.Limt
Mean/Interc.	2,83843	0,029821	95,1809	0,000000	2,77522	2,901652	2,838434	0,029821	2,775215	2,901652
(1)R	-1,08690	0,059643	-18,2235	0,000000	-1,21334	-0,960465	-0,543451	0,029821	-0,606670	-0,480233
(2)r	0,62372	0,059643	10,4576	0,000000	0,49729	0,750160	0,311861	0,029821	0,248643	0,375080
(3)I	0,56769	0,059643	9,5181	0,000000	0,44125	0,694125	0,283844	0,029821	0,220625	0,347062

ANOVA; Var.:Φ; R-sqr=,9559; Adj:,94929 (R
2**(3-0) design; MS Pure Error=,0213437
DV: Φ

Factor	SS	df	MS	F	p
(1)R	7,08814	1	7,088142	332,0957	0,000000
(2)r	2,33418	1	2,334179	109,3616	0,000000
(3)I	1,93361	1	1,933615	90,5943	0,000000
Lack of Fit	0,18236	4	0,045590	2,1360	0,123437
Pure Error	0,34150	16	0,021344		
Total SS	11,87980	23			

Figure 3. Regression model of the first order of current transformer based on Rogowski coil.

The adequacy of the model is tested using the Fisher criterion. For this, it is necessary to compare the critical F_{cr} and the calculated F_{calc} values of Fisher criterion. The critical value is selected from the Fisher distribution tables using the numbers of degrees of freedom f_1 and f_2, in this case $F_{cr} = 3.01$. The estimated value of the Fisher criterion, according to Figure 3, is 2.136. As a result of the fulfillment of the condition $F_{cr} > F_{calc}$, the obtained model is adequate:

$$F_{cr} > F_{calc};\ 3.01 > 2.136.$$

$$f_1 = N - L,$$

$$f_2 = N\,(n - 1),$$

where $N = 2^3 = 8$ – number of experiments;

L - the number of significant coefficients of the regression equation – 4;

n – number of repetitions experiments – 3.

$$f_1 = 8 - 4 = 4.$$

$$f_2 = 8(3\text{-}1) = 16.$$

However, the adequacy of this model was obtained with narrow ranges of variation of factors. This may indicate the nonlinear nature of the factors under study. As a result of additional studies, it was detected that the dependence I = *f* (Φ) is linear (Figure 4), and the dependence R = *f* (Φ) is non-linear, which indicates the need to develop a second-order model for an adequate description of the process (Figure 5).

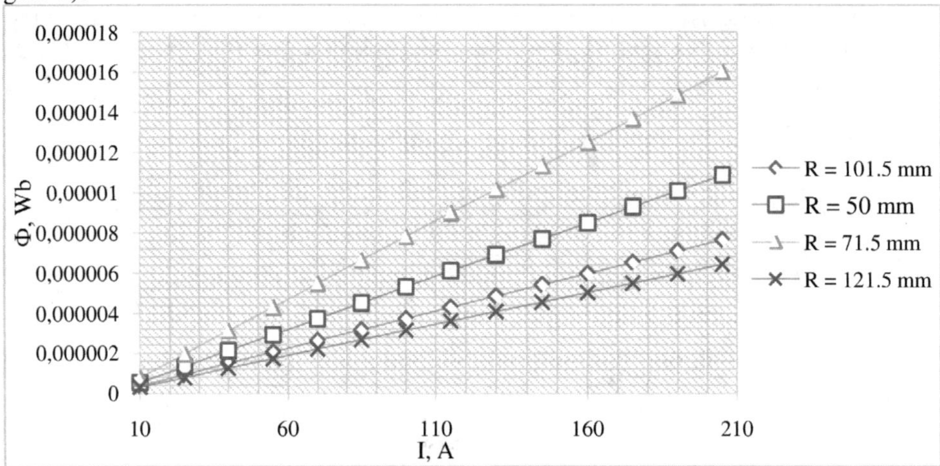

Figure 4. Dependence I = *f* (Φ).

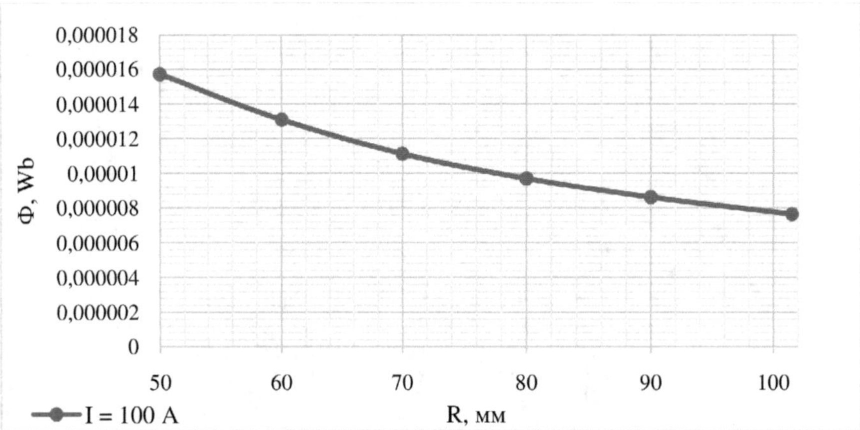

Figure 5. Dependence R = *f* (Φ).

The factors taken are d, R, r, I. Interval of variation d = 31.104 ± 20.583 mm; r = 15 ± 3.133 mm; R = 90.529 ± 12.2 mm; I = 80 ± 28.288 A. The intervals of variation are taken in such a way that the smallest value of R is the largest radius of the round bus d. As a result of the experiment, an adequate model was obtained with 3% deviations between repeated experiments:

$$F_{cr} > F_{calc};\ 1.85 > 0.45.$$

Regression model in coded values:

$$\Phi = 2.97592 - 0.40608 \cdot X2 + 0.0577 \cdot R2^2 + 0.56926 \cdot X3 +$$
$$+1.06825 \cdot X4 - 0.07719 \cdot X2 \cdot X3 - 0.14259 \cdot X2 \cdot X4 + 0.20221 \cdot X3 \cdot X4.$$

Regression model in named values:

$$\Phi = 0.42127 - 0.0401 \cdot \mathrm{R} + 0.00039 \cdot \mathrm{R}^2 + 0.182 \cdot \mathrm{r} + \\ +0.04094 \cdot \mathrm{I} - 0.00202 \cdot \mathrm{R} \cdot \mathrm{r} - 0.00041 \cdot \mathrm{R} \cdot \mathrm{I} + 0.00228 \cdot \mathrm{r} \cdot \mathrm{I}.$$

Rewrite (4) for a round bus:

$$\begin{cases} e_a = \omega \cdot n \cdot \psi_a \\ \psi_a = 0.42127 - 0.0401 \cdot R + 0.00039 \cdot R^2 + 0.182 \cdot r + 0.04094 \cdot I - . \\ -0.00202 \cdot R \cdot r - 0.00041 \cdot R \cdot I + 0.00228 \cdot r \cdot I \end{cases}$$

4.2. Mathematical model of measuring current transformer based on a Rogowski coil with a rectangular bus

For a rectangular bus (Figure 2, b) a similar experiment was conducted. The factors taken are the length and width of the bus, instead of the diameter. Considering the previous study, a second-order model is used with the following intervals of variation of factors: a = 62.5 ± 36.027 mm; b = 5.5 ± 2.819 mm; r = 15 ± 3.133 mm; R = 95.529 ± 15.333 mm; I = 80 ± 31.328 A, with N = 43. The intervals of variation are chosen in the same way as in the previous experiment. The number of repeated experiments - 3, with a deviation between them equal to 3%.

As a result of the experiment, the resulting model is adequate:

$$F_{cr} > F_{calc};\ 1.58 > 0.8.$$

Rewrite (4) for a rectangular bus:

$$\begin{cases} \mathrm{e_a} = \omega \cdot \mathrm{n} \cdot \psi_\mathrm{a} \\ \psi_\mathrm{a} = -0.1527 + 0.01007 \cdot \mathrm{a} - 0.0348 \cdot \mathrm{R} + 0.00036 \cdot \mathrm{R}^2 + 0.18111 \cdot \mathrm{r} + \\ +0.04068 \cdot \mathrm{I} - 0.00191 \cdot \mathrm{R} \cdot \mathrm{r} - 0.00039 \cdot \mathrm{R} \cdot \mathrm{I} + 0.00221 \cdot \mathrm{r} \cdot \mathrm{I} - 0.00009 \cdot \mathrm{R} \cdot \mathrm{a} \end{cases}$$

The number of turns n is determined from the formula [15]:

$$\mathrm{n} = \frac{\mathrm{l}}{\mathrm{d_w}}\mathrm{k},$$

where l – coil length, d_w – winding wire diameter, k – fill factor.

5. Conclusion

As a result of the conducted research, the influence of the displacement of the position of the primary bus relative to the center of the Rogowski coil on the measurement of the magnetic flux, and as a result, the EMF in the secondary winding, was revealed. The relative error at offset S1 = 20 mm, S2 = 20 mm is 2.56%, which indicate the importance of centering the primary winding. At the same time, the importance of taking into account the shape of the primary winding of the measuring current transformer on the basis of the Rogowski coil was revealed - the difference between the measurement results (using an example of a round and rectangular bus) is 6.5%. Two mathematical models were obtained that take into account both the shape of the primary bus and the dimensions of the measuring equipment for determining the EMF in the secondary winding of the current transformer based on the Rogowski coil.

This article is based on results obtained by using equipment of shared facility "Diagnosis and energy- efficient electrical equipment" (NPI).

References

[1] Oganyan R G Narakidze N D Shaykhutdinov D V 2017 *Inter. Conf. on Indust. Engin., Applic.s and Manufact (ICIEAM)* Available at: http://ieeexplore.ieee.org/document/8076429/
[2] Dubrov V I Oganyan R G Narakidze N D Aleksanyan G K 2017 *J.Eng.Aplliend Sci.* **12(2)** 276-282

IOP Conf. Series: Materials Science and Engineering **680** (2019) 012032 doi:10.1088/1757-899X/680/1/012032

[3] Oganyan R G Narakidze N D Shaykhutdinov D V Gorbatenko N I Maksuta S O 2017 *IOP Conf. Series: Materials Science and Engineering* **177**

[4] Dubrov V I Oganyan R G Shaykhutdinov D V Kirievsky E V Kruglova T N Narakidze N D 2016 *Fundamental Research* **9** pp 16-20

[5] Rykovanov S N, Ukhov V I, Slabospitssky S R, Kovtsova I O 2014 *J. Releyshchik* **1** pp 32-39

[6] Drozdova T V Elov N E Morozov A P 2016 *J. Energy of a Uniform Network* **3** pp 54-61

[7] In Skolkovo opened digital substation 110 kV Medvedevsky. Available at: http://www.en-prom.ru/2018/06/05/v-skolkovo-otkry-ta-tsifrovaya-ps-110-kv-medvedevskaya/

[8] Oganyan R G Shaykhutdinov D V 2018 *Fund. princip., theory, methods and means of measurem., control and diagnos.: mat. of the 19th Inter. young scientific-practical conf.* pp 145-152

[9] Rezaee M Heydari H 2010 *Inter. J. of electr. engin. & technol. (IJEET)* **6** pp 232–237

[10] Chatrefou D Ponchon P Duplan D Osbourne M P 2006 *Inter.Conf. on Large High Voltage Electric Systems* p 22

[11] Kojovie L Beresh R Bishop M Javora R Magruder B McLaren P Mugalian B Offner A 2010 *IEEE PSRC Speial Report* p 72

[12] Kojovic L 2008 *IEEE Guide for the Application of Rogowski Coils Used for Protective Relaying Purposes* p 55

[13] Usanov V 2019 *Bible electrician: EIC, IJSR, OR* p 752 (Moscow: Exmo)

[14] Zhilkin V A 2013 *Introduction to the finite element method* (Chelyabinsk: ChGAA) p 296

[15] Novikov Yu N *Теория и расчет электрических аппаратов* 1970 (Leningrad: Energy) p 327

IOP Conf. Series: Materials Science and Engineering **680** (2019) 012033 doi:10.1088/1757-899X/680/1/012033

Calculation of Stress-strain State of the Microtunneling Shield Housing

D A Grinko[1], A S Khoroshev[1] and M A Zemlyanoy[2]

[1]Platov South-Russian State Polytechnic University (NPI), Research Institute of Electromechanics str. Prosvesheniya 132, Novocherkassk 346428, Russia
[2]Project building company «GIDROSTROY» Society Limited, Novorossiysk, Russia

E-mail: vskych@gmail.com

Abstract. The work reviews the strength calculation of the microtunneling shield housing shell and bed plates of hydraulic cylinders of the mobile set. Based on the requirements for maximum loads on the shield, the analytical calculation of the housing shell thickness and basic design parameters of the tunneling shield bed plate was made. The bending moments and normal forces effecting on the housing were defined in analytical way. The calculation method was offered for the required solid shell of the shield housing considering the allowed values of its deformation due to the necessary construction clearance between the inner surface of the housing shell and mounted support. The numerical modelling of the stress-strain state of the tunneling shield housing was made using the finite-element method on the basis of solid model with due regard to the design peculiarities of the tunneling shield. The numerical modelling confirms the results of analytical calculation of design parameters of the tunneling shield.

1. Introduction

Currently, the majority of works on underground facilities construction in the densely populated cities and megalopolises are performed by underground method using the specialized machinery to provide for the underground facilities construction (wells, tunnels, manifold, etc.) by trenchless method that makes it possible to overcome the artificial and natural obstacles (main transport routes, railway roads, embankments, dams, rives, as well as overground and underground structures and utilities) encountered during the works, without violation of their functioning mode [1,2]. To construct the underground facilities of small diameter and relatively small length using the trenchless method, such methods as horizontally inclined drilling [3], puncturing or squeezing are used, and for medium and large diameters with large length, the tunneling shields of various modifications are used. The main advantage of the trenchless method of construction consists in possibility of construction of complex underground utilities with rather high economic efficiency in the conditions of the densely populated cities and various mining and technical conditions [4,5]. This means that the machinery designated for such works will actively develop and improve in future, as its market demand will increase and the area of application will expand [4].

The shield tunneling for construction of such facilities as tunnels, manifolds and subways has been widely used as this method has significant advantages over the alternative ones. These advantages can include the possibility of accident-free construction of facilities in complex mining and technical conditions, high work safety, high tunneling speeds, high mechanization levels and low labour capacity [6,7]. The listed advantages of the shield tunneling makes it possible to state that this method is preferred for construction of large and medium diameter underground facilities, but is also useful for construction of small diameter facilities when the suitable equipment is available [8,9].

One of the prospective samples of the equipment for laying the utilities of 1,420 mm in diameter and over 100 m long is the tunneling shield of the microtunneling complex. This equipment provides for tunneling the mine openings of round section in stable soft rocks with the hardness of up to 3 points according to the Protodyakonov scale of hardness and dip up to 12 degrees.

2. Analytical Model

One of the basic design parameters of any tunneling shield is its outer diameter. The outer diameter of the shield provides not only for the final diameter of the development made, but also such parameters as the mass, maneuverability factor and reliability level of the machinery, in general [10,11]. The shield housing is the main load-bearing structure that perceived the loads of the rock pressure and movement efforts and on which all main mechanisms are mounted. That is why the outer diameter values are to be thoroughly calculated and substantiated.

The outer diameter of the tunneling shield is used to be determined based on the following function [12]:

$$D_{sh} = D_{out} + t + 2\delta, \tag{1}$$

where D_{out} – outer diameter of finishing, mm; t – construction clearance value, mm; δ – shield shell thickness, mm.

The outer diameter of the finishing and the construction clearance value are used from the initial data and recommendations depending on the diameter of the constructed working. The required shield shell thickness considering the regulated safety factor amounts to the system calculation, where the number of unknown supporting responses is larger than the number of independent statics equation used for the structure under consideration, i.e. to solving the statically indefinite system.

The static calculation of the shield shell is performed according to general calculation techniques of the statically indefinite system using the method of forces with transfer of unknown quantities to the elastic centre of the system and using the symmetry of the shield and external loads relative to the vertical axis. When selecting the shield load scheme, the most disadvantageous combination is used, as a rule, when the whole column of the overlaying soft rocks effects the shield and there is no lateral pressure. Such load combination is possible when the shield leaves the stable rock zone and goes to the overlaying unstable soft rock zone with the long stay [12].

The design system of the shield shell and its main system are shown in Fig. 1.

The symmetric equations for solving the statically indefinite framed system in accordance with Fig. 1 provided that there are no additional bandages have the form of a set of equations:

$$\begin{cases} x_1\delta_{11} + \Delta_{1p} = 0, \\ x_2\delta_{22} + \Delta_{2p} = 0, \end{cases} \tag{2}$$

where x_1, x_2 – unit undefined forces acting in the main system; δ_{11}, δ_{22} – unit displacements in the x_1 and x_2 directions under the action of unit forces x_1 and x_2, respectively; Δ_{1p}, Δ_{2p} – displacements in the x_1 and x_2 directions under the action of a given load, respectively.

Fig. 2 and 3, respectively, show the distribution diagrams of bending moments from single forces and normal forces from the acting loads on the tensile fibre in the transverse section of the tunneling shield shell.

Single shifts are determined in accordance with the ratios:

$$\delta_{11} = \frac{2R}{EI}\int_0^{\pi} M_1 d\varphi, \tag{3}$$

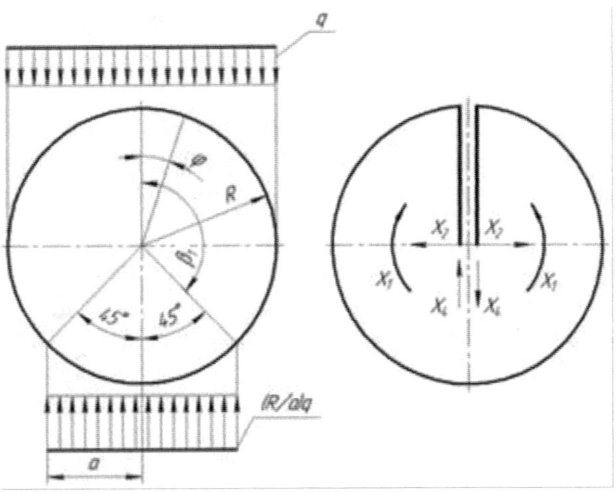

Figure 1. Transverse section of the 3unnelling shield shell: *a* – design scheme; *b* – main system of forces.

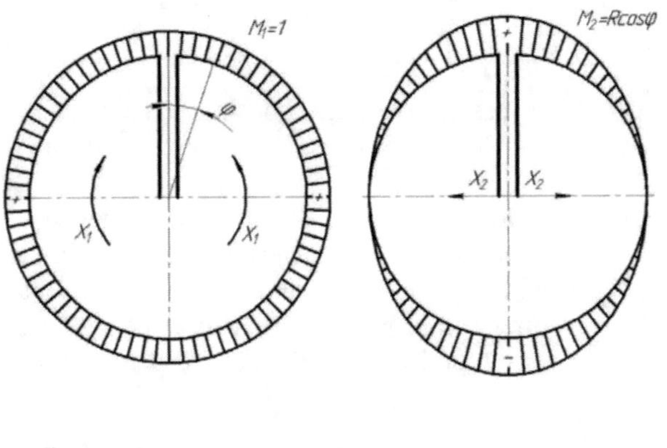

Figure 2. Distribution diagrams of bending moments from single forces.

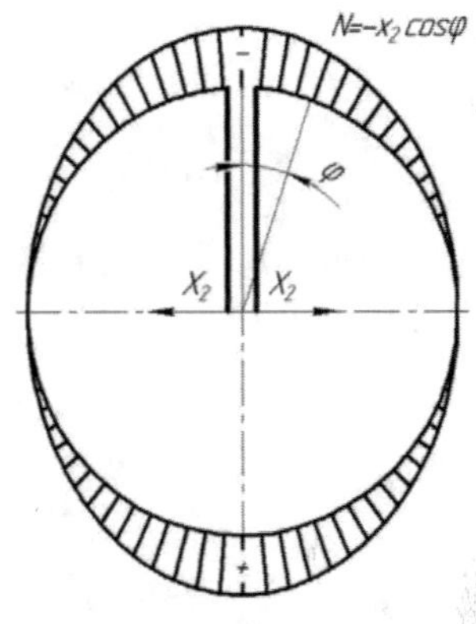

Figure 3. Distribution diagrams of normal forces from single forces.

$$\delta_{22} = \frac{2R}{EI}\int_0^{\frac{\pi}{2}} R\cos(\varphi)d\varphi\,, \tag{4}$$

$$\Delta_{1p} = 2R\left[\int_0^{\frac{\pi}{2}} \frac{qR^2}{2}\sin^2(\varphi)d\varphi + \int_{\frac{\pi}{2}}^{\beta_1} qR^2(\sin(\varphi)-0{,}5)d\varphi + \int_{\beta_1}^{\pi}\left[qR^2\sin(\varphi-0{,}5)+\frac{qR}{2a}(a-R\sin(\varphi))^2\right]d\varphi\right], \tag{5}$$

$$\Delta_{2p} = 2R\left[\int_0^{\frac{\pi}{2}} R\cos(\varphi)\frac{qR^2\sin^2(\varphi)}{2}d\varphi + \int_{\frac{\pi}{2}}^{\beta_1} R\cos(\varphi)qR^2(\sin(\varphi)-0{,}5)d\varphi + \right.$$
$$+$$
$$\left. + \int_{\beta_1}^{\pi}\left[qR^2\sin(\varphi-0{,}5)+\frac{qR}{2a}(a-R\sin(\varphi))^2\right]d\varphi\right], \tag{6}$$

where R – the radius of the tunneling shield housing shell; E – elasticity module; I – moment of inertia; M_I – bending moment on the component x_I, q – vertical load intensity per a meter of horizontal diameter of the tunneling shield housing.

The vertical load intensity per a meter of horizontal diameter of the tunneling shield housing q depends on the design shield depth, average volumetric mass of the overlaying rocks, design length of the shield housing and shall be determined according to the formulae:

$$q = H\gamma L_{des} g, \tag{7}$$

where H – shield depth; γ – average volumetric mass of the overlaying rocks; L_{des} – design length of the tunneling shield housing.

Bending moments effecting the tunneling shield housing are determined in accordance with the functions:

- from 0 to $\pi/2$:

$$M = x_1 + x_2 R\cos(\varphi) + \frac{qR^2}{2}\sin^2(\varphi), \tag{8}$$

- from $\pi/2$ to β_1:

$$M = x_1 + x_2 R\cos(\varphi) + qR^2(\sin(\varphi) - 0{,}5), \tag{9}$$

- from β_1 to π:

$$M = x_1 + x_2 R\cos(\varphi) + qR^2(\sin(\varphi) - 0{,}5) + \frac{qR}{2a}(a - R\sin(\varphi))^2 \tag{10}$$

To determine the bending moments and normal forces effecting the housing and the shell of the tunneling shield, respectively, a section with the most acting stress is selected and the future calculation is performed for this section.

To determine the main design parameters of the support ring of the tunneling shield designated for supporting and securing the shield jacks, as well as for the supply effort transfer from the shield jacks to the cutter-base plate, it is necessary to make assumption that a separate ring segment restricted by the strengthening ribs (kneepieces) acts to clear bending and represents a bar on two fixed supports. The design schemes of the support ring and its segment are given in Fig. 4 and 5, respectively.

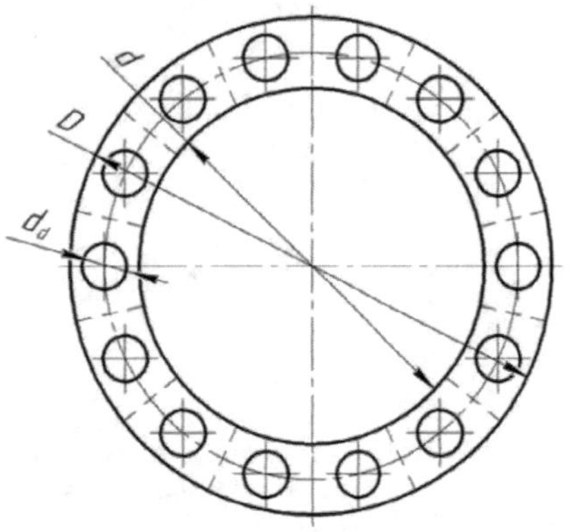

Figure 4. Design scheme of the support ring.

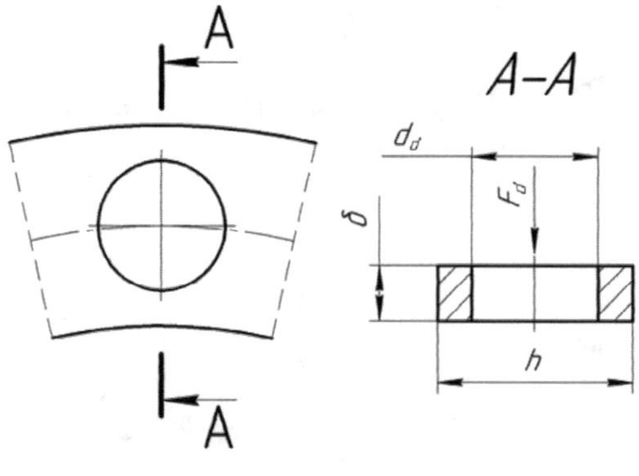

Figure 5. Support ring segment with dangerous section.

The dangerous section in the support ring segment is A-A section. To determine the acting stresses in the A-A support ring segment section, the following function should be used:

$$\sigma = \frac{M_{bend}}{W}, \tag{11}$$

where M_{bend} – bending moment in the A-A section; W – A-A section modulus.

The section modulus W is determined according to the formulae:

$$W = \frac{\delta\left(h^3 - d_d^3\right)}{6h}, \tag{12}$$

where δ – support ring thickness; h – support ring segment width; d_d – diameter of support ring segment opening for the jack.

This means that the support ring thickness can be defined from the following:

$$\delta = \frac{6h^2 F_d n}{2\sigma_T\left(h^3 - d_d^3\right)}, \tag{13}$$

where F_d – point loading acting in the dangerous section; n – specified safety factor of the support ring; σ_T – yield point of the material of support ring production.

3. Numerical Modelling

The offered technique does consider the tunneling shield as a complex object, with some inner design components directly effect the stress-strain state of the housing shell. To assess the stress-strain state of the shield housing shell, the numerical calculation was performed using the finite-element method in the numerical modelling environment SALOME-MECA v2018 [13,14].

The geometrical model of the microtunneling shield housing (Fig. 6) was created using the Geometry module tools of the numerical modelling platform SALOME-MECA. The housing model contains the internal power design elements: two support rings of thrust cylinders, cutter-base plate with the bearing housing and main drive shaft of the control, ratchet gear levellers of the main drive shaft, support plates of the brackets of the main drive shaft hydraulic cylinders.

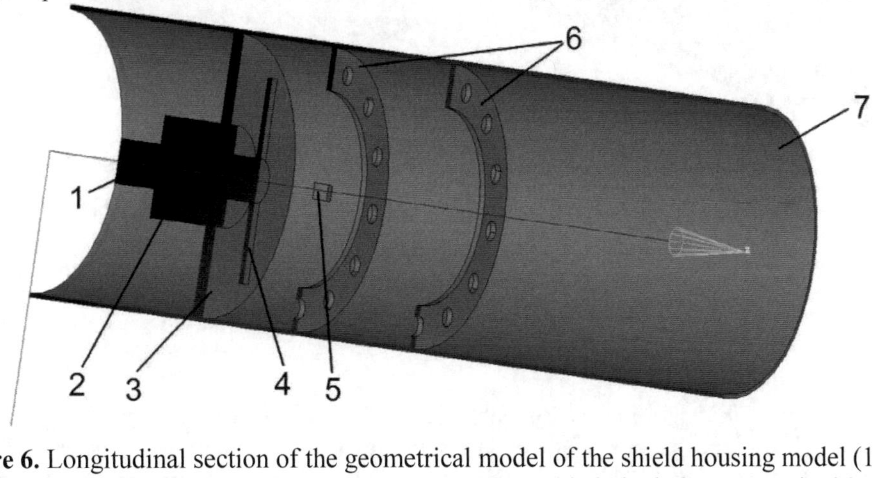

Figure 6. Longitudinal section of the geometrical model of the shield housing model (1 - main drive shaft, 2 - bearing housing of the main drive shaft, 3 - blade-bed plate, 4 - main drive link, 5 - bracket support of the main shaft hydraulic cylinder, 6 - support rings, 7 - housing shell).

The finite-element mesh is created using the Mesh module tools of the numerical modelling platform SALOME-MECA.

The basic characteristics of the finite-element length are limited by the value range of 5-10 mm. The degree of the finite-element optimization by quality is set as "VeryFine". The automatic conversion of the tetrahedral mesh into the quadrahedral one and adjustment of the matching units were performed. The finite-element mesh contains 1606090 nodes and 8451766 elements, 8012006 of which are three-dimensional.

The three-dimensional finite-element mesh has the separated surfaces of the forces applied that effect the design elements of the tunneling shield housing in the form of two-dimensional finite-element groups. The forces applied to such surfaces are uniformly distributed. The modelling of the forces applied was performed by using the operator FORCE_FACE, function AFFE_CHAR_MECA, "AsterStudy" module of the numerical modelling platform SALOME-MECA. The ground friction forces (459 kN) and vertical rock pressure forces (based on the rock pressure value of 0.85 MPa) are applied to the external surfaces (except for the end surfaces) of the shield housing shell. The force of resistance to intrusion of the shield housing edge into the rock was applied to the front end surface of the shield housing shell (1,340 kN). The basic surfaces relative to which the calculation of the stress-strain state of the shield housing is performed are the joint faces of the support rings of thrust cylinder. The linear static model "LinearStaticAnalisys" (function MECA_STATIQUE), "AsterStudy" module, was used for calculation of the stress-strain state. The calculation results of the stress-strain state of the microtunneling shield housing are given in the form of charts of equivalent stresses and strains in the shield housing. Post-processing and visualization of the calculation results of equivalent stresses and strains were performed using the integrated software tools of the "ParaVis" module [15-17] of the numerical modelling platform SALOME-MECA.

The calculation results established that the most strains of the microtunneling shield housing are observed on termination of the tail part. The strain value does not exceed 15 mm (Fig. 7) that is 4.7 times higher than the maximum constriction clearance (not more than 70 mm). The peak stress values in the shield housing do not exceed 530 MPa (Fig. 8) that makes it possible to use steel 14CrMn2SiVCu.

Figure 7. Chart of the shield housing strain.

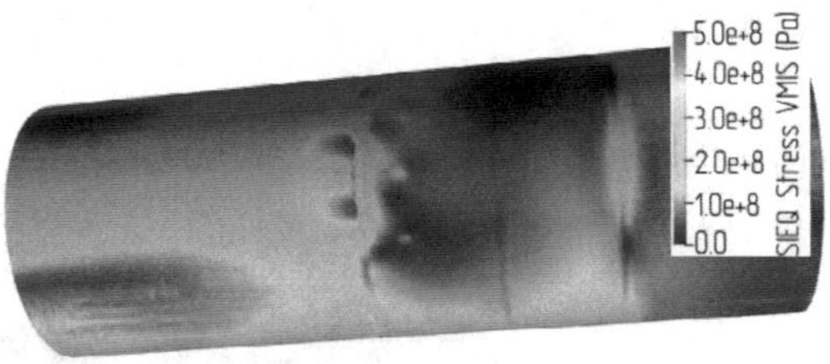

Figure 8. Chart of stress distribution in the tunneling shield housing.

Considering the obtained results, the reasonability of reduction in the tunneling shield shell thickness was assessed.

4. Conclusion

The given technique for determination of such design parameters of the tunneling shield as the housing shell thickness and support ring thickness provides for calculation of the specified design parameters considering the shield mass reduction provided that the regulated safety factor is preserved that will make it possible to create its balanced structure from the strength point of view in future.

Acknowledgments

Research and development carried out at the SRSPU(NPI) with the financial support of the Ministry of Education and Science of the Russian Federation - Agreement No. 074-11-2018-010 dated 5 June 2018 on the topic: “Creation of a high-tech production of import-substituting mining equipment, tooled with intelligent control systems, for the development of the underground space of cities”.

References

[1] G. Hemphill, Practical Tunnel Construction. Hoboken, New Jersey: John Wiley & Sons, 2013.
[2] B. Maidl, L. Schmid, W. Ritz, and M. Herrenknecht, Hardrock Tunnel Boring Machines. Berlin: Ernst & Sohn, 2008.
[3] B. Stack, Encyclopaedia of Tunnelling, Mining, and Drilling Equipment. Hobart, Tas.: Muden Publishing, 1995.
[4] G. West, Innovation and the Rise of the Tunnelling Industry. Cambridge, England: Cambridge

University Press, 1988.

[5] R Goel and B. Singh. Engineering Rock Mass Classification: Tunnelling, Foundations and Landslides, 1st ed. Woburn, United States: Elsevier, 2011.

[6] N. Barton, TBM Tunneling in Jointed and Faulted Rock. Rotterdam: Balkema, 2000.

[7] Z. Bieniawski, Rock Mechanics Design in Mining and Tunneling. Rotterdam: Balkema, 1984.

[8] B. Brady and E. Brown, Rock Mechanics: for Underground Mining, 3rd ed. Dordrecht: Kluwer Academic Publishers, 2004.

[9] V. Guglielmetti, P. Grasso, A. Mahtab, and S. Xu, Mechanized Tunnelling in Urban Areas: Design Methodology and Construction Control. UK: T & F Books, 2008.

[10] D. Chapman, N. Metje, and A. Stark, Introduction to Tunnel Construction. New York: Spon Press, 2010.

[11] C. Kirkland, Engineering the Channel Tunnel. Boca Raton, FL: CRC Press, 1995.

[12] V. Brenner, A. Zhabin, M. Shchegolevskij, Al. Polyakov and An. Polyakov, Complexes of Tunnelling Shields: Study Guide. Moskow: Gornaja kniga, 2009.

[13] J.-P. Aubry, Beginning with Code_Aster. A Practical Introduction to Finite Element Method Using Code_Aster, Gmsh and Salome. Paris: Framasoft, 2013.

[14] D. Thakore, Finite Element Analysis using Open Source Software, 2nd ed. Brisbane, Australia: Moonish Ent. Pty. Ltd., 2014.

[15] J. Ahrens, B. Geveci, and C. Law, ParaView: An End-User Tool for Large Data Visualization. Energy, vol. 836, pp. 717–732, 2005.

[16] U. Ayachit, The Para View Guide: A Parallel Visualization Application. USA: Kitware, 2015.

[17] D. DeMarle, B. Geveci, J. Ahrens, and J. Woodring, Streaming and Out-of-Core methods, High Performance Visualization: Enabling Extreme Scale Scientific Insight. Boca Raton, FL: CRC Press, 2013.

IOP Conf. Series: Materials Science and Engineering **680** (2019) 012034 doi:10.1088/1757-899X/680/1/012034

Modeling of competition between agents of the same level in a multi-level system with information asymmetry

A Gurtuev[1], M Dovletmurzaeva[2], M Chazhaev [2]

[1] Institute of computer science and problems of regional management, KBSC RAS, 360000, Nalchik, Russia,
[2] Chechen State University, 366007, Groznyi, Russia,

E-mail: alemao@mail.ru

Abstract. We propose a game-theoretic model of resource-based competition between agents of the same level in a multi-level system with information asymmetry. It is shown that under conditions of uncertainty, the optimism of agents leads to non-optimal solutions and an increase in the loss of the total added value in the system due to resource waste. The mechanisms of optimistic shifts in agent estimates and agent's Bayesian adjustment are described. Equilibrium conditions are obtained in the proposed model.

1. Introduction

Problems of estimating uncertainty are often named among the most significant factors that impede decision making in systems with partially independent agents, both technical and social. Among such systems, we could distinguish multilevel systems in which this problem is combined with the problem of agent opportunism [3, 11, 17]. For example, consider venture capital funds with some project filtering mechanisms, where reviewing experts, managers, decision makers, fund managers and shareholders are all agents of different levels; distributed hierarchical management systems, such as a regional management system; social network structures with agents of various levels of access and influence, including markets with imperfect competition, etc. In the models of such systems, the heterogeneity of individual agents is usually taken into account by means of various utility functions, and the differences in levels are taken into account by means of principal-agent models in modeling of relationships between agents [11, 15]. For such systems there is a need to model decision-making taking into account uncertainty, and the most common approach is to reduce uncertainty to risk where it is possible [7, 8]. The problem arises when such reduce is impossible without a significant loss of information.

A closer look allows us to distinguish two components of the problem of estimating uncertainty – rational decision-making mechanisms and management of risk attitude mechanisms. If a large number of works were the first to be devoted [2, 13, 16], then risk propensity management issues were addressed much less frequently [5], most often risk propensity in modern models is accepted by the immanent characterization of an agent that is not subject to changes [2, 6].

At the same time, the question of the optimal risk appetite or the optimal ratio of agents with different risk appetites seems to be important for the study of distributed decision-making systems (both technical and social) that are faced with the uncertainty of the future. Such systems are based on the aggregation of implicit knowledge, the carriers of which are various agents, for making decisions based on the forecast of the future state of the external environment. An example of such a system is the innovation investment system, when the evaluation of innovative projects is difficult due to their uniqueness, and the proportion of potentially successful projects is small enough so that a naive approach based on the initial screening of pseudo-innovators [9, 10] and the subsequent financing of all projects was irrational. In reality, the supply side in the innovation investment market is represented by many investors, differing in institutional characteristics and decision-making models. For some of them, risk appetite is normatively limited (investment funds), for others (business angels) - there are no such restrictions. This means that an effective mechanism for managing the system of innovative investment (and more broadly – systems with heterogeneous agents in an uncertain environment, both social and artificial) must take into account the existence of persistent deviations from risk neutrality in the behavior of agents. In this regard, it is of interest to study the effectiveness of various economic mechanisms in a multi-agent system with uncertainty at different risk appetite of agents.

While modeling social and economic systems the risk attitude for is usually represented as utility functions of risky gains. A convex utility function is characteristic for risk-prone agents, concave - for risk-avoiding agents. Arrow-Pratt's risk aversion measure, which characterizes the elasticity of the marginal utility of money in terms of consumption, is often used as the basis for assessing risk appetite [1, 14, 18]. For situations of symmetric information and risk (i.e., the known probabilistic distribution of uncertain future payments), this representation makes it possible to very well take into account the individual characteristics of agents in modeling interactions (insurance market, lotteries, stock market, etc.). In the case of information asymmetry or uncertainty, when the probability distribution parameters are unknown, this approach does not give the desired result, since in addition to the individual risk appetite, one must also take into account deviations of the estimates of the probability distribution of future incomes and expenses from the true values.

Thus, the question arises of how to model the risk appetite of an agent under uncertainty. Assume that agent wins are cashable. Then deviation from risk-neutral behavior is not optimal. The literature shows that this is true both for conditions of complete information and for risk conditions [4, 18]. Consider the uncertainty situation as a risk situation with true distribution parameters unknown to the agents (expectation, variance, asymmetry, shift, etc.). Let us imagine the risk appetite of the agent in the form of the combined effect of several irrational deviations of the estimate, which shift the subjective assessment of the probability of the occurrence of the event from some true probability. For competing innovators, this may be optimism about their performance, for business angels, optimism about their ability to assess the prospects of an innovative project. In the event of a dumping war, this may be the underestimation of the competitors' budget capabilities. For regional agents in a distributed hierarchical system, this is optimism regarding their ability to attract external and internal resources, etc.

2. The Model

Suppose that in a multi-level open system there are several agents with different risk appetite, competing for one long-term contract with the principal.

Consider a model of competition between two agents $\boldsymbol{i}_1$ and $\boldsymbol{i}_2$, each of which has a certain budget $\boldsymbol{R}_i(\boldsymbol{t})$ at time $\boldsymbol{t}>=0$. For simplicity, we take the initial value $\boldsymbol{R}_i(0)$ per unit. $\boldsymbol{R}$ reflects the agent's ability to bear the costs at the bidding stage for the subject of the contract, for example, the cost of participation in exhibitions, the manufacture and presentation of prototypes, and the preparation of marketing materials.

The game begins with a preliminary period, followed by a depletion game, if the participant decides to take part in it. In the preliminary period, each participant chooses to participate in the fight or not, and if either of the two chooses to participate, then a war of attrition type game begins [12], otherwise the status quo is saved and the budgets of the agents are not wasted. If the war of attrition began, then in each time period $\boldsymbol{t}>=0$ both agents simultaneously decide to continue to fight or to leave the game. Continuing the fight entails costs for both participants. These costs can be represented as a decrease in the agent's budget with a certain speed v_i, so that the costs of agent i over time t will be as follows:

$$\frac{dR}{dt} = -v_i R_i(t) \quad (1)$$

So, we can deduce that $R_i(t) = e^{-v_i t}$

Suppose that v_i is independently taken from the distribution $F_i(\hat{v}, \gamma^2)$, known to agents, where any v is in the range from 0 to 1. Agent i knows its v_i value (its type) exactly and has an estimate about the type of the other agent (assumption regarding the probability distribution).

Let us introduce the concept of optimistic bias in agents' assessments of the types of competing agents. Suppose that the expected distribution of competitor type F_{-i} can be represented as $\tilde{F}_{-i}\left(\hat{v} + K_i, \frac{\gamma^2}{\sigma_i^2}\right)$, where $\hat{v} + K_i$ is the estimate of mathematical expectation, and $\frac{\gamma^2}{\sigma_i^2}$ is the variance estimate. We now consider two types of optimistic biases in the assessment – "underestimating the opponent" ($K_i > 0$) and "re-evaluating the accuracy of the forecast" ($\sigma_i^2 > 1$). Underestimating an opponent is the average error in assessing the likelihood of an early exit of an opponent from the game. In the model, it takes the form of increasing the mathematical expectation of costs in each period of the game, which reduces the opponent's incentives to continue the rivalry in assessing player i. The second type of optimistic bias is a reassessment of the accuracy of your forecast. The higher σ_i, the more confident player i is in its assessment of opponent's costs.

Two circumstances should be noted. Firstly, underestimating an opponent in our model is a deviation from optimal behavior, because it fully explains the discrepancy between the expected distribution $\tilde{F}_{-i}$ with the true distribution F_{-i}. Secondly, underestimating the opponent leads to asymmetric configurations, because in its absence, the expected distributions of the players coincide.

For illustration purposes, Fig. 1 shows all four possible types of the deviations $\tilde{F}_{-i}$ from F_{-i} - a) underestimation of the opponent, b) overestimation of accuracy, c) both types of bias combined and d) no bias (zero deviation).

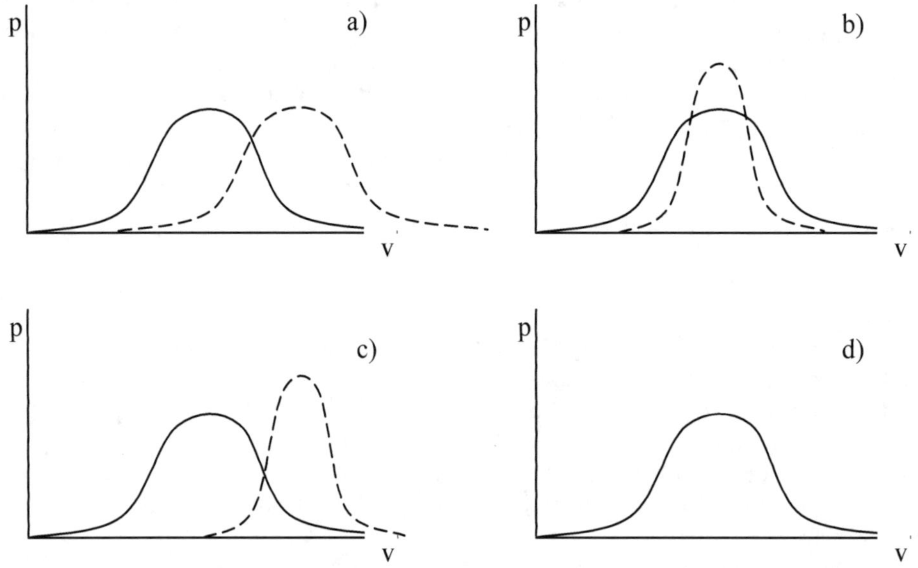

Figure 1. Types of deviations caused by differences in subjective ratings $\tilde{F}_{-i}$ (shown by a dotted line) from the true distribution of $\boldsymbol{F}_{-i}$ (shown by a solid line).

In each period of the game, agents choose whether to continue to compete or to leave the game. The game ends when at least one of them leaves the game. If one agent refused to continue the competition, and the second did not, then the winner receives the resources of the loser. If both refused to continue in the same round, they remain with their sources of resources.

Thus, if agent $\boldsymbol{i}$ intends to leave the game at step $\boldsymbol{T_i}$, its expected gain depends on the actions of the opponent and can be described as follows:

$$U_i(T_i, T_{-i}, v_i) = \begin{cases} -T_i, \; T_i < T_{-i} \\ -T_i + V_i^S(T_i), \; T_i = T_{-i} \\ -T_i + V_i^W(T_i), \; T_i > T_{-i} \end{cases} \qquad (2)$$

where $V_i^S(T_i), V_i^W(T_i)$ – expected final payouts, respectively, in case of simultaneous completion of the game and victory in the game

Denoting the distribution of expectations of agent $\boldsymbol{i}$ with respect to the future exit point of the opponent from the game by $\tilde{G}_{-i}(\cdot)$ and the corresponding probability density function by $\tilde{g}_{-i}(\cdot)$ we obtain the expected payout of the agent

$$V_i(T_i, v_i) = -T_i[1 - \tilde{G}_{-i}(T_i)] + \int_0^{T_i}[V_i^W(x) - x]\tilde{g}_{-i}(x)dx \qquad (3)$$

The equilibrium in this game is found as the Bayesian Nash equilibrium [8]. The strategy of agent $\boldsymbol{i}$ is described by a pair $(\boldsymbol{s}_i, \boldsymbol{T}_i(\boldsymbol{v}_i))$, where the second element is a function that determines the optimal predicted exit time for each possible value of $\boldsymbol{v}_i$.

3. Conclusion

Denoting the distribution of expectations of agent $\boldsymbol{i}$ with respect to the future exit point of the opponent from the game by $\tilde{G}_{-i}(\cdot)$ and the corresponding probability density function by $\tilde{g}_{-i}(\cdot)$ we obtain the expected payout of the agent

$$V_i(T_i, v_i) = -T_i[1 - \tilde{G}_{-i}(T_i)] + \int_0^{T_i}[V_i^W(x) - x]\tilde{g}_{-i}(x)dx \qquad (4)$$

The equilibrium in this game is found as the Bayesian Nash equilibrium [8]. The strategy of agent $\boldsymbol{i}$ is described by a pair $(\boldsymbol{s}_i, \boldsymbol{T}_i(\boldsymbol{v}_i))$, where the second element is a function that determines the optimal predicted exit time for each possible value of $\boldsymbol{v}_i$.

Acknowledgements

The research was supported by Russian Foundation of Basic Research (RFBR), research projects No. 19-010-00376 A, 19-010-00577 A.

References

[1] Arrow, K. J. (1965). Aspects of the Theory of Risk Bearing. The Theory of Risk Aversion. Helsinki: Yrjo Jahnssonin Saatio. Reprinted in: *Essays in the Theory of Risk Bearing*, Markham Publ. Co., Chicago, 1971, 90–109.

[2] Arve, M., Martimort, D. (2016). Dynamic Procurement under Uncertainty: Optimal Design and Implications for Incomplete Contracts. *Am. Econ. Rev.* **106**(11), 3238-74.

[3] Chassang, S., Padró I Miquel. G., Snowberg, E. (2012). Selective Trials: A Principal-Agent Approach to Randomized Controlled Experiments. *Am. Econ. Rev.* **102**(4), 1279-1309.

[4] Chemmanur, T., Simonyan, K. (2010). What Drives the Issuance of Putable Convertibles: Risk-Shifting, Asymmetric Information, or Taxes? *Financial Management*. **39**, 1027-1067.

[5] de Haan, T., Offerman, T., Sloof, R. (2011). Noisy signaling: Theory and experiment. *Games Econ Behav*. **73**(2), 402-428.

[6] De Luca G., Sekeris, P. (2013). Deterrence in contests. *Economica*, **80**(317), 171–189.

[7] Ehrlinger, J., Mitchum, A., Dweck, C. (2016). Understanding overconfidence: Theories of intelligence, preferential attention, and distorted self-assessment. *Journal of Experimental Social Psychology*, **63**:94–100.

[8] Fudenberg, D., Tirole, J. (1986). A theory of exit in duopoly. *Econometrica*, **54**:943–960.

[9] Gurtuev A. (2018). Direct incentive-compatible mechanism for innovator-investor bargain in an innovation investment system. XIV International Scientific-Technical Conference "Dynamic

of Technical Systems" (DTS-2018). *MATEC Web Conf.*, **226** (2018) 04032.

[10] Gurtuev A. O., Ivanov Z. Z., Derkach E. G. , Kazancheva H. K. , Dumanova A. Kh. (2018). Investor-expert interaction model in the innovation investment system with knowledge asymmetry. *Opcion*, Vol. **34**, Núm. 14: 1549-1574.

[11] Laffont, J.-J., Martimort, D. (2002). The Theory of Incentives I: the Principal-Agent Model. Princeton University Press.

[12] Menuet, M., Sekeris, P. Uncertainty, Overconfidence, and War. 2019. *WP hal-02155286v2.*

[13] Netzer, N., Scheuer, F. (2012). A game theoretic foundation of competitive equilibria with adverse selection. *NBER Working paper* No.w18471.

[14] Pratt, J. W. (1964). Risk Aversion in the Small and in the Large. *Econometrica*. **32**(1-2), 122-136.

[15] Riley, J. (2001). Silver Signals: Twenty-Five Years of Screening and Signaling. *J. Econ. Lit.* **39**, 432-78.

[16] Slantchev, B., Tarar, A. (2011). Mutual optimism as a rationalist explanation of war. *American Journal of Political Science*, **55**(1):135–148.

[17] Gurtuev A.O., Derkach E.G., Ivanov Z.Z. (2016). Sovremennoe sostoyanie issledovaniy v oblasti obespecheniya ispolneniya kontraktov v usloviyah neophedelennosti v sistemah innovatsionnogo investirovaniya. *Izvestiya KBNC RAN*, **5**(73): 62-68.

[18] Tarasova V.V., Tarasov V.E. (2017). Nepriyatie riska dlya investorov s pamyatyu: ereditarnye obobscheniya mery Errou-Pratta. *Nauchno-issledovatelskiy finansovyi institut. Finansovyi zhurnal*, **2**(36): 46-63.

IOP Conf. Series: Materials Science and Engineering **680** (2019) 012035 doi:10.1088/1757-899X/680/1/012035

Some specifics of modeling multi-level agent systems with uncertainty using Groves mechanisms

A Gurtuev[1], M Dovletmurzaeva[2], R Yalmaev[2]

[1] Institute of computer science and problems of regional management, KBSC RAS, 360000, Nalchik, Russia,

[2] Chechen State University, 366007, Groznyi, Russia,

E-mail: alemao@mail.ru

Abstract. Uncertainty is often named among the most significant factors impeding decision making in systems with partially independent agents, both technical and social. Among such systems, multilevel systems can be distinguished as a hard case, as uncertainty problem there is combined with the problem of agent opportunism. A special class of mechanisms — the Groves mechanisms that satisfy the requirements of truthfulness in dominant strategies and efficiency — has been well studied in the literature. The article shows how, under conditions of multilevel multi-agent systems with uncertainty, the implementation of Groves mechanisms can solve the problem of systematic distortion of the agents behaviour.

1. Introduction

The problems of developing effective mechanisms of interaction between independent or semi-independent entities of the system are among the most significant factors that affect overall decision making in technical and social complex multi-agent systems. In multi-level multi-agent systems, like, for example, venture funds, distributed hierarchical management systems, social network structures with agents of various levels, this problem is combined with the problem of agent opportunism [1-3]. In the models of such systems, the heterogeneity of individual agents is usually taken into account by means of various utility functions, and the differences in levels are taken into account by means of principal-agent models when modeling interagency relations [2, 4]. Moreover, an effective mechanism for managing the multi-level multi-agent systems (and more broadly - systems with heterogeneous agents in an uncertain environment, both social and artificial) should take into account the existence of persistent deviations from risk neutrality in the behavior of agents. In this regard, the study of effectiveness of particular decision-making mechanisms for a multi-agent system with uncertainty is very interesting.

At the present stage, one of the significant features of the socio-economic systems of the Russian regions is the existence of so-called interest groups, each of which seeks to realize its goals by influencing potential investors, both public and private. The problems of studying the mechanism for coordinating group interests and the influence of such groups on the effectiveness of the regional investment process began to be studied relatively recently, and the problems of the influence of groups on the processes of managing investment projects in the regions were not considered in detail [5, 6].

One of the most important factors underlying the phenomenon of the influence of regional interest groups on investor decisions is that decision-making mechanisms at all levels of the regional socio-economic system are in constant motion, they constantly collide, compete and conflict with each other.

At the same time, investment project managers are notable for their adaptability, opinions, positions, likes, dislikes, etc., therefore they are accessible to the influence of various interest groups [7].

If managers of investment projects can be considered to a greater extent as direct executors of managerial decisions, then strategic management is carried out by representatives of the groups financing the project (both private and state) - which will be called "investors" later in the paper. They offer shareholders certain strategies, promising a certain sequence of their future property management activities. It is important for investors that shareholders believe that they will carry out certain managerial influences and, most importantly, that they can carry out these influences (that is, that shareholders believe that they have certain resources).

In this case, another problem arises - an interest in managerial decisions being made, because the interests of shareholders on the one hand and individual managers or interest groups on the other do not always coincide [8, 9]. The success of interest groups in lobbying is largely determined by their access to useful information, as well as the low costs of collective action in such groups. The situation when the investment project manager makes managerial decisions guided by group interests can be demonstrated on the basis of formalized models.

Consider a situation in which a group of interests with the aim of resolving issues in relation to a certain investment object affects the target function of the investor making decisions on managing this object. The described situation is best represented in the form of an exchange model. We apply the approach proposed by Denzau and Munger [10] to the model and assume that interest groups may be interested in receiving from the investor something other than his support, because the investor spends his limited time and efforts to solve a number of specific problems.

At first glance, the official interests of the investor managing the investment project should coincide with the target settings of his company, so the target function of the manager should be determined by his nominal role as the representative of the interests of the investment company, and should not take into account his own target needs. However, we should note that behind a formal legal entity (an investment company, in this case) are real people with their inherent utility functions, and the influence of interest groups that affect investing managers is determined by the level of accessibility of information.

2. Model of agent behaviour in multi-level system with hidden information

Consider the activities of the investor, who was elected CEO of a large investment company, whose interests in the field of managing its property, he must represent. His goal is to be re-elected to this post again, so he will do everything to maximize the number of shareholders who will vote for his decisions on managing the company's property in the next election. However, the investor's activity is limited by his capabilities (worktime, natural capabilities, influence). It has some limited resource $\boldsymbol{E}$, which is given exogenously and which it cannot influence in any way. Then he distributes $\boldsymbol{E}$ into three types of activity in such a way as to maximize his function:

1. to actions directly useful to the shareholders of the company;
2. to actions related to the ownership of the company, which are primarily useful to interest groups that are not shareholders, but that can positively or negatively affect the shareholders of the company;
3. to actions that advertise his activities.

He spends some resources on activities that benefit interest groups, because in exchange for implementing such a policy he receives resources for his PR campaign. His promotion, as a good manager, by a group of interests can be measured in monetary terms. Such an objective function of the investor's welfare level W has the form:

$$W[P_u(E_u), P_i(E_i), R_i] \rightarrow max E_u, \{E_i, R_i\} \quad (1)$$

The investor spends his resource E on activities that increase the likelihood of his re-election. Spending a share of its E_u resource, the investor brings P_u of utility to the company. Other share of the investor's resource E_i, is spent in favor of the i-th group of interests. The resource E_i spent on a certain policy brings the investor P_i efficiency level (he invests E_i and receives P_i). His utility function also depends on R_i, which he receives in exchange for P_i, because he invested some efforts of E_i (E_1, E_2, ..., E_n) in promoting various interest groups and received the resource R_i.

Thus, the investor maximizes his utility function. Consider the border cases:

• a situation where shareholders have complete information about decisions made by the investor;

• a situation where shareholders are uninformed.

In a situation where shareholders have complete information, the real activities of the investor are known to all, and advertising is unproductive. Then the objective function takes the form:

$$W = W(P_u, P), P = (P_1, P_2, \dots, P_n) \tag{2}$$

The welfare of the investor will depend only on the results of the company. And if the P_i decision, beneficial to a certain group of interests, will negatively affect the company's results, the investor simply will not accept it. This means that in a situation of complete information, the investor will only pursue a policy that positively affects the shareholders of his company.

In a situation where shareholders are uninformed, they are not able to assess the benefits of investor actions, because they receive all relevant information only from external sources. Then the function will look like this:

$$W = W[R(P_i(E_i))] \tag{3}$$

Thus, the activities of the investor for the benefit of the company are not directly taken into account by the shareholders, since they are unobservable for them. In this case, the investor depends entirely on interest groups – only they can "advertise" it by informing shareholders. Obviously, real situations are something in between, but these extreme cases are very obvious.

$$\begin{cases} W[P_u(E_u), P_i(E_i), R_i] \to max E_u, \{E_i, R_i\} \\ W = \begin{cases} W(P_u, P), P = (P_1, P_2, \dots, P_n), & E_i = 0 \\ W[R(P_i(E_i))], & E_i > 0 \end{cases} \end{cases} \tag{4}$$

In this model, the interests of the investor are corrected from the outside, influencing its objective function. Assume the opposite situation – the interests of the influence group and the investor coincide, i.e. the investor acts in the direction of resolving issues within the scope of interests of the group, therefore it is possible to consider the case when the impact is on the budgetary constraints of the investor.

Consider a strategy aimed not at changing the position and preferences of the investor (since his preferences already coincided with the group of interests), but at changing his resources, i.e. in this case, there is an impact on the budget line of the investor, and not on its utility function.

A model appropriate to this situation will be based on three assumptions:

1. An investor, in order to be able to make decisions, must actively participate in management, i.e. spend certain resources on the implementation of the decision-making process and monitoring their implementation during the implementation of the investment project.

2. In any period, the investor simultaneously solves several problems related to the management of investment projects, and he spends more of his resources on some issues than on others. Choosing a project for implementation, he will coordinate his preferences about positive results in various areas with other people who have an influence on the company's policy (other directors and managers).

3. Investor's resources (time, staff, and information) are limited. Therefore, he cannot simultaneously carry out all types of activities that should ensure success in resolving the issues that he supports. Therefore, he sets priorities.

Practically this model reflects the relationship of a group of interests and an investor who are simultaneously interested in solving the same issue (implementation or refusal to implement an investment project). To solve the problem in the context of the model means that the investor brings the current policy in relation to the project closer to the policy preferred by the group of interests, or at least as a result of joint actions, the likelihood of making an appropriate decision increases. As an illustration, we use the case of a decision to sell a share of shares $\boldsymbol{D}$ of a certain enterprise $\boldsymbol{A}$. The effectiveness of solution to this issue is indicated on the horizontal axis (Fig. 1) as $\boldsymbol{D_a}$.

The vertical axis $\boldsymbol{D_o}$ reflects the changes made by the investor in the framework of other issues to be resolved (selling and buying other property, making decisions on the distribution of profits or additional investments in specific projects, etc.). But since the investor has a limited amount of resources, which is represented by budget line *B* in Fig. 1, his efforts to implement all the issues of interest to him are limited.

In the model, a group of interests subsidizes the information and labor costs for the investor and helps him to take independent actions to change the policy regarding this investment project or protect an existing solution. In fact, the interest group acts as the "additional staff" of the investor.

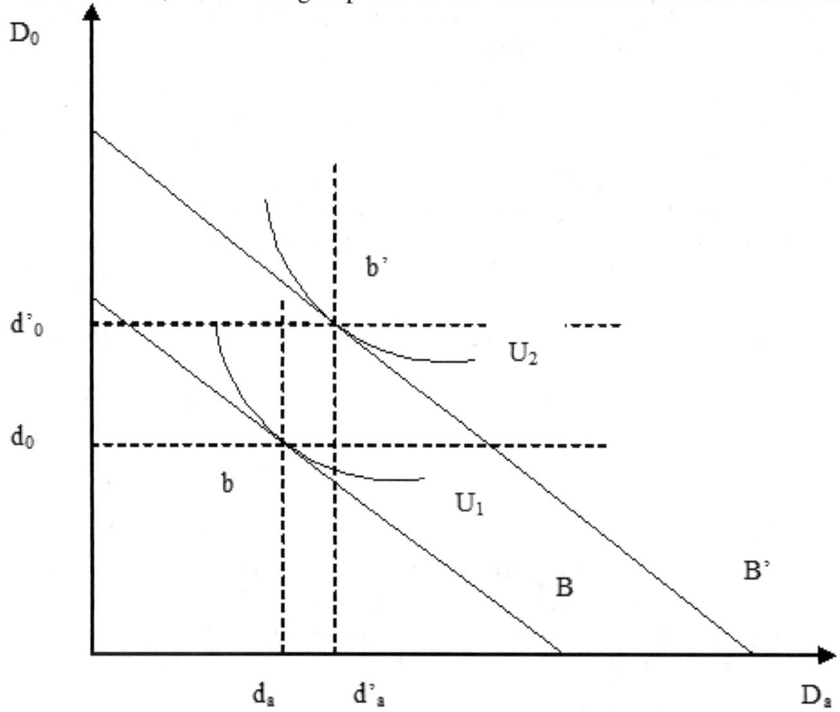

Figure 1. The influence of interest groups on investment decisions.

Investor preferences *i* in this model are reflected by their willingness to bear the costs (in time and effort) of selling a share in project *A*. The indifference curve U_i is a function of the interests and opinions of shareholders. Suppose that they are informed about the decisions made by the investor - then he will be interested in making decisions that best meet the interests of shareholders. The point of contact of the budget line (d_a, d_o) and U_i is the optimal distribution of investor time. He will not be able to do anything more while maintaining the available resources.

A group of interests differs from the investor in its interest in solving one problem of interest to it and has a significant resource for solving it. Since a group of interests cannot directly participate in decision-making in the field of managing the property of an investment company, the group's resource (for example, in the form of a "grant", change in income) must be realized through those who have the opportunity to participate in making these decisions. To this end, the group of interests provides the investor associated with the management of its resource, and thereby shifts its budget line to the right. The investor's optimum thereby increases ($U_2 > U_1$).

Figure 1 reflects the impact of the resource provided by the investor to make the necessary decisions. An additional resource shifts the investor's budget line in parallel from *B* to *B'*. As a result, part of the time and effort is redistributed to additional support for the decision to sell a share in the investment project ($d_a' - d_a$), and most of the attracted resources ($d_o' - d_o$) are used to achieve other investor tasks. Obviously, the use by the investor of the resources provided to solve other problems is completely ineffective for the group of interests.

Therefore, the interest group needs to limit the investor's ability to use the provided resources. When establishing control over the intended use of the provided resource, its value to the investor becomes different depending on how it is used. The investor can fully use the resource to solve the problem. But with the cost of the resource for other purposes, its value for the investor increases, since the provided resource is used not only to achieve other goals, but also to hide the very fact of such activities.

If the restriction works properly (i.e., the interest group can be sure that all the information and work carried out by the investor will be additional to what has already been spent), the new budget line will leave the point on the ordinate axis (curve *B*), but with a kink at point *b*, where it shifts horizontally to the right to point *b''* (Fig. 2). However, in reality it is difficult to avoid the substitution effect: if the investor organizes the process of making the necessary decisions on project A and further intervention is not necessary, then he will reduce the cost of his resource on this problem and will pay more attention to other investment projects.

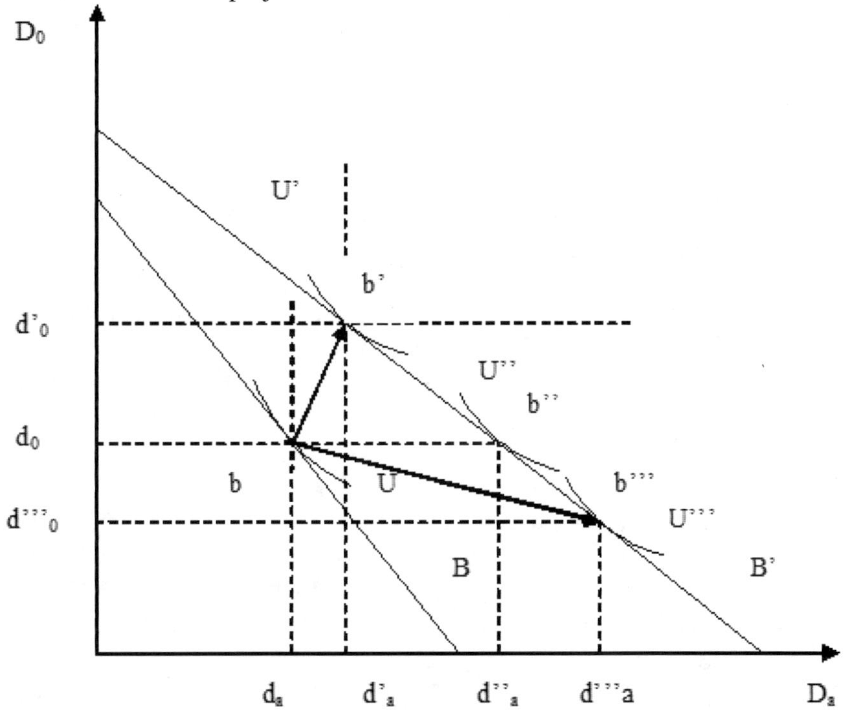

Figure 2. Investor's agent behavior

A way to minimize the effect of substitution is to limit and balance the use of the resource. An investor may try to convince him that he has sufficient resources, but, unfortunately for a group of interests the effectiveness of maintaining a balance when using a resource can also be affected by the substitution effect. Suppose that an investor allocated a large share of his resources to the solution of the da problem, compared with the one he planned to spend on the solution of the problem on his own. As a result, the distribution of costs of investor resources shifts from point *b* to *b''*, and as a result, greater results have been achieved in resolving the issue of the size of the project's share ($d_a''' - d_a > 0$), but less effort is spent on achieving alternative goals ($d_a''' - d_a < 0$).

Ultimately, with the interaction of income and substitution effects, the distribution can be any point on the line [***b***; ***b'''***]. Point ***b''*** will be the new distribution point if resource constraints work, but a balance of resources does not.

Three decision-making options are similar in one characteristic – they ensure the sale of a larger share of the investment project as a result of the impact of a group of interests on the budgetary constraint of the investor.

IOP Conf. Series: Materials Science and Engineering **680** (2019) 012035 doi:10.1088/1757-899X/680/1/012035

3. Conclusion and Discussion

Based on the situations considered affecting the investor, from manipulating information to direct bribery, the following conclusions can be drawn:

1. Regional interest groups provide the investor with the opportunity to carry out their investment project management functions, achieving a greater level of their own usefulness, increasing the resources that the investor can use, including for the benefit of the company (in particular, in situations where the information advantage is on the side investor). In this regard, regional interest groups are beneficial to the investor, as increase the effectiveness of its activities.
2. The resource provided by the group of interests is used by the investor not only for the implementation of those investment goals for which it was provided, but also for other tasks.

Summing up, we can conclude that in the considered system, the possibility of establishing adequate compensations is associated with the problem of identifying the true demand of the economic agents. One of the ways to identify demand and establish an effective volume of public goods production is set by the Groves-Clark mechanism.

Such a mechanism consists from 3 mandatory rules:

- Each agent makes a "bid" for the particular action, bi. It can be either truthful or deceitful.
- The action is performed if $\sum_i b_i \geq 0$.
- Each agent i has a payout equal to the sum of other agents' bids, $\sum_{j\neq i} b_j$ if the action is performed (payouts may be negative).

Let's show that the truthfulness is optimal for each agent in our system. Consider n agents, each with true vi value and declared bi value. Truthfulness means that for each agent it is optimal to declare bi = vi regardless of what the other agents behavior. Consider payout for agent i:

$$\text{Payout}_i = \begin{cases} v_i + \sum_{j\neq i} b_j\,, & \text{if } \left(v_i + \sum_{j\neq i} b_j\right) \geq 0 \\ 0, & \text{otherwise} \end{cases} \quad (5)$$

Assume that $\left(v_i + \sum_{j\neq i} b_j\right) > 0$. Then agent i can guarantee that the action will be performed by declaring bi = vi. Now, let's consider the case when $\left(v_i + \sum_{j\neq i} b_j\right) < 0$. Then agent i can guarantee that the action will not be performed by declaring bi = vi. In any case, it's beneficial for the agent to tell the truth. There is no incentive to reveal distorted preferences regardless of the behavior of other agents. As a result, the mechanism for comparing information is modified in such a way that each agent is faced with the problem of a social solution rather than an individual one, and thus everyone has an incentive to identify their preferences correctly.

The described preference identification scheme has a significant and well known drawback – total payments can potentially be very large and the mechanism can be extremely expensive. Unfortunately, in the general case, it's impossible to make a zero-budget Groves mechanism, but it is possible to establish a mechanism by which payments are always negative.

In such mechanism agents may be required to pay "tax", but they never will receive payments. Due to these "lost" tax payments, overall solution will not be Pareto-efficient, but the decision of performing a particular action by the multi-level system will be made if and only if it would be efficient in terms of aggregate utility. Let's describe how this can be done. The essence of the idea is as follows: we can require each agent i to make an additional payment, which depends only on what other agents do and does not affect incentives i.

Let b-i be the vector of all bids without the bid of agent i, and let hi(b-i) be the additional payment required from agent i.

Then the payout of agent i is equal to:

$$\text{Payout}_i = \begin{cases} v_i + \sum_{j\neq i} b_j - h_i(b_{-j})\,, & \text{if } \left(b_i + \sum_{j\neq i} b_j\right) \geq 0 \\ -h_i(b_{-j}), & \text{otherwise} \end{cases} \quad (6)$$

If the hi functions are reasonably selected, the amount of payments to the agent can be significantly reduced. One of the possible options for choosing a function is as follows:

$$h_i(b_{-j}) = \begin{cases} \sum_{j\neq i} b_j\,, & \text{if } \sum_{j\neq i} b_j \geq 0 \\ 0, & \text{otherwise} \end{cases} \quad (7)$$

Payout of agent i then will be as following:

$$\mathrm{Payout_i} = \begin{cases} v_i,\ \text{if } \sum_i b_j \geq 0 \text{ and } \sum_{j\neq i} b_j \geq 0 \\ v_i + \sum_{j\neq i} b_j,\ \text{if } \sum_i b_j \geq 0 \text{ and } \sum_{j\neq i} b_j < 0 \\ -\sum_{j\neq i} b_j,\ \text{if } \sum_i b_j < 0 \ and\ \sum_{j\neq i} b_j < 0 \\ 0,\ \text{otherwise} \end{cases} \qquad (8)$$

Note that agent i will never receive positive payout; he may be taxed but will never receive subsidies. Supplementary payments have an effect on the utility of agent i only if he changes the whole system action. Note that for the agent this mechanism is also preferable, since he is never taxed to a greater extent than the system action is valuable to him.

Acknowledgements
The research was supported by Russian Foundation of Basic Research (RFBR), research projects No. 19-010-00376 A, 19-010-00577 A.

References

[1] Chassang S, Padró I Miquel G. Snowberg E. 2012 Selective Trials: A Principal-Agent Approach to Randomized Controlled Experiments. *Am. Econ. Rev.* **102**(4), 1279-1309.
[2] Laffont, J.-J., Martimort, D. (2002). The Theory of Incentives I: the Principal-Agent Model. Princeton University Press.
[3] Gurtuev A.O., Derkach E.G., Ivanov Z.Z. (2016). Sovremennoe sostoyanie issledovaniy v oblasti obespecheniya ispolneniya kontraktov v usloviyah neophedelennosti v sistemah innovatsionnogo investirovaniya. *Izvestiya KBNC RAN*, **5**(73): 62-68.
[4] Riley, J. (2001). Silver Signals: Twenty-Five Years of Screening and Signaling. *J. Econ. Lit.* **39**, 432-78.
[5] De Figueiredo J, and Cameron, Charles M., Endogenous Cost Lobbying: Theory and Evidence (August 3, 2009). *CELS 2009 4th Annual Conference on Empirical Legal Studies Paper.*
[6] Mitchell, W.C. and Munger, M.C. (1991) Economics models of interest groups: an introductory survey. *American Journal of Political Science*, Vol.**35**, No. 2, pp. 512-46.
[7] Ehrlinger, J., Mitchum, A., Dweck, C. (2016). Understanding overconfidence: Theories of intelligence, preferential attention, and distorted self-assessment. *Journal of Experimental Social Psychology*, **63**:94–100.
[8] Hackmann, Martin B., Jonathan T. Kolstad and Amanda E. Kowalski. 2015. Adverse Selection and an Individual Mandate: When Theory Meets Practice. *Am. Econ. Rev*, **105**(3): 1030-66
[9] Arve, M., Martimort, D. (2016). Dynamic Procurement under Uncertainty: Optimal Design and Implications for Incomplete Contracts. *Am. Econ. Rev.* **106**(11), 3238-74.
[10] Denzau A., Munger M. Legislators and Interest Groups: How Unorganized Interests Get Represented. *American Political Science Review*, 1986, **80**: 89-106.
[11] Holmstrom, B., and Paul Milgrom. 1991. Multitask Principal-Agent Analyses: Incentive Contracts, Asset Ownership, and Job Design. *Journal of Law, Economics, and Organization* **7**:24-52.
[12] Niskanen, William A. (May 1968). "Nonmarket Decision Making: The Peculiar Economics of Bureaucracy". *Am. Econ. Rev.* **58** (2): 293–305

IOP Conf. Series: Materials Science and Engineering **680** (2019) 012036 doi:10.1088/1757-899X/680/1/012036

Algorithm of linear discrete filtering with fuzzy modification of structure

A Eliseev and O Sokolova

Don State Technical University, Gagarin square 1, Rostov-on-Don 34400, Russia

Abstract. The paper considers a task of processing measurements to determine the state vector of a dynamic object, for example, an aircraft. A linear discrete filtering algorithm based on the Kalman filter is proposed. It is capable of modifying its structure using fuzzy expert system depending on the intensity of the object maneuver. The performance and efficiency of the algorithm are confirmed by the results of numerical simulation.

1. Introduction

When solving problems of controlling dynamic objects, for example, aircraft of various classes, it is necessary to determine their state vector based on the measurement vector available to the observer. The vector can contain errors of various types. To reduce the influence of measurement errors on the resulting estimate, in practice the processing algorithms such as classical and advanced Kalman filters, nonlinear filter, $\alpha - \beta$ -filter, etc. are widely used [1-25]. The quality of estimates generated by these filters is largely determined by the adequacy of the state vector models used The adequacy of models, as a rule, is degraded by a random (from the point of view of the observer) maneuver of the aircraft. Using an inadequate model in the filter leads to its divergence. To prevent this, the following methods of filter adaptation to the maneuver [1-3, 5, 6] are used in the theory and practice of dynamic filtering: periodically adjusting the filter gain to the initial value, increasing the dispersion of the shaping noise, increasing the covariance matrix of the state vector estimation errors, expanding the state vector, alternate using the multiple filters, using the multi-model filters. The use of the considered methods, with the exception of the latter, involves solving the problem to detect a maneuver. For this reason, the adaptive filtering algorithms contain three main modules: a maneuver detection module; filter settings module; filter module [1-3]. The existing variety of dynamic filtering algorithms is due to the features of the implementation of these modules. For example, ref. [6] proposes an algorithm to adjust the Kalman filter by changing the values of the elements of the matrix of the intensities of the forming noises. A feature of this algorithm is the iteration adjustment, causing the dynamic filtering errors to arise. Thus, the problem to develop an filtering algorithm adaptive to maneuver intensity is still relevant.

The purpose of the work is to reduce the dynamic filtering error of trajectory measurements under the conditions of the aircraft maneuver.

The problem to be solved is to develop a linear discrete filtering algorithm capable to modify its structure depending on the intensity of maneuver using a fuzzy expert system.

2. Mathematical formulation of the problem

Let the motion of a dynamic object on the time interval $[t_0, T]$ is described by the difference equation [1]

$$X_j = \Phi_j X_{j-1} + \Gamma_j \left(A_{xj} + N_{xj}\right), j = 1,2,... \quad (1)$$

and the observation equation has the form

$$Z_j = H_j X_j + N_{zj}, j = 1,2,... \quad (2)$$

where $X_j = X(t_j) = \left[x_{sj}, s = \overline{1,q}\right]^T$ is the vector of the object motion parameters, $Z_j = [z_{sj}, s =$ 1,pT is the vector of measurements, Φj=ϕlsj,⇌l,s=1,q, Γj=γskj,s=1,q,k=1,m, Hj=bksj,k=1,p,s=1,q are the known functional matrices, $A_{xj} = \left[a_{xsj}, s = \overline{1,m}\right]^T$ is the maneuver intensity vector, which elements belong to a priori unknown ranges $a_{xs_j} \in [a_{x\,min\,s}, a_{x\,max\,s}]$, $s = \overline{1,m}$ and represent the object accelerations with respect to the corresponding coordinate, $\mathbf{N}_{xj} = \left[n_{xsj}, s = \overline{1,m}\right]^T$, $\boldsymbol{N}_{zj} = \left[n_{zsj}, s = \overline{1,p}\right]^T$ are random noises of the object (1) and the observation channel (2), respectively, which have zero mathematical expectations and correlation matrices $Q_j = diag\ \left[q_{ssj}, s = \overline{1,m}\right]$,

$R_j = diag\ \left[r_{ssj}, s = \overline{1,p}\right]$.

It is required: to obtain based on the results of current observations Z_j the optimal (in terms of the mean-square) estimate $\hat{X}_j$ of the filtration of the state vector (1) under conditions of a priori uncertainty with respect to the values of the elements of the acceleration vector A_{xj}.

2 Solution to the problem

Suppose that to form an estimate $\hat{X}_j$ of the vector of motion parameters of the dynamic object a finite set of operators $\{\Re_k,\ k = \overline{1,M}\}$ is used, and each operator is the Kalman filter [1-6] adjusted to a specific value of the acceleration vector $A_{xj}^k \in \{A_x^1, A_x^2, ..., A_x^M\}$:

$$\hat{X}_j^k = \hat{X}_{j|j-1}^k + K_j\left[Z_j - H_j \hat{X}_{j|j-1}^k\right], \quad (3)$$

$$\hat{X}_{j|j-1}^k = \Phi_j \hat{X}_{j-1}^k + \Gamma_j A_{xj}^k, \quad (4)$$

$$P_{j|j-1} = \Phi_j P_{j-1} \Phi_j^T + \Gamma_j Q_j \Gamma_j^T, \quad (5)$$

$$K_j = P_{j|j-1} H_j^T \left[H_j P_{j|j-1} H_j^T + R_j\right]^{-1}, \quad (6)$$

$$P_j = \left[I - K_j H_j\right] P_{j|j-1}, \quad (7)$$

where $\hat{X}_{j|j-1}^k$ is the estimate of the prediction of the state vector at the time j, $P_{j|j-1}$ is the symmetric prediction error matrix, P_j is the covariance matrix of filtering errors $X_j - \hat{X}_j$, K_j is the filter gain, I is the identity matrix.

At any given time only one filter is used, this means that the filters are switched depending on the information about the value of A_{xj}^k.

It can be seen that for the case when in model (1) $A_{xj} \notin \{A_x^1, A_x^2, ..., A_x^M\}$ the estimate at the output of the filter $\hat{X}_j^k$ contains a significant dynamic error due to the inadequacy of the model (4) to the real process.

To ensure the sustainability of the estimation $\hat{X}_j$ it is necessary to develop an adaptive filtering algorithm that takes into account the uncertainty of the value A_{xj}.

This algorithm should detect the divergence of the estimate $\hat{X}_j$ and modify the filter structure.

Let us successively consider the possible solutions to these problems.

The problem to detect the divergence. To solve this problem, we need to introduce the divergence index. To increase the reliability of the solution to the problem of the divergence detection, we will simultaneously use two particular indices.

The first partial indicator. Let the updating process (residual) $\varepsilon_j = Z_j - H_j \hat{X}_{j|j-1}$ is used as the decisive statistics, and for the threshold S_0 – the value

$$S_{0lj} = c\sqrt{\Psi_{llj}}, \tag{8}$$

is used [1], where Ψ_{llj}, $l = \overline{1,p}$ is the diagonal elements of the matrix $\Psi_j = H_j P_{j|j-1} H_j^T + R_j$; c is a positive constant which value is determined by the probability $P_Д$ of the value ε_{lj} to fall into the interval $[-S_{0lj}, S_{0lj}]$. For example, when $P_Д = 0.9973$ we have $c = 3$.
In view of (8), the divergence can be detected by the condition [1]

$$|\varepsilon_{lj}| > S_{0lj}. \tag{9}$$

From (9) it follows that it is reasonable to use the relative value of the residual of the form

$$|\delta\varepsilon_{1j}| = \frac{|\varepsilon_{lj}|}{S_{0lj}}. \tag{10}$$

The second partial indicator. Indicator (10) takes into account both the abnormal random error and the dynamic error due to the inadequacy of the model (4) to the real process. To better take into account the dynamic error we introduce, by analogy with ref. [6], the modulus of the arithmetic mean of the residual $|\bar{\varepsilon}_j|$ (for simplicity, without loss of generality, we assume that the residual is scalar):

$$|\bar{\varepsilon}_j| = \frac{|\sum_{i=0}^{n-1} \varepsilon_{j-i}|}{n}, \tag{11}$$

where n is the number of consecutive measurements used to detect the divergence of the filter.

The calculation of indices (11) is possible when $j \geq n$, i.e. when no less than n observations of the form (2) is performed.

By analogy with (10) we transform (11) to the form:

$$|\delta\bar{\varepsilon}_{2j}| = \frac{|\sum_{i=0}^{n-1} \varepsilon_{j-i}|}{n\sigma_{prg}}, \tag{12}$$

$\sigma_{prg}$is some admissible value of the standard deviation (MSE) of the error ε_j, $j \geq n$.

The problem to modify the filter structure. For example, let the evaluation system contain two filters $\{\Re_k,\ k = \overline{1,2}\}$, the first filter ($k = 1$) is set to the lowest acceleration value $A_{xj}^1 = A_{xmin}$ and the second filter ($k = 2$) is set to the highest acceleration value $A_{xj}^2 = A_{xmax}$. In this case, the actual value of acceleration in model (1) can take any value from the known interval $A_{xj} \in [A_{xmin}^{xmax}]$.

To modify the filter structure, we use the operatively advising expert system (OAES), which is based on the use of fuzzy logical inference [26]. The use of OAES is reasonable due to the uncertainty of setting the values of the acceleration vector A_{xj}. We represent the process of the filter operation in the form of a tuple of some problem situations (PrS) [26-28]. Any PrS can be described by a situational vector $sv = [sv_\xi,\ \xi = \overline{1,\Xi}]^T$, and each coordinate sv_ξ of the vector is a linguistic variable with a given set of terms $SV_\xi = \{SV_\xi^l,\ l = \overline{1,\ m_\xi}\}$. We believe that for some particular implementations of the situational vector sv^* there are precedents of successful solving the current PrS, characterized by a certain precedent vector $pv = \{pv_m,\ m = \overline{1,\ m_{PV}}\}$, and each coordinate pv_m of the vector is also a linguistic variable with a given set of terms $\{PV_m^p,\ p = \overline{1,\ n_m}\}$. Let us introduce for the considered filtering system consisting of the finite set of operators $\{\Re_k,\ k = \overline{1,2}\}$ of the form (3)–(7) a situational vector $sv = [sv_\xi,\ \xi = \overline{1,4}]^T$ with elements: sv_1 – "relative value of the residual of the first filter $|\delta\varepsilon_{1j}^1|$", sv_2 – is "value of the relative error modulus of the first filter $|\delta\bar{\varepsilon}_{2j}^1|$", sv_3 – is "relative value of the residual of the second filter $|\delta\varepsilon_{1j}^2|$", sv_4 – is "value of the relative error modulus of the second filter $|\delta\bar{\varepsilon}_{2j}^2|$".

Let the variables of the situational vector are described by the following term-sets:

$$SV_1 = SV_3 = \left\{ \begin{array}{l} \text{"}Very\ low\ \left(OH\left(\left|\delta\varepsilon_{1j}\right|\right)\right)\text{",} \\ \text{"}Low\left(H\left(\left|\delta\varepsilon_{1j}\right|\right)\right)\text{",} \\ \text{"}Average\ \left(CP\left(\left|\delta\varepsilon_{1j}\right|\right)\right)\text{",} \\ \text{"}\ High\ \left(B\left(\left|\delta\varepsilon_{1j}\right|\right)\right)\text{"} \end{array} \right\},$$

$$SV_2 = SV_4 = \left\{ \begin{array}{l} \text{"}Very\ low\ \left(OH\left(\left|\delta\bar{\varepsilon}_{2j}\right|\right)\right)\text{",} \\ \text{"}Low\left(H\left(\left|\delta\bar{\varepsilon}_{2j}\right|\right)\right)\text{",} \\ \text{"}Average\ \left(CP\left(\left|\delta\bar{\varepsilon}_{2j}\right|\right)\right)\text{",} \\ \text{"}\ High\ \left(B\left(\left|\delta\bar{\varepsilon}_{2j}\right|\right)\right)\text{"} \end{array} \right\}.$$

We assume that linguistic variables sv_ξ, $\xi = \overline{1,\ 4}$ are given on the universe $E = [0, \infty]$, and the terms are described by the membership functions $\mu_{SV_\xi} \in \left\{\mu_{SV_\xi^l},\ l = \overline{1,\ m_{L\xi}}\right\}$, $\xi = \overline{1,\ \Xi}$, presented in Figures 1 and 2.

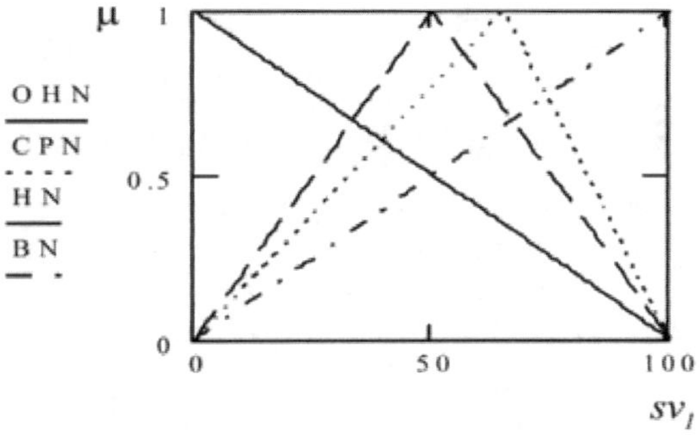

Figure 1. Membership functions of term set SV_1 and SV_3: OHN – "very low"; CPN – "medium"; HN – "low"; BN – "high".

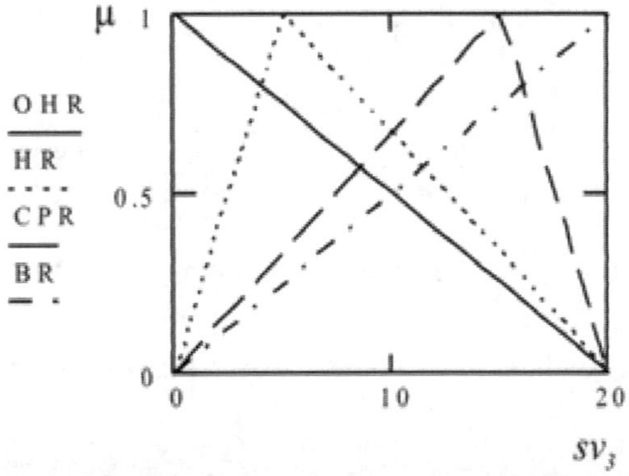

Figure 2. Membership functions of term set SV_2 and SV_4: OHR – "very low"; CPR – "medium"; HR – "low"; BR – "high".

Let the following precedents be known for this class of PrS: pv_1 – the first model of the form (4) $(k = 1)$ is "very close" to the actual process, it is reasonable to use the estimation formed by the first filter, that is: $\hat{\boldsymbol{X}}_j = \hat{\boldsymbol{X}}_j^1$; pv_2 – the first model of the form (4) $(k = 1)$ is "closer" to the actual process, compared to the second model $(k = 2)$, it is reasonable to use a weighted estimate $\hat{\boldsymbol{X}}_j = 0{,}7\hat{\boldsymbol{X}}_j^1 + 0{,}3\hat{\boldsymbol{X}}_j^2$; pv_3 – the second model of the form (4) $(k = 2)$ is "closer" to the actual process, compared to the first model $(k = 1)$, it is reasonable to use a weighted estimate $\hat{\boldsymbol{X}}_j = 0{,}3\hat{\boldsymbol{X}}_j^1 + 0{,}7\hat{\boldsymbol{X}}_j^2$; pv_4 – the second model of the form (4) $(k = 2)$ is "very close" to the actual process; it is reasonable to use the evaluation formed by the second filter, that is: $\hat{\boldsymbol{X}}_j = \hat{\boldsymbol{X}}_j^2$.

Taking into account the previously introduced situational vector and the known precedents, the system of rules R^{mr_m}, $m = \overline{1,\ m_{PV}}$, $r_m = \overline{1,\ N_m}$, describing the mechanism to solve the current PrS, takes the form:

$$R^{mr_m} = 1{,}1\text{: if} \begin{pmatrix} \left(sv_1 = OH\left(\left|\delta\varepsilon_{1j}^1\right|\right)\right) \\ \text{and } \left(sv_2 = OH\left(\left|\delta\bar{\varepsilon}_{2j}^1\right|\right)\right) \\ \text{and } \left(sv_3 = B\left(\left|\delta\varepsilon_{1j}^2\right|\right)\right) \\ \text{and } \left(sv_4 = B\left(\left|\delta\bar{\varepsilon}_{2j}^2\right|\right)\right) \end{pmatrix}, \text{ then } (pv_1),$$

$$R^{mr_m} = 2{,}1\text{: if} \begin{pmatrix} \left(sv_1 = H\left(\left|\delta\varepsilon_{1j}^1\right|\right)\right) \\ \text{and } \left(sv_2 = H\left(\left|\delta\bar{\varepsilon}_{2j}^1\right|\right)\right) \\ \text{and } \left(sv_3 = CP\left(\left|\delta\varepsilon_{1j}^2\right|\right)\right) \\ \text{and } \left(sv_4 = CP\left(\left|\delta\bar{\varepsilon}_{2j}^2\right|\right)\right) \end{pmatrix}, \text{ then } (pv_2),$$

$$R^{mr_m} = 3{,}1\text{: if} \begin{pmatrix} \left(sv_1 = CP\left(\left|\delta\varepsilon_{1j}^1\right|\right)\right) \\ \text{and } \left(sv_2 = CP\left(\left|\delta\bar{\varepsilon}_{2j}^1\right|\right)\right) \\ \text{and } \left(sv_3 = H\left(\left|\delta\varepsilon_{1j}^2\right|\right)\right) \\ \text{and } \left(sv_4 = H\left(\left|\delta\bar{\varepsilon}_{2j}^2\right|\right)\right) \end{pmatrix}, \text{ then } (pv_3),$$

$$R^{mr_m} = 4{,}1\text{: if} \begin{pmatrix} \left(sv_1 = B\left(\left|\delta\varepsilon_{1j}^1\right|\right)\right) \\ \text{and } \left(sv_2 = B\left(\left|\delta\bar{\varepsilon}_{2j}^1\right|\right)\right) \\ \text{and } \left(sv_3 = OH\left(\left|\delta\varepsilon_{1j}^2\right|\right)\right) \\ \text{and } \left(sv_4 = OH\left(\left|\delta\bar{\varepsilon}_{2j}^2\right|\right)\right) \end{pmatrix}, \text{ then } (pv_4).$$

To calculate the membership function of a precedent pv_m we use the Mamdani-Zade rule [26-28]:

$$\mu_{pv_m}(sv_1, sv_2, \ldots,\ sv_{\Xi},) = \max_{r_m}\min_{\xi}\mu_{SV_{\xi}(mr_m)}(sv_{\xi}), \tag{13}$$

where $\mu_{SV_{\xi}(mr_m)} \in \left\{\mu_{SV_{\xi}^l},\ l = \overline{1,\ m_{L\xi}}\right\}$ is the membership function of the linguistic variable sv_{ξ}, which is part of the production rule R^{mr_m}.

Taking into account in the equation (13), the most preferable precedent to solve observed PrS can be defined as follows:

$$pv_m^* = \underset{m}{\text{argmax}}\mu_{pv_m}(sv_1, sv_2, \ldots,\ sv_K,). \tag{14}$$

The scheme OAES to modify the structure of the filter, is presented in Figure 3.

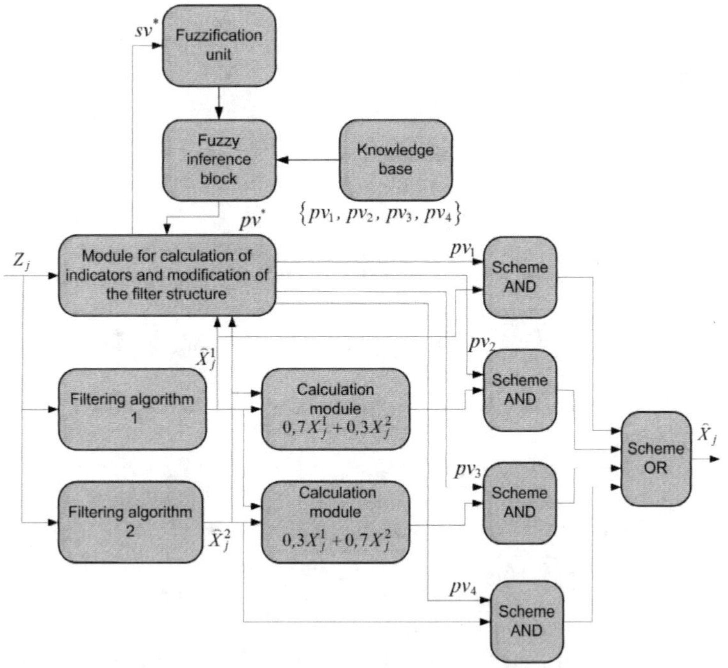

Figure 3. Structure of the intelligent measurement processing system.

The system functions as follows: the specific values of the elements of the situational vector $\boldsymbol{sv}^*$, formed in the module for calculating indices and modifying the filter structure (MCIMFS), are fed to the phasification block, where they are converted into fuzzy sets. The obtained data are input to the fuzzy inference block that implements the algorithm of choice of the most preferable precedent $pv^* \in \{pv_1, pv_2, pv_3, pv_4\}$ based on expressions (13), (14), using information from the knowledge base containing fuzzy production rules, as well as the form and parameters of the membership functions (see Figure 1 and Figure 2). The number of the most preferred precedent chosen to solve the observed PrS is used in the MCIMFS to switch one of the four calculated arrays of estimates $\{X_j^1\}$, $\{0{,}7X_j^1 + 0{,}3X_j^2\}$, $\{0{,}3X_j^1 + 0{,}7X_j^2\}$, $\{X_j^2\}$ to the output of the filtering device.

Thus, the OAES provides a modification of the filter structure depending on the change in the model of the estimated information process of the form (1).

To confirm the performance of the proposed algorithm, mathematical modeling was performed. It is assumed that the model of movement of the maneuvering aircraft has the form:

$$X_j = \begin{cases} \Phi X_{j-1} + \Gamma n_{x\,j-1}, 0 \le j < 200, \\ \Phi X_{j-1} + \Gamma\left(2g + n_{x\,j-1}\right), \ 200 \le j < 400, \\ \Phi X_{j-1} + \Gamma\left(4g + n_{x\,j-1}\right), 400 \le j < 600, \end{cases} \tag{15}$$

where

$$\Phi = \begin{bmatrix} 1 & \tau \\ 0 & 1 \end{bmatrix}, \Gamma = \begin{bmatrix} \frac{\tau^2}{2} \\ \tau \end{bmatrix}, \tau = t_{j+1} - t_j = 1c, j = \overline{0,600}.$$

To estimate the state vector $\hat{X}_j$ we use the system containing two filters $\{\Re_k, \ k = \overline{1,2}\}$. The first filter ($k = 1$) is set to the lowest acceleration $A_{xj}^1 = 0$, and the second filter ($k = 2$) is set to the highest acceleration $A_{xj}^2 = 4g$, $g = 9.8$ m/s^2. The actual value of acceleration in model (1), as can be seen from (15), take three different values $A_{xj} \in [0, \ 2g, \ 4g]$.

IOP Conf. Series: Materials Science and Engineering **680** (2019) 012036 doi:10.1088/1757-899X/680/1/012036

As a result of the simulation, we find that the application of the developed algorithm allows to reduce the dynamic estimation error by 27% compared with the multi-structured filtering algorithm based on the choice of only one particular filter at a time [1].

References

[1] Farina A, Studer F, 1993 *Digital Processing of Radar Information. Targets Tracking*, Transl. from English, M.: Radio and communications
[2] Sorenson H W, 1985 *Kalman Filtering: Theory and Application* (New York: IEEE Press)
[3] Zarchan P, 2005 *Fundamentals of Kalman Filtering - A Practical Approach* (Reston: American Institute of Aeronautics and Astronautics)
[4] Gibbs B P, 2011 *Advanced Kalman Filtering, Least-Squares and Modeling: a Practical Handbook* (Hoboken: John Wiley & Son)
[5] Petersen I R, Savkin A V, 1999 *Robust Kalman Filtering for Signals and Systems with Large Uncertainties* (Boston: Birkhauser)
[6] Rodkin M, 2012 *Adaptive Method for Adjusting the Kalman Filter* (VI All-Russian Conference Radiolocation and radio communication, Kotelnikov IRE of RAS) 125–128
[7] Bulychev Y G, Eliseev A V, 1998 Computational aspects of the nonlinear filtration problem, *Avtomatika i Vychislitel'naya Tekhnika*, 32 (2), 26-35
[8] Eliseev A V, 2006 Estimation of the state vector of an object based on a filter with fuzzy logic, *Aviakosmicheskoe priborostroenie*, v.4. pp. 30-38
[9] Yeliseyev A V, 2005 A nonlinear filtering algorithm immune to singular errors, *Radioelectronics and Communications Systems*, 48, 10, P. 13-19
[10] Ramachandra K V, 2000 *Kalman Filtering Techniques for Radar Tracking* (New York: Marcel Dekker)
[11] Gomez V, Maravall A, *1993 Initializing the Kalman Filter with Incompletely Specified Initial Conditions, Approximate Kalman Filtering* (Singapore: World Scientific) 39-62
[12] Verhaegen M, Van Dooren P, 1986 Numerical aspects of different Kalman filter implementations, *IEEE Trans, on Automatic Control*, 10, 907-917
[13] Xia Q, Rao M, Ying Y and Shen X, 1994 Adaptive fading Kalman filter with an application, *Automatica*, 8, 1333-1338
[14] Geng Y, Wang J, 2004 Adaptive fading Kalman filter with application in integrated navigation system, *J. of Beijing University of Aeronautics & Astronautics*, **5**, 434-437
[15] Oussalah M, Schutter J, 2000 Adaptive Kalman filter for noise identification, Proc. of the 25th Int. Conf. on Noise and Vibration Engineering (Leuven, Belgium) 1225-1232
[16] Singer R A, 1970 Estimating optimal tracking filter performance for manned maneuvering targets, *IEEE Trans, on AES*, 4, 473-483
[17] Simeonova T, Semerdjiev T, 2002 Specific features of IMM tracking filter design, *Inform. & Security*, 9, 154-165
[18] Best R A, Norton J P, 1997 A new model and efficient tracker for a target with curvilinear motion, *IEEE Trans, on AES*, 3, 1030-1037
[19] Wang T C, Varshney P K, 1993 A tracking algorithm for maneuvering targets, *IEEE Trans, on AES*, **3**, 910-924
[20] Chan Y T, Couture F, 1993 Manoeuvre detection and track correction by input estimation, *IEEE Proc. on Radar and Signal Processing*, 1, 21-28
[21] Bar-Shalom Y, Birmiwal K, 1982 Variable dimension filter for maneuvering target tracking, *IEEE Trans, on AES*, 5, 621-629
[22] Sui H B, Fang X Y and Wu Y, 2008 A modified adaptive tracking algorithm based on current statistic model, *Radar Science and Technology*, 3, 202-205
[23] Sun F M, Wu X Q and Qi K, 2008 Adaptive strong tracking algorithm based on current statistical mode, *J. of Data Acquisition & Processing*, 23 (2), 191-195

IOP Conf. Series: Materials Science and Engineering **680** (2019) 012036 doi:10.1088/1757-899X/680/1/012036

[24] Li X R, Jilkov V P, 2010 Survey of Maneuvering Target Tracking. Part II: Motion Models of Ballistic Targets, *IEEE Trans, on AES*, 46 (1), 96-119

[25] Li X R, Jilkov V P, 2010 A Survey of Maneuvering Target Tracking. Part VI: Density-Based Exact Nonlinear Filtering, *Proc. SPIE*, 6-8

[26] Gostev V, 2011 Designing fuzzy logic controls for automatic control systems (SPb: BHV-Petersburg)

[27] Eliseev A V, 2005 Fuzzy controller to control a multistructure object, *Izvestiya Vysshikh Uchebnykh Zavedenij. Aviatsionnaya Tekhnika*, 3, 20-25

[28] Eliseev A V, 2005 Identification of fuzzy model in the task of regulator synthesis, *Avtomatizatsiya i Sovremennye Tekhnologii*, 11, 3-12

Model of deformation of a feather quill when rolling with a ribbed roller on the flat surface

Y E Chertov, E Y Fisunova

Institute of service and business (branch) of Don state technical university in town Shakhty, 147 str. Shevchenko, Shakhty 346500, Russian Federation

E-mail: Chertov-6112@mail.ru

Abstract. Mathematical modeling is an effective and cost-effective alternative to a trial physical experiment in solving problems associated with the development of new technologies and equipment. The paper is devoted to modeling the deformation process of a feather quill when rolling by a ribbed roller on a flat surface to form thorough kneading on its concave side, providing additional bending. The technology is used to enhance the Executive ability ("Fill Power" measure) of feather stock of land birds, which are used for manufacturing down-feather products, as an alternative to increasingly scarce and expensive feahers of waterfowl. This is especially true for products with a short service life, intended for use in the rectifying natural disasters and the consequences of accidents.

The developed mathematical model allows us to investigate the dependence of the resistance force of the roller on the design parameters of the working body: the knurling pitch; the angular size of the flute; the diameter of the riffle vertices.

Stresses arising in the structure of the quill under the action of compressive load, when loading and removing the load in an arbitrary cross-section of a feather quill have different values and are described by the corresponding regression models. The solution of the system of regression equations together with the kinematic equations of motion of the working body allows to determine the strength of resistance and power requirement of the deformation of feather quill of land birds to improve the "Fill Power" of feather stock.

The dependence of the total rolling resistance force for the working body and the chicken feather quill of the given parameters obtained on the model is given.

1. Introduction

Environmental problems of modern civilization make us develop technologies that implement full cycles of processing of raw materials. At the enterprises of the poultry industry and at poultry farms the renewable product is feather raw product. The implementation of various production processes makes it possible to use it in the production of consumer goods, including feather-down products [1-4].

Research in the field of technological advancement of processing of feather - down raw materials and development of equipment for manufacturing feather - down products have been carried out by many scientists: B. N. Nikitin, I. S. Mitrofanov, E. V. Gaevoy, L. A. Bekmurzaev, I.Y. Brink, H. Bauer, E. Benz, R. Gurtler, L. H. Lorch, C. Monjarret, etc.

Down and feather of waterfowl is of higher quality, but more expensive at the same time. Feathers of land birds are cheaper, but its use for products is limited due to low quality indicators.

Recent studies [5-8] have shown that the filling capacity of the feather stock of land birds can be increased by 20% by reducing the radius of the quill. The technology of deformation of the quill with a ribbed roller allows to increase the filling capacity ("Fill Power" measure [9]) of feather stock of land birds for use in the feather - down products, significantly reducing their cost, instead of more expensive raw of waterfowl (see figure 1).

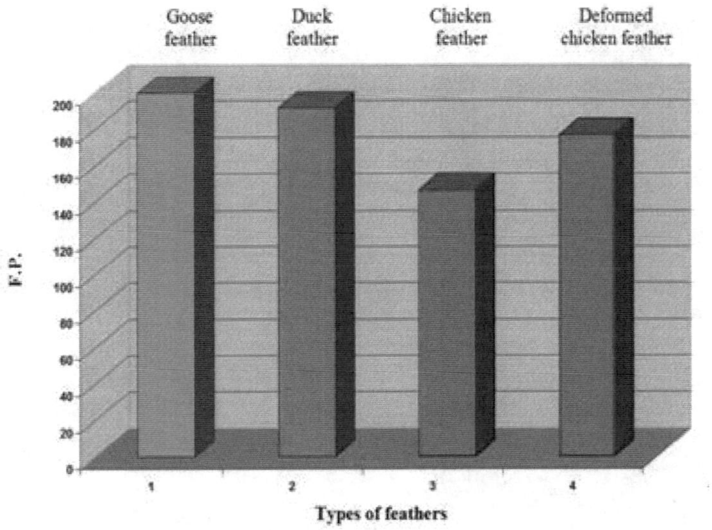

Figure 1. Filling capacity "Fill Power" measure for different types of feathers.

This is especially important for products with a short service life, intended for intended for use in the rectifying natural disasters and the consequences of accidents.

Various technical solutions are known to increase the "Fill-Power" measure of the feather stock by mechanical deformation [10-19], including those that implement the process of deformation of the feather quill when rolling with a ribbed roller on a flat surface [14-17]. Increasing the efficiency of such devices is achieved by improving the design and is an urgent task. To this end, the paper considers a model of deformation of a bird feather quill when rolling with a ribbed roller on a flat surface.

Mathematical modeling is an effective and cost-effective alternative to the trial physical experiment in solving problems associated with the development of new technologies and equipment. In this case, the law of motion of the working body, the ribbed roller, is known, it is required to find the strength of the deformation resistance of the feather quill during the rolling. The tasks of this type refer to the inverse [20].

2. Research materials and methods

The structure of the feather quill in the place of thorough kneading of the working body determines the magnitude of the deformation resistance force.

The structure of poultry feathers was studied in many works, e.g. [1,2,4,21,22]. As we know, the feather consists of keratin, water and fats. Keratin is a typical representative of the class of fibrous proteins with mechanical strength. Sectional shape of the quill changes from a staff to a web from almost round to elliptical. Figure 2 shows the cross-section of the quill in the feather web area, where the bast layer 1 and the core 2 can be distinguished. The bast layer is formed by keratinized flat epithelial and spindle-shaped cells. The core is formed by keratinized cells of various shapes and sizes. These cells are filled with air, so that the resulting porous structure provides a low weight and heat-shielding properties.

To simplify the model of the feather quill, it is advisable to introduce the concept of conditional diameter of a feather d_n, mm Parameters in this case will be the diameter of a staff, the length of the feather quill L_c, mm, and the polynomial characterizing the change of conditional quill diameter d_n on the length l_c

$$d_n(L_c) = a_1 {l_c}^2 + b_1 l_c + c_1, \tag{1}$$

IOP Conf. Series: Materials Science and Engineering **680** (2019) 012037 doi:10.1088/1757-899X/680/1/012037

where a_1, b_1, c_1 – polynomial coefficients;

l_c – the current value of the length of the feather quill, mm.

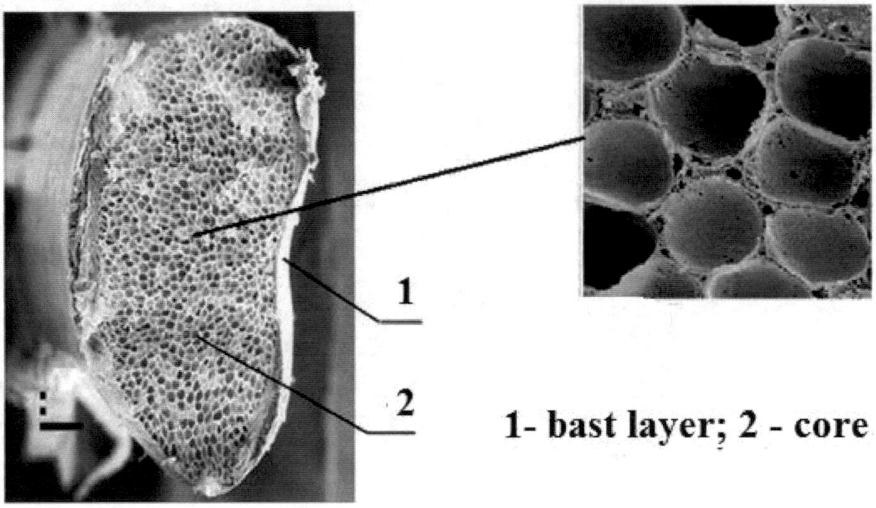

Figure 2. Structure of the feather quill.

Figure 3, for example, shows the polynomial dependence of the change in the conditional diameter of the chicken feather quill of the average fraction along the length, which is obtained as a result of experimental studies.

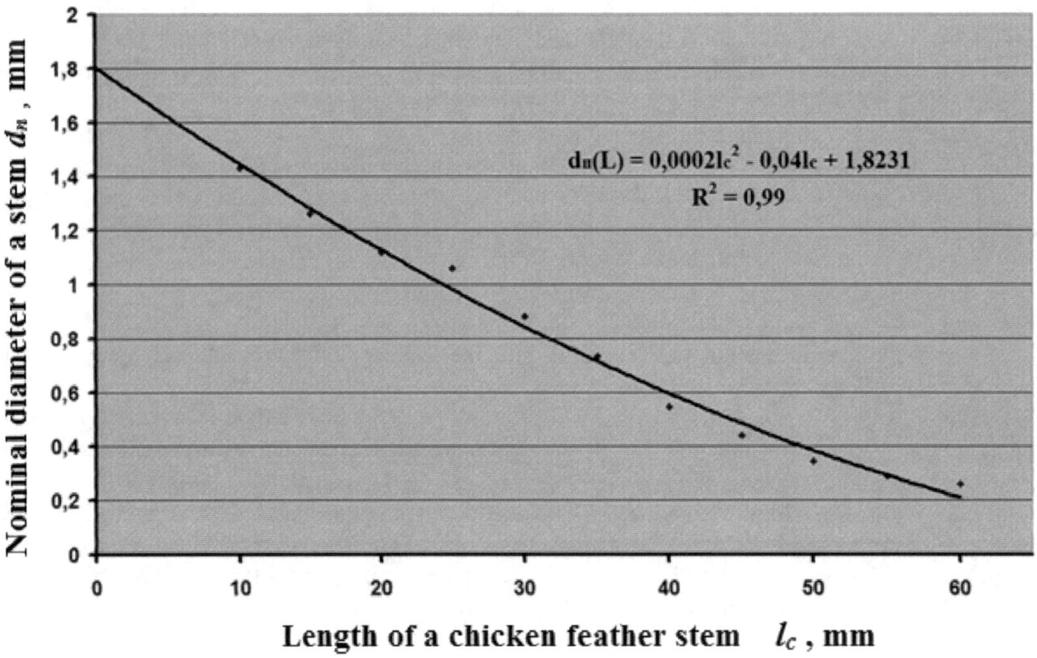

Figure 3. Change of the nominal diameter of the chicken feather quill along the length.

As a working body in the node of mechanical deformation, a ribbed roller acts, which carries out making of thorough kneading on the concave surface of the quill to reduce its radius.

The term groove in the technical literature has two meanings. From one hand, it denotes a notch, flute, rifle on any surface.

From another hand, when describing the structural elements of flax processing equipment, the term "flute" means a tooth, a work tool that causes thorough kneading on the flax stem [23].

Numerous design parameters of ribbed roller and their influence on the deformation process of the quill are considered in [7, 8, 24]. The developed model uses the basic parameters: the diameter of the peaks of the flutes, the pitch of fluting, the width of the tops of the flutes, the angular size of the peaks of the flutes, the angle of fluting.

Deformation of feather quills by rolling with the ribbed roller on a flat surface (it is assumed that the surface and the roller are the rigid bodies) occurs in the gap, the size of which must not exceed the admissible value.

Otherwise, there will be a significant destruction of the fibers of the quill. The size of the minimum allowable gap is determined from the experimental diagram of the deformation of the feather quill under compressive loading. Conditional work area of the flute during rolling of the quill is shown in figure 4*a*. The range of variation of the angle φ is from minus α to γ.

The ribbed roller when rolling rotates with angular velocity ω, knowing its diameter we can define a circumferential speed V_r. The projection of the circumferential speed V_{ry} on the Y-axis characterizes the rate of deformation of the feather quill, and the projection of V_{rx} on the X axis characterizes the speed of displacement of the quill on a horizontal surface (see figure 4*b*). The strain rate of the quill in the process of rolling changes from the maximum value at the point of contact of the quill with the flute of the working body, to zero when an acceptable gap h_{add}, and depends on frequency of rotation of the roller, its design parameters, dimensions of a feather quill.

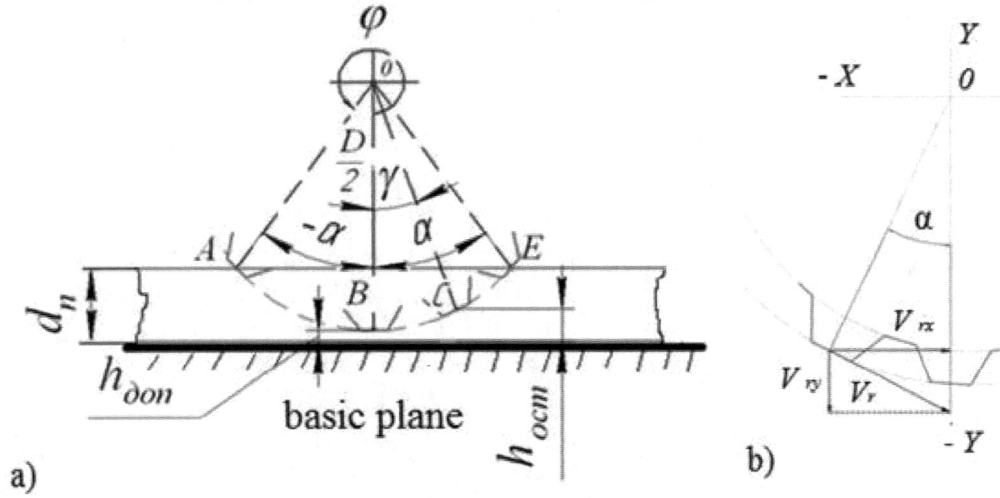

Figure 4. The kinematic scheme of rolling a quill:
a) to determine the degree of deformation;
b) to determine the rate of deformation

Its value will determine the quality of the process of formation of thorough kneading with local structure disturbance on the concave surface of the feather quill.

The dependence of the gap h, mm, from the angle of rotation of the ribbed roller is determined by the expression

$$h = \frac{D}{2}\left(1 + \frac{2h_{add}}{D} - \cos\phi\right), \tag{2}$$

where h_{add}. - minimum allowable clearance during rolling, mm;

D - diameter of the peaks of the flutes of the roller, mm;

φ - angle of rotation of the ribbed roller, rad.

The degree of deformation of the quill corresponding to the gap

$$\varepsilon(\phi) = \frac{d_n - \frac{D}{2}\left(1 + \frac{2h_{add}}{D} - \cos\phi\right)}{d_n}, \tag{3}$$

The strength of the deformation resistance when rolling the quill for one flute *F*, H, is determined in accordance with figure 3 by the expression

$$F = N(\varphi)(1+f) = (1+f)\frac{n(\varphi)S_n}{\cos\varphi}, \tag{4}$$

where *N* is the normal pressure force of one flute, N;

f - coefficient of friction of the stem sliding along the surface;

$n(\phi)$ - normal pressure on the quill when rolling, H/м^2;

$S_n = \frac{d_n}{2} D \int_{-\frac{\lambda}{2}}^{+\frac{\lambda}{2}} d\theta$ – the area of the deformation of the flute, mm^2;

λ-angular size of the peak of the flute, rad;

θ - the current value of the angular size of the peak of the flute. glad.

The normal pressure on the quill of one flute of the roller during rolling $n(\phi)$ is distributed over the contact surface, the position of which depends on the angle φ. Assuming that the structure of the quill is uniform, when rolling on a surface, it is possible to accept

$$n(\phi) = \begin{cases} K\sigma_n(\varepsilon, d_n), \text{ if } -\alpha \le \varphi \le 0; \\ K\sigma_n(\varepsilon, d_n), \text{ if } 0 < \varphi \le \gamma; \\ 0, \text{ if } \gamma \le \varphi \le (2\pi - \alpha), \end{cases} \tag{5}$$

where $\sigma_н$ - the stress in the structure of the chicken feather quill under compression, MPa;

ε -degree of deformation of the chicken feather quill;

α - the angle of grasping a feather, rad;

σ_p - voltage in the structure of the feather quill with decreasing compression load, MPa;

γ - the angular size of the zone to reduce the forces of resistance, rad;

K - the correlation coefficient between static and dynamic stress states during deformation.

The angle of grasping feathers α, rad, depends on the size of the feather quill and is defined by the following expression

$$\alpha = \frac{\pi}{180} \cdot \arccos\left(1 + \frac{2(h_{под} - d_n)}{D}\right). \tag{6}$$

The angular size of the zone of decreasing the resistance of deformation γ, rad (see figure 4a) is determined from the experimental dependence of residual deformation of the quill δ from the nominal size of the feather in contact with a flute [25] from the expression

$$\gamma = \frac{\pi}{180} \cdot \arccos\left(1 + \frac{2(h_{под} - d_n(1-\delta))}{D}\right). \tag{7}$$

A feather quill of medium fraction has a length L_c from 27 to 75 mm. Therefore, for roller of diameter from 30 to 90 mm with a pitch of corrugations from 2 to 4 mm rolling of feathers will be less than one turn. The number of flutes of the roller *M* involved in the rolling process of one quill determines its length

$$M = \frac{2L_c}{D \cdot \psi}, \tag{8}$$

where ψ is the angular pitch of the rolling of the roller, rad.

The rolling resistance force for each of the *M* flutes F_i H, can be calculated using the expression (3-7) and the parameters of the model of the feather quill can be calculated by piecewise linear approximation.

The resulting rolling resistance force ΣF, H, is defined as the sum of the resistance forces of each flute taking into account their shift by the corresponding angle multiple of the rolling step ψ

$$\sum F = \sum_{i=1}^{M} F_i. \tag{9}$$

This approach allows us to investigate the process of rolling of a feather quill with a known law of change in the conditional diameter along the length without the use of specialized software.

Figure 5 shows the dependence of total force of resistance to the rolling from the angle of the ribbed roller constructed by the graphic-analytical method using a standard package of computer programs. As a model of the feather quill we used with polynomial dependence (1) characterizing the change of the conditional diameter of the stem along the length, the coefficients $a_1=0,0002;\ b_1=-0,04;\ c_1=1,8231.$

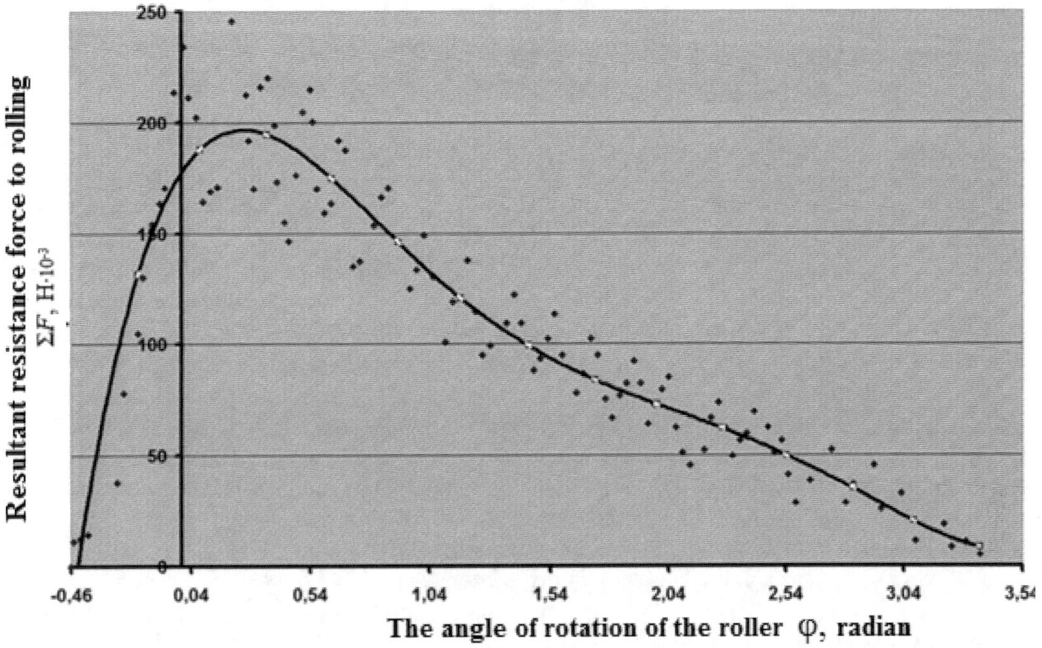

Figure 5. Dependence of the resultant resistance force on the angle of rotation of the roller

The following design parameters of the roller were taken: D=30 mm; λ = 0,02 rad; ψ = 0,195 rad. To determine the stresses in the structure of the chicken feather quill under compression and reduction of the compressive load, the experimental dependences given in [25] were used.

3. Conclusion

The mathematical model of deformation of a feather quill when rolling with a ribbed roller on the surface allows to study the dependence of the resistance force to rolling from structural parameters of the working body; knurling pitch, the angular size of the flutes; the diameter of the peaks of flutes.

Stresses arising in the structure of a quill under the action of compressive load, under loading and unloading in an arbitrary section of a quill have different values and are described by the corresponding regression models. The solution of the system of regression equations together with the kinematic equations of motion of the working body allows to determine the resistance force and the required deformation power of a quill of a land bird to increase the index of "Fill Power" of the feather stock.

In the framework of further studies of the problem discussed in the paper, it is planned to improve:

- the methodology for the analysis of the stress state of the feather quill of land birds, to clarify the correlation between the stresses of the deformed structure of static and dynamic nature;
- the model of the feather quill.

References

[1] Mitrofanov, N. S. Collection and processing of feather down raw materials at the enterprises of the poultry industry and poultry farms / N. S. Mitrofanov. – M.: Myasomolprom, 1989. – 196 p.

[2] Mitrofanov, I. S. Processing of poultry / I. S. Mitrofanov, Y. A. Plyasov, E. G. Shumkov. – M.: Agroprom-Izdat, 1990. – 303 p: il.

[3] *Directory of a technologist of the poultry industry 2-nd revised and enlarged edition. - M.: Light and food industry*, 1981. – 320 p.
[4] Gaevoy, E. V. *Processing of feather-down raw materials* / E. V. Gaevoy. - M.: Food industry. – 1978. – 126 p.
[5] Saprykina, T.N. *Prospects of application of chicken feathers in the products of the clothing industry* / T. N. Soprikina, I. Y. Brink, Y. E. Chertov // Sewing industry. 2007. No. 4 . P. 53–54.
[6] *The effect of mechanical modification with "FillPower" of chicken feathers* [Text] / I. Y. Brink, T. N. Saprykina, Y. E. Chertov // News of higher educational institutions. North Caucasus region. Engineering science: collection of scientific works. . – No. 3 (139). - Southern Federal University (Rostov-on-Don), 2007. – P. 123 -125.
[7] Chertov Y. E., Galatov K. S. Molev M. D. *Regression analysis of the deformation process of the feather quill with a ribbed roller* // Modern problems of science and education. – 2015. – No. 2; access mode: http://www.science-education.ru/131-23819 (date of access: 03.12.2015).
[8] *Improving the efficiency of the modification of the feather quill* / Y.E.. Chertov, K. S. Galatov// Poultry and poultry products. – 2012. – No. 6. P. 60-62
[9] European Standard EN 12934: 1999 has the status of a DIN Standard. *Composition labelling of processed feathers and down for use as. English version ICS* 59.040; 1999 -09- 05.
[10] *Vorrichtung zum Quetschen und Knicken von Federn*: pat. De 130468 МПК B 68G 3/08/ Robert Pee; patentee Robert Pee; app. 25.08.1901; publ. 09.05.1902.
[11] *Bettfedernschleis- und sortiermachine*: pat. De 519318: МПК1B68G 3/08, B68G 3/12/ Hermann Bauer; patentee Hermann Bauer; app. 24.01.1930; publ. 05.02.1931.
[12] *Device for separating the feather from the down*: pat. SU 2024233 A41G 9/00 IPC/ Melikov, E. N., Smirnov, V. P.; applicant and patentee Melikov, E. N. No 4937867/12; Appl. 22.05.1991; publ. 15.12.1994.
[13] Pat. 2404914 Russian Federation, МПК8 B68G 3/08. *Method of increasing Fill Power of chicken feathers and a device for its implementation* / Y. E. Chertov, I.Y. Brink, T. N. Soprikina; applicant and patentee of the GOU VPO YuRGUES. No. 2009125854/21 (036020); Appl. 06.07.2009; Publ. 27.11.2010. – Bull. No. 33. – 2 p..
[14] Pat. 2346885 Russian Federation, МПК8 B68G 3/08. *Device for mechanical modification of chicken feather* /I. Y. Brink, T. N. Soprikina, Y. E. Chertov; applicant and patent of the GOU VPO YuRGUES. No. 2007108628/12; Appl. 21.05.2007; Publ. 20.02.2009. –Bull. No. 5. – 2 p. I
[15] Pat. 2366767 Russian Federation, МПК8 B68G 3/08. *Device to sort of chicken feathers*/ Y/E. Chertov, K. S. Galatov. applicant and patentee of the GOU VPO YuRGUES. No. 2008113670/12; Appl. 07.04.2008; Publ. 10.09.2009. –Bull. No. 25. – 2 p.
[16] Pat. 2424384 Russian Federation, МПК8 B68G 3/08. *The device for separation feather down raw materials into fractions* / Y. E. Chertov, K.S. Galatov; applicant and patentee of the GOU VPO YuRGUES. No. 2010102611/05; Appl. 26.01.2011; Publ. 20.07.2011. – Bull. No. 20. – 2p.
[17] Pat.2467948 МПК8 B68G 3/08, B07B 1/20. *The device for separation feather down raw materials into fractions in a liquid medium* / K. S. Galatov; Y. E. Chertov; applicant and patentee of the GOU VPO YuRGUES. No. 2011105570/02; Appl. 14.02.2011; Publ. 27.11.2012. Bull. No. 33. – 1p.
[18] Pat. 2468987 МПК8 B68G 3/08. *Device for processing feathers into chopped feathers*/ Y. E. Chertov; K. S. Galatov; applicant and patentee of the GOU VPO YuRGUES No. 2011133439/12; Appl. 09.08.2011; Publ. 10.12.2012. Bull. No. 34. – 1p.
[19] Classification of methods of modification of the bird feather / K. S. Galatov, Y. E. Chertov, // Poultry and poultry products. – 2013. – No. 6. P. 52-56.

[20] *Mathematical modeling of technological processes and the method of inverse problems in mechanical engineering* /A. N. Tikhonov, V. D. Kalner, V. B. Glasko. – M: Mechanical Engineering, 1990. – 264 p.

[21] Kock, J.W. *Physical and mechanical properties of chicken feather materials: a thesis in partial fulfillment of the requirements for the degree master of science in the school of civil and environmental engineering* / Jeffrey W Kock. – Georgia Institute of Technology, May, 2006. – 101 p.

[22] Wilde, T.P. *An energy based model for the compressive behavior of goose down: a thesis in partial fulfillment of the requirements for the degree masters of mechanical engineering* / Timothy Philip Wilde. – Georgia Institute of Technology, november, 2004.

[23] Kruchinin, M. I. Flax / M. I. Kruchinin, L. I. Petrov, M. E. Egorov / monograph / under the editorship of A. R. Rogash: Moscow: Kolos. – 1967. – 583 p.

[24] *Determination of parameters of the working body of the modification arrangement of the feather sorting machines* / K. C. Galatov, Y. E. Chertov // Fundamental and applied problems of technics and technology. – Orel: Orel State University named after I. S. Turgenev, 2012. No. 1. P. 48-51.

[25] Galatov, K. S. *Improvement of equipment for separating feather down raw materials into fractions with the development of a host of mechanical deformation of the chicken feather stem*: extended abstract of dissertation. Shakhty: 2013. – 22 p

IOP Conf. Series: Materials Science and Engineering **680** (2019) 012038 doi:10.1088/1757-899X/680/1/012038

Development of a new method of management of ergo-technical system on the basis of its dynamics monitoring

S O Kramarov[1], A R Groshev[1], V V Khramov[2], AV Belyaev[3]

[1] Surgut State University, 1 str. Lenina, Surgut 628412, Russia
[2] Southern University (IMBL), 105/42 Bolshaya Sadovaya Str., Rostov-on-Don 344006, Russia
[3] Don State Technical University, Gagarin's square 1, Rostov-on-Don 344000, Russia

E-mail: maoovo@yandex.ru

Abstract. This paper explores the state assessment and management of complex human-machine systems that ensure that results are quickly obtained with the desired accuracy. The problem is solved by monitoring the real parameters of the process and comparing them with the results of the simulation, taking into account their fuzziness, incompleteness and lack of definition. The proposed method of solving control problems can be used in a number of subject areas (robotics, aero-space monitoring, etc.).

1. Introduction

In the modern world of digitalization of many technical and socio-economic processes ergatic systems are dominant in both public and private life. The effectiveness of such systems depends not only on their technical perfection and human professionalism, but also on the architecture of the system of their interaction, integrity, self-organization, level of synergy.

One of the necessary conditions of creation of effective system of information support of state management of ergatic systems (further - ES) is the unity of methodological approaches to the synthesis of such systems, giving the opportunity to form the theoretical basis of this synthesis and makes it possible to further standardize the mathematical, information and software processes assessment of the current state of ES from different application fields of engineering, technology, management, etc.

It should be noted that, depending on the specific situation, the features of digital transformation, it can be not only (and even not so much) about the synthesis of a new ES, but about the harmonization and adaptation of the functioning of individual components of the so – called (SoS) - System of Systems [1-3].

Currently, there is no unified methodology for creating systems for assessing the state and functioning of ES, but there is a theoretical basis for its development in the form of general systems theory, artificial intelligence theory, fuzzy logic systems, combinatorial topology, classical control theory [4-5].

2. Problem statement

The purpose of this research is to develop the foundations of adequate modeling within Soft Computing, the development of theory and creation of mathematical models for the assessment of the state of ergatic systems, providing the construction of tools in the form of mathematical and software intelligent systems and formational support for the state of ES.

This goal can be achieved by solving the following main tasks:

- Analysis of the possibilities of a systematic analysis of the methods and tools used to evaluate ergatic systems, clarification and definition of their role, which will be aimed at solving specific developments;
- Determination of the system of components of the generalized system description of dynamic models of ergatic systems and its constituent parts, including the scientifically-based structural decomposition of ES;
- Carrying out theoretical evaluation of the state of ES and construction on the basis of their classification of new estimates of ergatic systems;
- Definition of structure, role and place of information support systems for ES state management and development of their classifications
- Development of generalized system-dynamic models and algorithms of human activity within ES.

3. Method of researches

The theoretical and experimental studies carried out by the authors are based on the application of methods of complex systems theory, set theory, combinatorial topology, control and regulation theory, artificial intelligence, ergatic systems theory, as well as engineering psychology and ergonomics.

In real situations, researchers find it convenient to act on the assumption of the chaotic nature of the observed process [6]. As shown in [6-8] the attractor of the system is quite simple to recover, there are also numerical methods for estimating its dimension. Methods of approximate solution of the inverse dynamics problem [2,9], i.e. restoration of the given type of differential equations from experimental data, are developed.

Let the time series be available for measurement characterizing the dynamics of ES $\{y_k\}$, k=1…n, it is assumed that this series is generated by a system whose discretized form looks like:

$$\begin{cases} x_{k+1} = f(x_k, \dots, x_0), \\ y_k = g(x_k) \end{cases}, \quad (1)$$

where $\boldsymbol{x}$ – this n - dimensional point in the state space;

$\boldsymbol{y}$ - observed one-dimensional process;

$\boldsymbol{k}$ - selectable time (number);

$\boldsymbol{f}, \boldsymbol{g}$ - function vectors.

The use of discrete models is conditioned to the features of a digital source of initial data about ES state and allows to operate with the data obtained as a result of the experiment.

Phase trajectory representing the state of $\boldsymbol{x}_k$ dynamic system in time $\boldsymbol{t} = \boldsymbol{k}\ \boldsymbol{T}$, where $\boldsymbol{T}$ is time between measurements. For stable systems under study, the attractor is a point, for vibrational ones is closed trajectories (cycles). For chaotic ones, there is an attractor called strange, in which case the trajectories are contracted into some subset of phase space.

There are several variants of classification of scenarios of transition to chaos:

- through cascade of bifurcation of the doubling period (Feigenbaum);
- through the destruction of a two-dimensional torus or through the destruction of a closed invariant curve (Ruel-Takens);
- alternation in time of almost regular oscillations with intervals of chaotic behavior (Pomo-Manneville).

Let's stay on one of them [8].

The research of scenarios of transition to chaos allows predicting the possibility of chaotic behavior of dynamic system in the control parameters change. To concretize our proposed approach we will use the Ruel-Takens scenario.

Note that the attractor is an invariant characteristic of the system, so it is preserved after admissible transformations. The task considered in the study is to build a model of type (1) $\{\boldsymbol{y}_k\}$.

4. Simulation of systems with chaotic dynamics based on the evaluation of weak symmetry breaking in the reconstructed attractor

Source data: time series $\{y_k\}$

Step 1. Lyapunov exponents are calculated by numerical methods [10]. The conditions for the existence of chaos are given in table 1.

Step 2. Recovering the attractor by the method of Packard (delay method) [2,8].

Step 3. If the assumption of chaotic dynamics is confirmed, the presence of symmetries (under conditions of weak symmetry breaking) is checked on the basis of the reconstructed attractor analysis [9-10].

Step 4. According to the accepted symmetries of the Hausdorff-Lee formula the type of equations is built in the minimal invariant diversity [1].

Step 5. The structure of the equations is parametrically identified [2].

Step 6. The qualitative difference between the dynamics of the model and the initial series is estimated.

Step 7. The output is a model in the form of finite difference equations.

The significant number of papers are devoted to the first two steps of the method, so their discussion is beyond the scope of this article. Let's stop on the fourth step.

The algorithm for finding symmetric sections of reconstructed trajectories under conditions of weak symmetry breaking solves the problem, which is NP-complete, was described in detail in [10].

The algorithm consists of five stages. The initial data of the algorithm are n-dimensional points belonging to the reconstructed attractor in the phase space. The whole algorithm is implemented in the form of a program in MATLAB.

Note that, in fact, there is a check whether the system allows the specified transformations, in conditions of weak symmetry breaking. The latter is verified as the existence of some small quantity that deviates slightly from the symmetric representation.

The algorithm of quantitative assessment of the state of the system model assumes the following sequence of actions:

- formation of Cauchy model based on reconstructed attractor;
- obtaining Jacobi matrix composed of partial derivatives of the right parts of the corresponding differential equations;
- substitution of initial conditions;
- finding own value of the resulting numerical matrix [8] for n=3 there are three: μ_1, μ_2, μ_3 ;
- finding Lyapunov exponents of the first order as the real part of own value λ_j^1=Reμ_j^1;
- finding the one-dimensional Lyapunov exponent $\lambda_1 = \max_j \lambda_j^1$;
- finding $\lambda_1^2 = \lambda_1^1 + \lambda_2^1$; $\lambda_2^2 = \lambda_1^1 + \lambda_3^1$; $\lambda_3^2 = \lambda_3^1 + \lambda_2^1$;
- finding $\lambda_2 = \max_j \lambda_j^2$;
- finding $\lambda_1^3 = \lambda_1^1 + \lambda_2^1 + \lambda_3^1$

Cauchy initial model Jacobian matrix of the model Initial conditions of the simulation

$$\begin{cases} \dot{x} = -10(x-y) \\ \dot{y} = -xz - y + 28x \\ \dot{z} = -26z + xy \end{cases} \quad \begin{bmatrix} -10 & 10 & 0 \\ 28-z & -1 & -x \\ y & z & -2{,}6 \end{bmatrix}$$

	a	b	c
X(0)	0	1	1
Y(0)	0	2	5
Z(0)	0	-1	-7

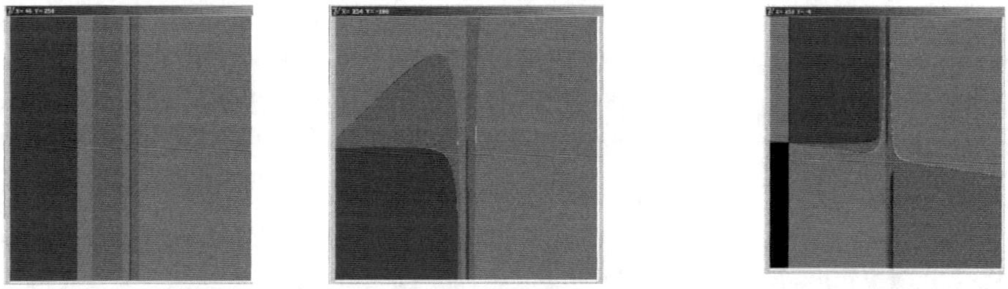

a *b* *c*

Figure 1. Numerical simulation of a chaotic system

Table 1. Attractor type depending on Lyapunov exponents

Attractor type	Phase space dimension	Signs of the Lyapunov exponents
Fixed point	1	(-)
Fixed point	2	(-,-)
Limit cycle	2	(0,-)
Fixed point	3	(-,-,-)
Limit cycle	3	(0,-,-)
Two-dimensional torus	3	(0,0,-)
Strange attractor	3	(+,0,-)

Note: in parentheses, respectively, (λ_1), (λ_2, λ_1), $(\lambda_3, \lambda_2, \lambda_1)$,

During the machine experiment, one parameter was fixed, initial conditions were introduced on other parameters and each point of the phase space was evaluated from the point of view of determining the attractor type. A variant of this study is shown in Fig.1. for three initial conditions.

5. Conclusion

The obtained models can be used to construct "robust chaos" [2], as well as to generate chaos with specified qualitative properties.

The research showed the principal possibility of analytical assessment of the stability of trends emerging in the process of economic activity. It turned out that this can be done quickly, using high-quality visual and cognitive evaluations and computer programs with a point-by-point evaluation of the entire phase space.

The use of dynamic models of chaotic processes allows predictive proactive management for dynamic resource allocation in complex socio-economic infrastructures. An important task is to choose the shape of the model. Using the proposed in this paper forms will allow to obtain reliable computational models that, on the one hand, reflect the essential qualitative properties of nonlinear socio-economic systems, on the other - have the necessary robustness - not sensitive to small changes of the parameters.

The work was carried out on the state task of the Khanty-Mansi Autonomous district of the Russian Federation

References

[1] Khramov V V 2002 Using the paradigm of self-organization to describe information processes in ergatic systems. In the collection: Transport-2002 *Proceedings of the scientific and theoretical conference of the teaching staff* (Rostov-on-Don: Rostov State Transport University) pp 25-27 URL: https://elibrary.ru/item.asp?id=3265311 (Date of application 17.07.2019)

IOP Conf. Series: Materials Science and Engineering **680** (2019) 012038 doi:10.1088/1757-899X/680/1/012038

[2] Khramov V V 2011 Theory of information processes and systems: methodical manual (Rostov-on-Don) URL:https://elibrary.ru/item.asp?id=32764234 (Date of application 17.07.2019)
[3] Kramarov S O and Khramov V V 2018 System-engineering approach to research of complex multidimensional systems on the basis of soft models *Intellectual resources to regional development* issue 4, №1 pp 211-218 URL:https://elibrary.ru/item.asp?id=36739538 (Date of application 17.07.2019)
[4] Kramarov S O, Smirnov Y A, Sokolov S V and Taran V N 2016 System methods of analysis and synthesis of intelligent adaptive control *Monography* (Moscow: RIOR: INFRA-M) p 238 URL:https://elibrary.ru/item.asp?id=35338793 (Date of application 17.07.2019)
[5] Sokolov S V, Kovalev S M, Kramarov S O 2018 Fuzzy logic optical processors *Monography* (Moscow: RIOR: INFRA-M) p 202 URL: https://elibrary.ru/author_items.asp?authorid=661471 (Date of application 17.07.2019)
[6] Bratishchev A V 2017 About the characteristic polynomial of the equilibrium state of autonomous system having an attracting invariant manifold *Differential equations and control processes* № 2, pp 11–19 URL:http://diffjournal.spbu.ru/RU/numbers/2017.2/article.1.2.html (Date of application 17.07.2019)
[7] Volkova V E 2010 Identification criteria for nonlinear models of dynamic systems *Metal construction* Vol. 16 № 3 pp 171–178 URL:http://donnasa.ru/publish_house/journals/mk/2010-3/03_volkova.pdf (Date of application 17.07.2019)
[8] Khramov V V 2000 Generation of models of objects of intellectual space. Theory and use for control of complex systems *Management in social, economic and technical systems. Proceedings of the inter-republican scientific conference* pp 67-68 URL:https://elibrary.ru/item.asp?id=32737843 (Date of application 17.07.2019)
[9] Nikulchev E V 2007 Geometric approach to modeling nonlinear systems from experimental data (Moscow: MGUP) URL:https://www.twirpx.com/file/19818 (Date of application 17.07.2019)
[10] Nikulchev E V 2014 Modeling of systems with chaotic dynamics based on evaluation of weak symmetry breaking in reconstructed attractors *Modern problems of science and education* issue №4 URL:https://www.science-education.ru/ru/article/view?id=14319 (Date of application 17.07.2019)

IOP Conf. Series: Materials Science and Engineering **680** (2019) 012039 doi:10.1088/1757-899X/680/1/012039

Geometric conditions of mathematical modeling of human heat exchange processes with the environment for CAD systems creating heat-shielding clothing

I Cherunova[1,2], S Tashpulatov[3] and Yu Davydova[1]

[1] Department of Design and technology Don State Technical University, Shevchenko street 147, Shakhty 346500, Russia Federation
[2] University of Rostock, Albert-Einstein-Straße, 2, Rostock, Germany
[3] Department of Technology and design of garments
Tashkent Institute of textile and light industry, SHohdzhahon, 5, Tashkent, Uzbekistan

Abstract. In the article research results are presented, which aim to provide of automation algorithm mathematical model of designing technology clothing of the cold protection. The method of calculating heat transfer in the "human - environment" system takes into account the variable thermal parameters and the geometric dimensions of the elements, the presence of internal heat sources and convective heat transfer with the heat-carrier flow along the layers and between the calculated elements . Then the problem is reduced to a one-dimensional heat conduction problem with a variable thermal conductivity coefficient. When modeling a human body, we assume that each element of the model is represented as a set of elliptical cylinders forming the inner layers. Four main concepts of the geometric representation of the outer and inner solutions of the human body model have been developed to simulate the problems of human heat exchange with the environment. An algorithm is proposed for determining the type and structure of a mathematical model for use in CAD systems of protective clothing designing. The research was made in Don State Technical University within the framework of State Assignment of the Ministry of education and science of Russia under the project 11.9194.2017/BCh.

1. Introduction

The modern development of mathematical modeling for technical systems of various types has become the basis of modern technology. The purpose of mathematical modeling is to withdraw the need for a significant number of experiments as part of the program for creating new technology objects or processes and their treatment (processing).

Mathematical models for designing clothes that protect a person from dangerous heat flow include the main object - the human. It is a complex system of biological, thermodynamic, and psychological interaction with the environment. It is almost impossible to consider all functioning parameters and features of person as an object of the model. For technical tasks, including the creation of effective protective clothing, thermodynamic processes of human interaction with the environment is of the top priority [1-3]. They have two components: inner and outer.inner and outer.

Human internal thermodynamics refers to a group of very complex tasks, since it includes various physiological reactions, often amenable to the combined laws not only in terms of human physics but the human psychology as well. And it is extremely difficult to model them as a set of mathematical dependencies with a high degree of adequacy. Therefore, it is necessary to determine the limitations and tolerances that form the geometry for the boundaries defining the provisions of the thermodynamics laws. The justified concept of determining the geometric conditions for modeling thermal processes in the "man-environment" system is a necessary component to development the mathematical software for CAD systems for designing individual means of thermal protection for a person.

2. Scientific relevancy

The problem of mathematical modeling of heat transfer processes in the human body has been considered in a number of works dedicated to the rationing, prediction and regulation of the thermal state of a person [4-7]. A number of works is based on the principles laid down by such authors as Stolvik, Hardy [8,9]. The basis of these models is the principle of idealization of the human body. These works are based on the concept of the human body as a set of geometric elements, most of which are cylinders. The human body is considered as a system of integral interacting components. Such components are the layers of outlined cylinders: core, muscles, fat, skin. These layers create the elements of the system, i.e. parts of the human body called segments. In the majority of works there are distinguished 6 such segments: head, body, arm, palm, leg, foot. The mathematical model of Stolvik [10] reviews the thermal state of a person as an isolated object, without clothes. This approach allows to represent the of heat exchange process of a person mathematically beyond the external factors affecting him, including thermal stimuli. In terms of the geometric concept, in other papers [11-12] the human body is treated in a similar way, where the body elements are cylinders of various sizes and the head is a sphere. In papers [5], a mathematical model of human thermoregulation considers it similarly to the above model, i.e. as an isolated system, and the components outlined in the system are the brain, scalp, internal organs, body muscles, body skin, limbs muscles, limbs skin, and blood. Human body ability to lose excessive heat and the regulating property of the blood flow is the basis of the thermostatic function. However, the model takes into account the standard amount of heat loss when there is no increased perspiration, and the heat transfer by evaporation is constant and equal to 20% of the total heat transfer [13].

The mathematical model describes the "thermal state of a person in clothing with heating, but does not take into account evaporative heat loss from the skin surface and when breathing, which can reach 27% of the total heat loss of the body. Heating is carried out by convective heat, which does not have a direct contact action on the surface of the human body and is not subject to limiting the ergonomic indicators of clothing design. The mathematical model is a description of human body parts having form of sphere (head) and cylinders (body, arms, legs), which, in its turn marks the areas covered by clothing packages. The model does not take into account the system of organism physiological thermoregulation, since a person is considered to be in a state of heat balance, which is a significant deficiency of this model.

Numerous human bioheat models have been developed with the aim to calculate the human thermal field and estimate the body heat loss in a stable and uniform thermal environment at various levels of activity in order to describe the human thermoregulatory apparatus. The development began with the analytical bio thermal models of Pennes [13] and Weinbaum and Jiji [14], containing two details "core" and "skin". With the work of Gagge [15], development has shifted to single-segment and multi-segment (multisite) models of the human body with thermoregulation solutions, which were originally developed by Stolwijk [10] and Wissler. Multisegmentalbioheat models are based on physiology theories, thermodynamics and transfer processes for modeling thermal distributions of the entire human body or its part. A more sophisticated 3D finite-element bio-thermal model of the element was also developed by Smith (known as the KSU model) [16], but this model computationally prevents the integration of thermal distribution and the calculation of the human thermal comfort state. Few of the developed multi-segmented bioheat models have dealt with a spatially inhomogeneous and unsteady environment. There is a need to develop effective bio-thermal models for a dressed person to formalize a number of design solutions for the creation of thermal protective means of life support.

The mathematical model UTCI-FIALA [17] of human temperature control is a multi-detail dynamic model of human physiology and thermal comfort. It is based on a new universal thermal climate index (UTC). After numerous verification, adaptation and expansion tests, such as the inclusion of an adaptive clothing model, this model was used to predict a person's temperature in open climatic conditions. The system inserted in this model (passive version) is a multi-segmental multilayered representation of the human body with a spatial division, which includes a detailed designation of the anatomical, thermal, and thermophysiological properties of the human body. It presents conditionally average person weighing 73.5 kg with a fat content of 14%. and body surface area of 1.82m2 [17]. The body is idealized as a set of spherical and cylindrical elements constructed of annular concentric layers of tissue with corresponding thermophysical properties and physiological functions. However, modern means of individual protection of a person against dangerous heat include not only common methods of heat transfer by convection and radiation, but also a significant role in a number of structures, which is achieved by conduction using special parts of different shapes. This requires expanding the possibilities of mathematical modeling for heat transfer processes in the "person-protective clothing against heat flux-environment" system and the creation of an improved concept to automate the adoption of a number of important decisions in CAD systems.

3. Theoretical part

The person comfort and the internal climate are designed in sub-models, for example such as: Outdoor Standard Effective Temperature [15]); the Munich energy model of human balance MEMI (1984), the ecological thermal model MENEX (1994, 2004), the perspiration rating approach (ISO 7933), AUB is a human multi-segmented bio-thermal model for transient and asymmetric process. “Multi-segmented human bioheat model for transient and asymmetris” is an improved multi-segmented bio thermal model of the human body and its integration with clothing to calculate the peripheral temperature change of the skin of undressed and dressed segments of a conditional person. In such models, the system is subjected to a complex transient process in a spatially inhomogeneous environment. The passive bioheat model implies the division of the human body into 15 cylindrical segments. Each body segment is divided into one primary detail, six corner details of the skin, one artery as a blood detail, and one detail of venous blood. The Avoliosubmodel is used to calculate blood circulation parameters. In such submodels, the unsteady heat conduction equations are simultaneously solved for the details of each body segment in order to predict the main indicators of skin temperature, rates for perspiration, absolute and latent heat loss. There is a conjugation principle of border conditions: the thermal model of the undressed body is combined with the clothing model, which takes into account the absorption of moisture by the fibers to make calculations at high temperature and with heat transfer through the clothing layers. The local temperature of the skin can be evaluated as a reaction to the influence of a heterogeneous environment.

The multi-segmented bioheat model is extended to two dimensions in each segment, relative to the standard AUB model, to represent heat transfer in an angular distribution in addition to the radial distribution [17]. The clothing layer is divided into six parts parallel to the skin details. The heat generated by the human body is dispersed into the environment through radiation, convection and evaporation to maintain the main temperature.

The method of calculating heat transfer in the "human - environment" system takes into account the variable thermal parameters and the geometric dimensions of the elements, the presence of internal heat sources and convective heat transfer with the heat-carrier flow along the layers and between the calculated elements [17]. However, in case of asymmetric structures formation with addressable elements of thermal impact, a more detailed concept of modeling is required.

The process of human heat exchange with the environment can be divided into a number of stages:

1. Heat exchange between limbs and the environment.
2. Heat exchange between the head and the environment.

3. Heat exchange between the human body and the environment.

The first two sections are sufficiently detailed in [10], where the classical approach to the geometric concept of human body elements is used.

In the case considered in this paper, we assume that there is no heat transfer in the cross section between sectors. In addition, we replace each sector of an ellipse with a sector of a circle whose radius passes through the middle of the arc of the corresponding elliptical sector (A…P), which is shown in figure 1.

Then the problem is reduced to a one-dimensional heat conduction problem with a variable thermal conductivity coefficient [5]. In this case, the variability of the heat conductivity coefficient is determined by two factors:

1. Body tissue, through which the heat flows (core, muscle, etc.);
2. Expansion with distance from the center of the cross section of the area through which the heat flows.

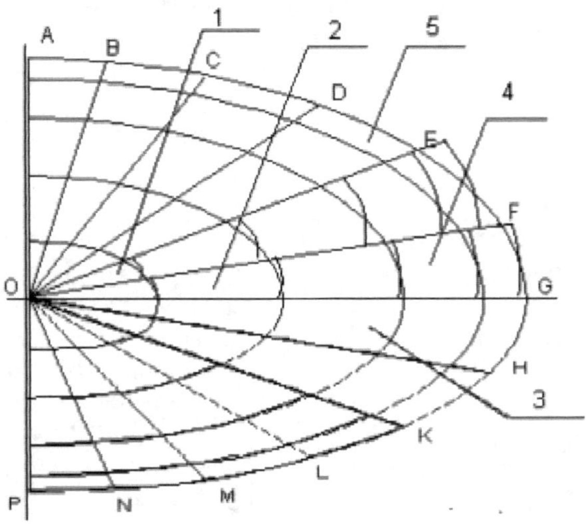

Figure 1. Model representation of human body section. 1,2 – core, 3 – muscles, 4 – fat, 5 – skin.

The process of heat transfer can be described in each of the horizontal section sectors, combined into a single unit. The common sector combines all the constituent layers that makes up the structure of the human body. In each layer of the horizontal section of the sector, the heat transfer process is described by the following equation:

$$dT/dt = \lambda(x,t)\frac{d^2T}{dx^2} + c(x,t)\frac{dT}{dx} + f(x,t) \quad (1)$$

Where x- distance from the centre of human body, m;

t - time, sec;

$T(x,t)$ - temperature,°C, depending on x and t;

$f(x,t)$ - the function of heat production process characteristics for human body;

$\lambda(x,t)$ and$c(x,t)$ - coefficients, characterizing heat transfer processes.

4. Practical relevance

When modeling a human body, we assume that each element of the model is represented as a set of elliptical cylinders forming the inner designated layers 1,2,3,4,5 (Fig.1). For the further development of mathematical models applicable in CAD systems for designing heat-protective clothing [18], 2 concepts of representing the inner structure of the body were proposed:

The inner layers of the body are distributed in accordance with the proportion of each of them to the total radius of curvature;

The inner layers of the body are distributed in accordance with the specified thickness of layers, and size of the core is formed depending on the radius of curvature in the cross-section area.

Thus, it is possible to form a set of basic concepts for the mathematical description of the outer and inner structure of the human body, shown in Figure 2

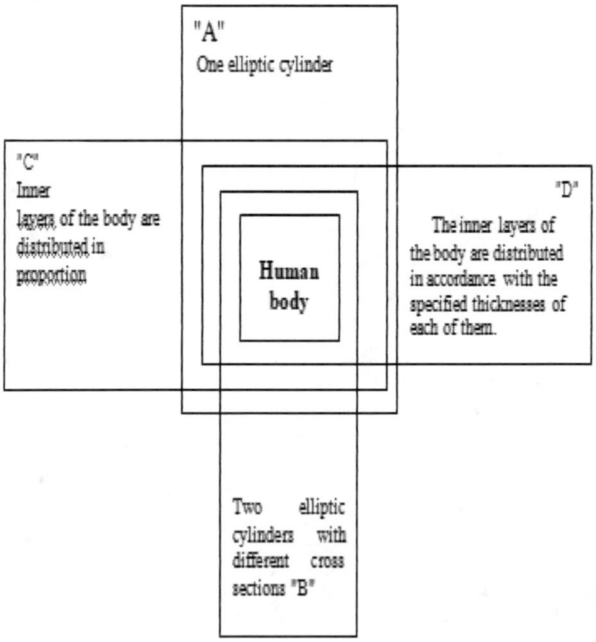

Figure 2. The structure of the concepts of the structure of human body.

On the basis of the proposed definition scheme, a concept variant can be chosen to describe the heat exchange processes for a particular scheme of geometric distribution of layers in the system [19].

All the studied approaches in mathematical modeling allow to form a specific algorithm for identifying the type and content of a mathematical model based on the criteria that are meaningful for the task, which is shown in Figure 3.

Based on the research and development of principles and limitations in mathematical models of "human-heat-protective clothing-environment", the geometric features typical for creating human models are systematized. Four main concepts of the geometric representation of the outer and inner solutions of the human body model have been developed to simulate the problems of human heat exchange with the environment.

IOP Conf. Series: Materials Science and Engineering **680** (2019) 012039 doi:10.1088/1757-899X/680/1/012039

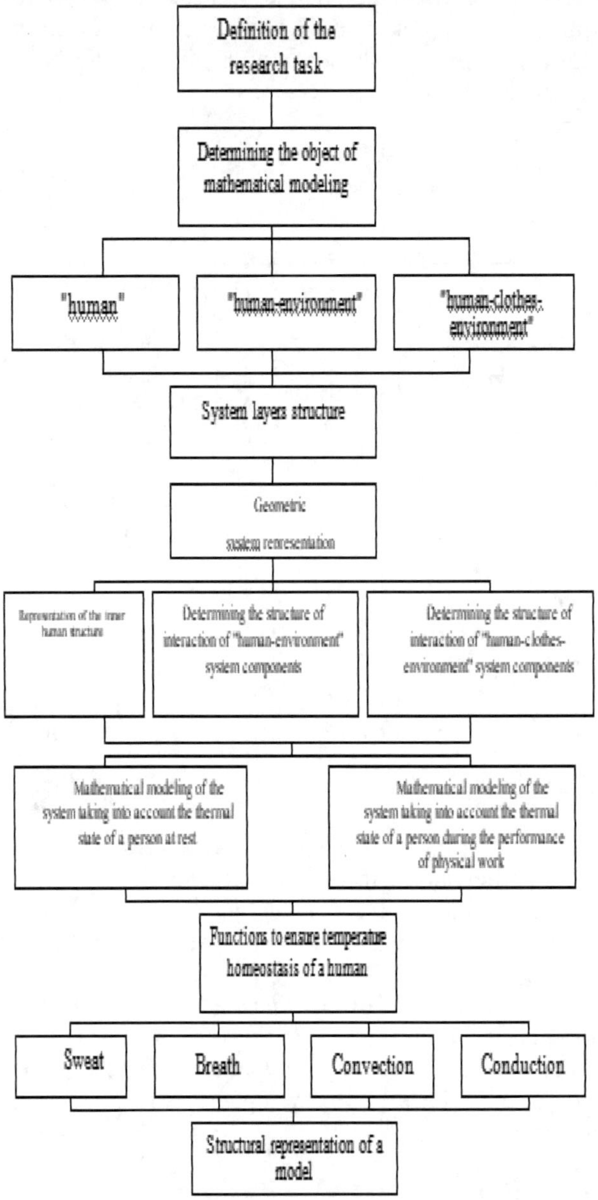

Figure 3. An algorithm for determining the type and structure of a mathematical model for use in CAD systems in designing the protective clothing.

An algorithm is proposed for determining the type and structure of a mathematical model for use in CAD systems of protective clothing designing.

The results obtained allow to expand the possibilities of using mathematical modeling resources for CAD systems in order to create heat-protective clothing.

5. Conclusion

An algorithm is proposed for determining the type and structure of a mathematical model for use in CAD systems of protective clothing designing.

IOP Conf. Series: Materials Science and Engineering **680** (2019) 012039 doi:10.1088/1757-899X/680/1/012039

The results obtained allow to expand the possibilities of using mathematical modeling resources for CAD systems in order to create heat-protective clothing.

References

[1] Gupta D. Design and engineering of functional clothing / Indian Journal of Fibre and Textile Research. 2011. Vol.36. – pp.327-335.

[2] Dabolina I., Vilumsone A. The Role of the Latest Clothing CAD/CAM System Applications in the . – Vol.Educational Process / Material Science. Textile and Clothing Technology. – 2012.- №7 – PP. 63-67.

[3] Zhang L., Zhang H., Li Y. The New Development of CAD in Clothing / Computer and Information Science. – 2008. – Vol.1. – No.8. – 104-106.

[4] Cherunova I., Dhone M., Kornev N. Coupled thermo-aero. - dynamical Problems in design of protection Cloth / Proc. VI International Conference on Computational Methods for Coupled Problems in Science and Engineering COUPLED PROBLEMS- 2015 (Venice, May 2015). – pp. 1303-1311.

[5] Cherunova I., Kornev N., Brink I.Mathematical model of the ice protection of a human body at high temperatures of surrounding medium . Forschung im Ingenieurwesen / Springer: Gernany. 2012. - Volume 76. - Issue 3-4. - pp 97-103

[6] Cherunova I., Kornev N., Jacobi G., Treshchun I., Groß A., Turnow J., Schreier S., Paschen M. Application of calculations of heat transfer and computational fluid mechanics to the design of protection clothes / Journal of Engineering Physics and Thermophysics. Journal of Engineering physics and Thermophysics. - 2014 . - Vol. 87 . - No. 4.

[7] Parsons K. C., Havenith G., Holmér I., Nilsson H. and Malchaire J. The effect of wind and human movement on the heat and vapour transfer properties of clothing / Annals of Occupational Hygiene. - 1999. - 43(5). - 347-352.

[8] Stolwijk J.A.J. Mathematical model of Thermoregulation / In: Physiological and behavioral Temperature Regulation / Springfield: Charles C. Thomas Publisher.-1970. (48) -703-721.

[9] Parsons, K. Human thermal environments: The effects of hot, moderate and cold environments on human health, comfort and performance, 2nd ed. London : Taylor & Francis, 2003 .

[10] Stolwijk JAJ, Hardy JD (1966) Temperature regulation in man - A theoretical study. Pflugers Arch Gesamte Physiol Menschen Tiere 291(2): 129-162.

[11] d'Ambrosio F. R. A., Palella B. I. & Riccio G. THERMODE 193: An Enhanced Stolwijk Thermoregulation Model of the Human Body / Proceedings of 7th International Thermal Manikin and Modelling Meeting. – 2008. - University of Coimbra Portugal .

[12] Olesen B. W Measurement of Thermal / Comfort in Vehicle, from Application notes: Evaluation of the Thermal Environment in Vehicles. - Bruel&Kjaer, 1989.

[13] Thake C.D., Simons C. Physiological, perceptual and cognitive responses to conducting nuclear decommissioning related activities when wearing a pressurised breathing air suit (PBAS) in moderate and hot conditions / 13th International Conference on Environmental Ergonomics. BOSTON, MASSACHUSETTS, USA, 2009. - pp. 189 – 194,.

[13] Pennes, H. H. Analysis of Tissue and Arterial Blood Temperature in Resting Forarm / Journal of Applied Physiology, 1948. - Vol. 11 - pp. 93-122.

[14] Jiji, L.M., Weinbaum, S. & Lemons, D.E. Theory and experiment for the effect of vascular microstructure on surface tissue heat transfer - Part I1:Model formulation and solution. J. Biomed. Eng., 1984. - (106). - 331-341.

[15] Mangum B.W., J.E. HillThermal Analysis--human Comfort--indoor Environments: Proceedings of a Symposium Held at the National Bureau of Standards, Gaithersburg, Maryland. - 1977. – (11). – Vol. 13 . – 191 p.

[16] Smith CE. A transient, three-dimensional model of human thermal system. / PH.D. Dissertation, Kansas State University, Manhattan, Kansas. 1991.

[17] Fiala D., K.J. Lomas, and M. Stohrer First Principles Modelling of Thermal Sensation Responses

in Steady State and Transient Boundary Conditions / ASHRAE Transactions, . – 2003. – Vol.109(1). - 179-186

[18] Aihua M. , Yi L., Xiaonanb L., Ruomei W., Shuxiaoa W. A CAD system for multi-style thermal functional design of clothing / Computer-Aided Design, 2008. – No. 40. – pp.916–930.

[19] Cherunova, I., Samarbakhsh S., Kornev N. CFD simulation of thermo- aerodynamic interaction in a system humancloth-environment under very low temperature and wind conditions / ECCOMAS Congress 2016: Proceedings of the VII European Congress on Computational Methods in Applied Sciences and Engineering, Held in Crete, Greece , 5-10 June 2016. – Volume IV . - pp. 7703-7710

[20] Meng Y., Mok P.Y., Jin X. Interactive virtual try-on clothing design systems / Computer-Aided Design, 2008. – No.40. – pp.916–930.

[21] Cherunova I., Dhone M. , Kornev N. Coupled thermo-aerodynamical Problems in design of protection Cloth / Proc. VI International Conference on Computational Methods for Coupled Problems in Science and Engineering COUPLED PROBLEMS 2015, Venice, May 2015, 1303-1311.

IOP Conf. Series: Materials Science and Engineering **680** (2019) 012040 doi:10.1088/1757-899X/680/1/012040

Medical Image Inpainting Using Multi-Scale Patches and Neural Networks Concepts

N V Gapon[1], V V Voronin[1,2], R A Sizyakin[1], D Bakaev[3], A Skorikova[4]

[1]Don State Technical University, sq. Gagarin 1 Rostov-on-Don, 344000 Russia
[2]Moscow State University of Technology "STANKIN",ln. Vadkovsky 1 Moscow, 127055 Russia
[3]Rostov State Medical University, st. Pushkinskaya 212 Rostov-on-Don, 344000 Russia
[4]Voronezh State Medical University, st. Studencheskaya 10 Voronezh, 394036 Russia

E-mail: gapon.nv@gmail.com

Abstract. In this paper, we consider the problem of medical images inpainting, where the goal is to reconstruct missing or damaged parts of the image. This is a good tool for medical applications such as vascular restoration, removal of specular reflections for endoscopic images, removal of MRI artifacts, etc. The new method combines the search for patches of various sizes and the operation of a pre-trained neural network. The large patches are used to reconstruct homogeneous areas, and then small patches are used to reconstruct structure image details. As a result, the proposed method provides a plausible of medical images inpainting. Experimental results demonstrate the effectiveness of the proposed method in the tasks of medical images inpainting.

1. Introduction

Image inpainting is a technique to remove the undesirable or filling missing (repair damaged) section from an image, by selecting patches from the rest of the image, or the class of images and placing these patches in the region/section using some "optimality" criterion. Most image inpainting methods can be divided into the following groups: methods based on the solution of partial differential equations (PDE); methods based on orthogonal transformations; methods based on texture synthesis; methods based on the operation of a neural network. Each of them has own advantages and disadvantages.

We consider in this paper the problem of image inpainting in medical image analysis, where the objective is to reconstruct missing or deteriorated parts of an image. It is a good tool for such medical applications as vascular reconstruction, specular reflection removal for endoscopic images, MRI artifacts removing, etc. Most inpainting approaches require a good image model to infer the unknown pixels. Such examples are:

- MRI and CT metal artifact reduction (Fig. 1).

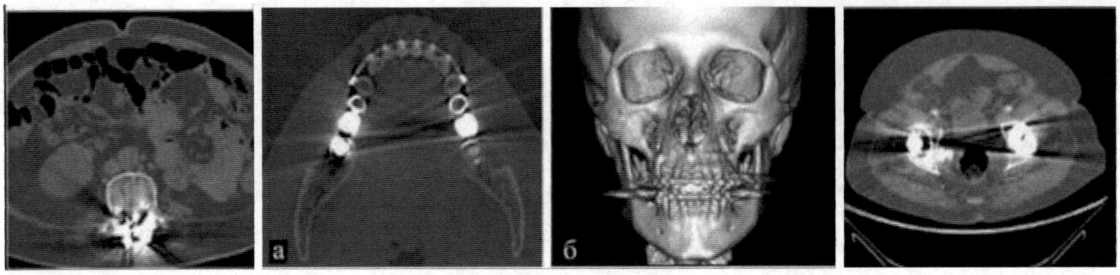

Figure 1. MRI and CT images.

- A Specular Reflection Suppression for Endoscopic and Cardiac Images (Fig. 2).

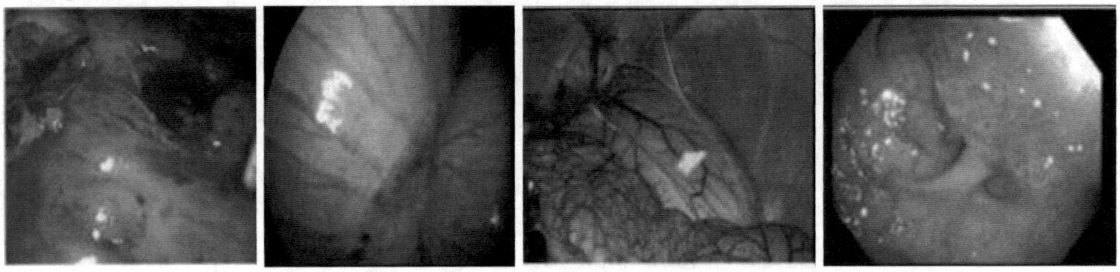

Figure 2. Endoscopic and Cardiac Images.

- Vessel detection in retinal images (Fig. 3).

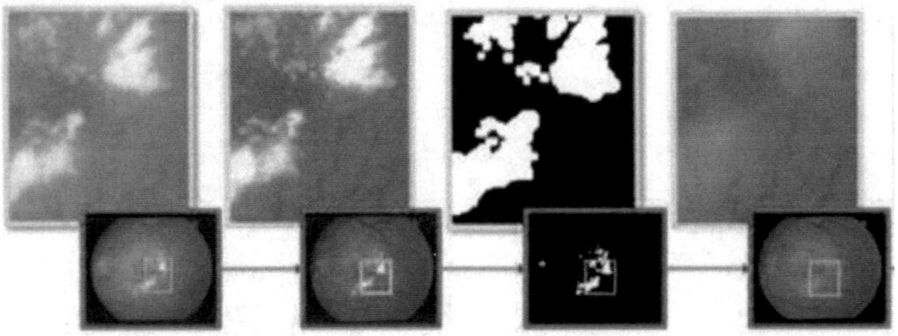

Figure 3. Vessels on retinal images.

2. Proposed method

2.1. Reconstruction method

At the first step, for each pixel of the boundary $\delta S_{i,j}$ using the inversion method the shape of the domain for similarity search is adaptively determined by combining two adjacent homogeneous sub regions in the direction of the maximum gradient [2-6].

The second step calculates the priority value $P(\delta S)$ for each pixel value of the boundary, which consists of two factors [3, 5]:

$$P(\delta S) = C(\delta S) \cdot D(\delta S) \tag{1}$$

$$C(\delta S) = \frac{\sum_{l \in \Psi_{\delta S}} C(l)}{|\Psi_{\delta S}|}, \; D(\delta S) = \frac{|\nabla I_{\delta S}^{\perp} \cdot n_{\delta S}|}{\alpha} \tag{2}$$

Using the method proposed in the work [1] the entire image is preliminarily divided into patches of 5×5, 7×7,..., 21×21, where large patches are used to reconstruct homogeneous areas, and then small patches are used to reconstruct structure image details (Fig. 4).

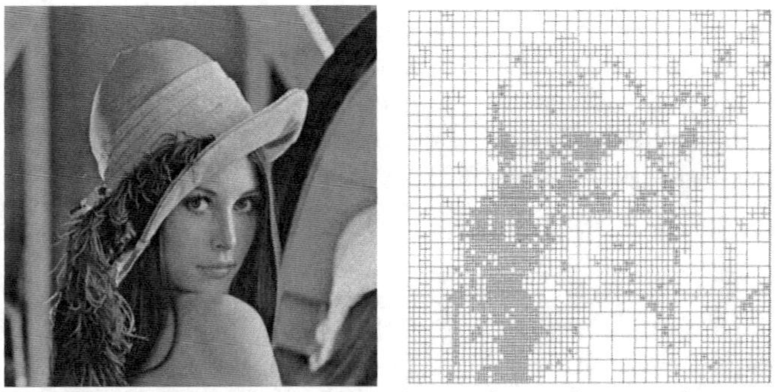

Figure 4. Example of splitting an image into patches

Next, in the third step, we determine blocks $\Psi_{q^{(h)}}$, $h=\overline{1,R}$ in the area of available pixels S for which the Euclidean metric is minimal [4, 6, 8]:

$$\sqrt{\sum(\Psi_p-\Psi_q)^2}\longrightarrow \min. \tag{3}$$

The pixel values in the region η adjacent to the pixel with the highest priority p are reconstructed by averaging the corresponding pixels from chosen areas $\Psi_{q^{(h)}}$ in the area of available pixels S using a neural network, in particular, a multi-layer perceptron[8].

The confidence coefficient C for restored pixels is assigned to the current value $C(p)$.

After that, the procedure of priority correction and search of similar areas with subsequent replacement is repeated.

2.2. Neural network

In this work, a neural network of direct signal propagation, which was trained using the algorithm for back propagation of the error was developed. The activation function which is used in this network is the sigmoidal nonlinear function (sigmoidal nonlinearity), namely the hyperbolic tangent function (4) [8-10].

$$f(x)=\tanh\left(\frac{ax}{2}\right)=\frac{1-e^{-ax}}{1+e^{-ax}} \tag{4}$$

a - is the slope parameter of the sigmoidal activation function.

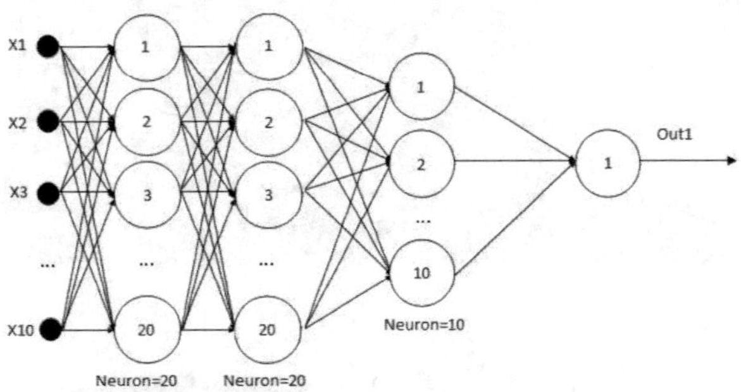

Figure 5. Architecture of the neural network

At the training stage of the neural network, pre-prepared data was fed to the input: a blocks with random coordinates 3×3, 5×5, 7×7,..., 21×21 in size was allocated to the image, then the central pixel was removed, and five most similar blocks were found on the whole image comparing them by MSE.

Then the procedure was repeated on 35 images, and 100,000 blocks were obtained, they were used as a training sequence for this network. This network contains three layers [9]: the first layer contains 20 neurons, the second layer also contains 20 neurons, the third layer contains ten neurons (Fig. 5). The network created ten inputs, 5 of them were fed only the central pixels in the blocks found, and the other five inputs fed the MSE of these blocks. As an output, there were the central pixels of the original blocks.

3.Experimental results

To test the effectiveness of the proposed reconstruction method when deleting objects in the image, four test images were selected. The peculiarity of these test images is that the areas of missing pixels are located at the intersection of several borders. Figures 6-9 show examples of image restoration (a) of the original image, b) image with a mask of distorted pixels, c) image restored by the proposed method). The proposed method of image inpainting allows to "correctly" restore borders.

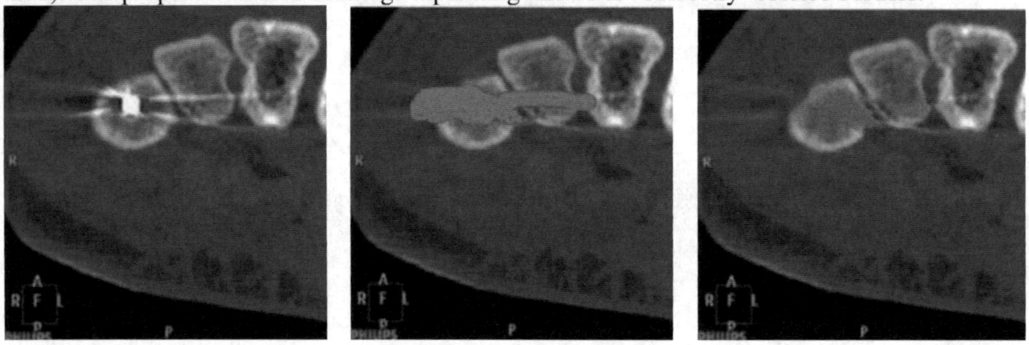

a) b) c)

Figure 6. Reconstruction example (a) the original image, b) image with a mask of distorted pixels, c) image restored by the proposed method.

It should be noted that the proposed method does not blur the texture and structure when restoring large areas with lost pixels. The proposed method allows you to correctly restore details on the image, and there is also the absence of artifacts when restoring lost blocks.

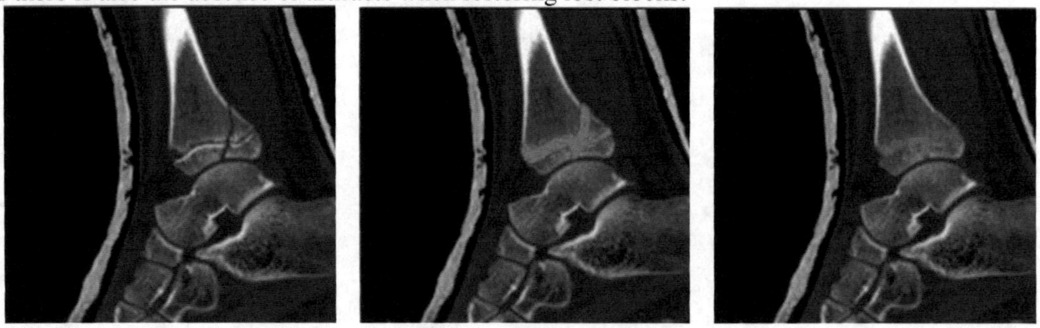

a) b) c)

Figure 7. Reconstruction example (a) the original image, b) image with a mask of distorted pixels, c) image restored by the proposed method.

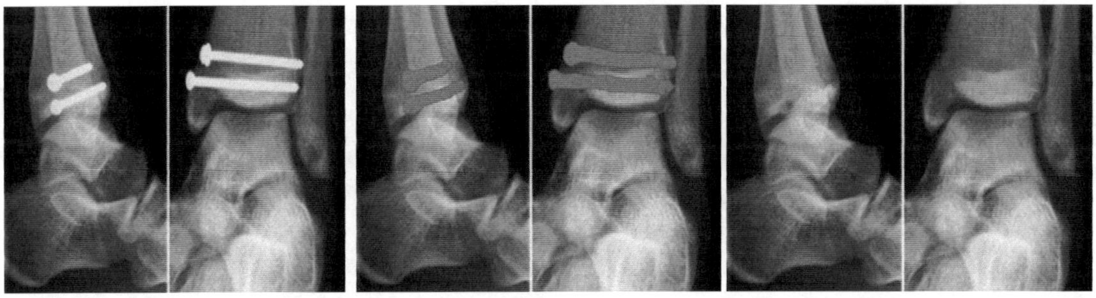

Figure 8. Reconstruction example (a) the original image, b) image with a mask of distorted pixels, c) image restored by the proposed method.

Figure 9 shows an example of image recovery (a) the original image, b) image with a mask of distorted pixels, c) image restored by the proposed method). A feature of this image is that the area for restoration is at the intersection of several borders. The proposed method allows you to correctly restore the boundaries of objects.

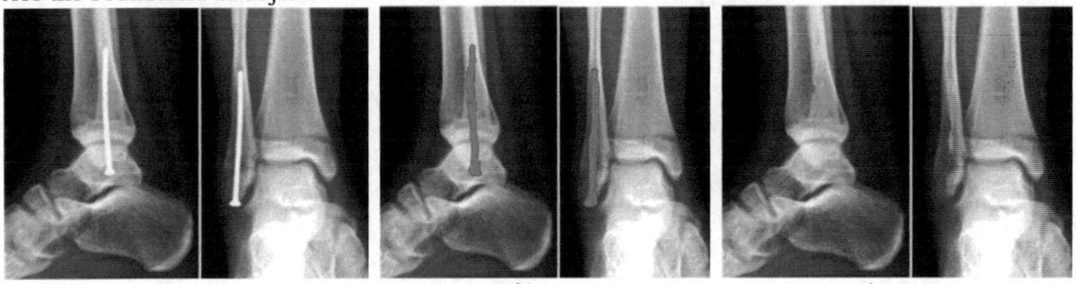

Figure 9. Reconstruction example (a) the original image, b) image with a mask of distorted pixels, c) image restored by the proposed method.

The processing error values for the proposed method, EBM, Navier-Stokes and Telea methods [2] are presented in Table 1. It should be noted that the quantitative values of the errors are calculated in the reconstructed areas. When using the developed image reconstruction method, the error values are on average 10–15% less than when processed by EBM, Navier Stokes and Telea.

Table 1. Processing error values.

Images		**Fig. 6**				**Fig. 7**		
Method	***PP***	***EBM***	***NS***	***Telea***	***PP***	***EBM***	***NS***	***Telea***
PSNR	30.73	27.37	25.98	24.62	25.36	24.64	25.42	23.52
RMSE	8.61	10.75	13.73	14.85	15.87	15.14	16.36	14.72
MAE	3.55	4.72	5.34	7.81	7.11	7.8	7.68	7.93
SNR	26.53	23.13	22.46	20.78	24.68	14.27	21.55	22.33
Images		**Fig. 8**				**Fig. 9**		
Method	***PP***	***EBM***	***NS***	***Telea***	***PP***	***EBM***	***NS***	***Telea***
PSNR	44.3	41.91	18.24	22.56	26.22	24.65	23.95	24.35
RMSE	3.56	7.36	8.67	12.92	16.18	16.98	17.54	16.33
MAE	1.5	2.35	3.87	4.68	6.18	6.95	6.25	5.95
SNR	41.18	36.79	33.26	31.54	28.58	28.27	32.18	31.88

4. Conclusions

In this paper, a patch-based inpainting algorithm for MRI and CT metal artifact reduction has been developed. We used the method of splitting the image into patches of different sizes and a trained neural network to select the "best similar" patch. Analysis of the results indicates that the proposed method correctly restores both details and background of the medical image in case of the presents metal artifacts and produces 10-15% lower reconstruction errors, than widely used state-of-the-art methods.

5. Acknowledgment

This work was supported by Russian Ministry of Education and Science in accordance to the Government Decree № 218 from April 9, 2010 (project number № 074-11-2018-013 from May 31, 2018 (03.G25.31.0284)).

References

[1] Ponomarenko N N, Lukin V V, Egiazaryan K O, Astola J T 2002 Modified horizontal vertical partitioning scheme for compressing fractal images *5th Northern Symposium on Signal Processing on board Hurtigrunen*

[2] Ballester C, Bertalmio M, Caselles V, Sapiro G, Verdera J, 2001 Filling-in by join interpolation of vector fields and grey levels *IEEE Trans. Image Process.*, **vol.** 10, no. 8, 1200–1211

[3] Bertalmio M, Sapiro G, Caselles V, Ballester C, 2000 Image inpaintin *Proc. Comput. Graph. (SIGGRAPH2000),* 417–424

[4] Bertalmio M, Bertozzi AL, Sapiro G, 2001 Navier–Stokes, fluid dynamics, and image and video inpainting *Proc. IEEE Conf.Comput. Vis. Pattern Recognition*, 355-362

[5] Paget R , Longstaff I D 1998 Texture Synthesis via a Noncausal Nonparametric Multiscale Markov Random Field *IEEE Transactions on Image Processing*, **vol.** 7, no. 6

[6] Shen Huanfeng, Ai Tinghua, Li Pingxiang, 2009 A MAP-Based Algorithm for Destriping and Inpainting of Remotely Sensed Images Geoscience and Remote Sensing, *IEEE Transactions on,* **Vol.** 47, Issue: 5, 1492 – 1502

[7] Criminisi A, Perez P, Toyama K 2003 Object removal by examplar-based image inpainting. *In Int. Conf. on Computer Vision and Pattern Recognition (CVPR)*, pp. 721–728

[8] Gapon N, Ponamorenko M, Pismenskova M, Tokareva O 2017 Image inpainting using a neural network. *In MATEC Web of Conferences* **132** 05015

[9] Voronin V V, Marchuk V I, Gapon N V, Sizyakin R A, Sherstobitov A I, Egiazarian K O 2014 Exemplar-based inpainting using local binary patterns. *Proceedings of SPIE - The International Society for Optical Engineering Сер. "Proceedings of SPIE-IS and T Electronic Imaging - Image Processing: Algorithms and Systems XII"*

[10] Gapon N, Voronin V, Sizyakin R, Pismenskova M, Nadykto A 2019 Video block error recovery using modified patch-based inpainting. *In Eleventh International Conference on Machine Vision (ICMV 2018)* **Vol. 11041**, p. 110410D

Image defect detection algorithm based on deep learning

R A Sizyakin[1,3], V V Voronin[1,2], N V Gapon[1], A A Zelensky[2], A Pižurica[3]

[1]Don State Technical University, sq. Gagarin 1 Rostov-on-Don, 344000 Russia
[2]Moscow State University of Technology "STANKIN",ln. Vadkovsky 1 Moscow, 127055 Russia
[3]Ghent University, st. Rozier 44 Ghent Belgium,

E-mail: roma_sz@mail.ru

Abstract. In this paper proposed a system for automatic defects detection in images. The solution to this problem is widely used in practice. Automatic detection is found in the challenge of detecting defects on the road surface, in the textile industry, as well as virtual restoration of archival photo images. The solution to this range of problems allows speeding up work in these areas, and in some cases, completely solving. To solve the first two problems (search for defects on the pavement and textiles), it is enough to create a mask that localizes defects in the image with maximum reliability, while photo restoration requires additional algorithms to restore the detected damaged areas. The proposed method is based on the latest achievements in the field of machine learning and allows solve the main disadvantages of traditional methods. Automatic defect detection is performed using a neural network with compound descriptor. A series of experiments confirmed the high efficiency of the proposed method in comparison with traditional methods for detecting defects.

1. Introduction

Solution automatic defect detection problem is widely used in practice. This problem occurs when searching for defects in the road surface, the textile industry, as well as virtual restoration of archival photo images. The decision of the range of tasks to speed up work in the direction of the data, and in some cases completely resolve. For defects occurring in the pictures are different kinds of stains, scratches, cracks and other foreign objects. Their appearance can be due to aging, physical impact, improper storage, or operation. Recovery of such defects in the present, in most cases carried out through manual processing, making it challenging to restore archived facsimiles.

In [1] for the detection of defects in the image using a method based on machine learning. For a mask with an estimated localization defects using morphological operations top and bottom hats to detect light and dark cracks. To reduce false marked areas on the pre-mask errors, using a method based on training. As the descriptor values are the hue, saturation, and brightness (HSV- Hue, Saturation, Value (Brightness)). For the construction of the separating hyperplane using a support vector machine (SVM) with a median radial basis function as the kernel (MRBF).

IOP Conf. Series: Materials Science and Engineering **680** (2019) 012041 doi:10.1088/1757-899X/680/1/012041

In [2] the same approach is used to detect defects that in [1]. The main distinguish this work from work [1] is the method of filling in damaged areas. Reconstruction of the damaged regions in this study occurs through search blocks similar to those damaged. As a similarity metric used the sum of squared differences (SSD). If the same unit was not found, then the bad block is filled with the average values. In [3] authors for detect defections proposed to use a morphological operation top hat, followed by binarization with the automatic selection of the threshold value [4]. Choice of the threshold value depends on the separability of the original image histogram. For optimal separation of the histogram iteratively selects threshold values that minimize the variance within a class, which is defined as a weighted sum of the variances of the two classes. In his work Otsu shown that minimizing the variance within a class is equivalent to maximize the variance between classes.

In [5] algorithm to search for defects in the photographs consists of three steps. The first is the improvement in pre-input image contrast. Then it calculates the convolution of the modified image from different directions Gaussian kernels. The second step is to obtain mask defects by applying morphological operations top hats with various sizes of structure-forming element. The third step is constructed for a pre-defect mask method using K-SVD. The point of this step is to train the algorithm, on pre-prepared templates, followed by the classification of areas on the original image. To reduce the number of false positives in this paper, the method to-medium. The handle acts as a color component, the length, orientation and eccentricity ratio. After that three-prepared masks are combined into one by adding. In those areas where there is the imposition of the alleged defects, the resultant mask takes the single value, or zero.

In this paper, we propose a new approach to detection defects on the photograph. The defect detection method includes two main stages: preliminary localization of defects and subsequent accurate classification of detected defects.

2. Proposed method

Damages encountered in the photo image include various kinds of spots, scratches, cracks and other foreign objects. Their appearance can be caused by aging, physical stress, improper storage or use. In this paper, we focus on crack detection. The method for detecting damage to an image consists of two main steps: the first step is to use morphological filtering as a pre-processing, the second step is to use the machine learning method, which is necessary to classify pixels that have received a large response in the preprocessing phase (see Fig.1).

Morphological filtering reduces the computational complexity of the algorithm, because there is no need to process each pixel of the image, but only those that received a higher response. Also, the use of this operation can reduce the number of false alarms [6].

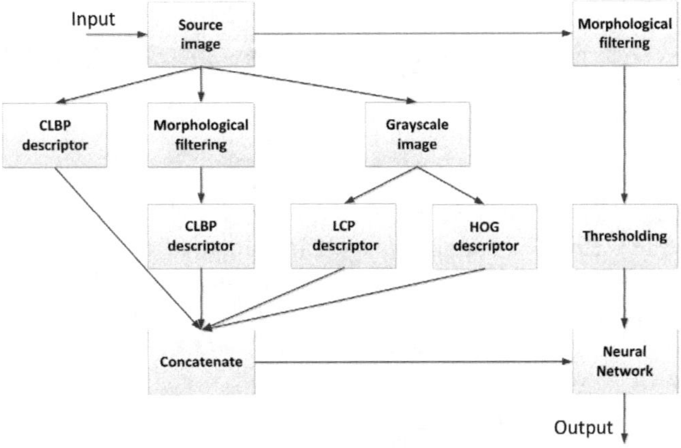

Figure 1 General scheme of the proposed method of defect detection.

In our work we perform morphological operation "bottom hat". The essence of this operation consists of subtraction from the original image $I_{r,c}$ the result of its "opening" with the structural element B. Use of this morphological operation allows locating the expected defects in the image.

$$Mask_{r,c} = Y_{r,c} - (Open_{r,c})^B \tag{1}$$

$$(Open_{r,c})^B = (Y_{r,c} \Theta B) \oplus B) \tag{2}$$

$$(Dilate_{r,c})^B = MAX_{(u,\upsilon)\in B}(Y_{r,c}(r+u, c+v) - B(u,v)) \tag{3}$$

$$(Erode_{r,c})^B = MIN_{(u,\upsilon)\in B}(Y_{r,c}(r+u, c+v) - B(u,v)) \tag{4}$$

where B is a structural element with size pixel size $u \times \upsilon$, Θ - erode, $\oplus$ - dilate.

After the procedure of morphological filtering, the preliminary mask with defects contains both correctly detected defects and a number of false positives. To reduce the number of false alarms, we use a neural network as a classifier.

The neural network allows reducing false alarms on the preliminary mask obtained by morphological filtering. The neural network architecture for defect detection is shown in figure 2.

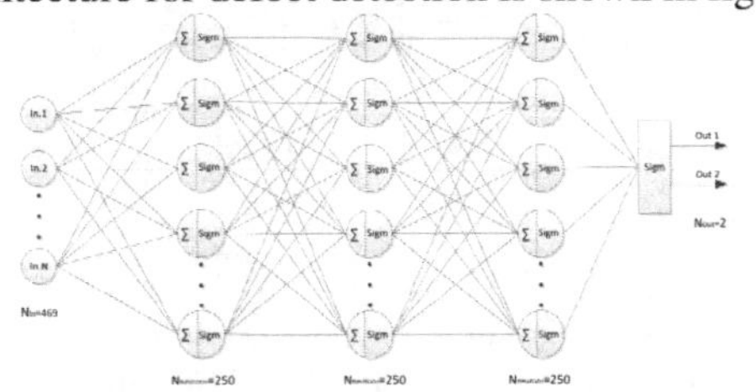

Figure 2. Neural network architecture for damage detection

The network consists of 3 hidden layers. Each layer includes 250 neurons. The logistic sigmoid is used as the activation function:

$$f(x) = \frac{2}{1+e^{-x}} - 1 \tag{5}$$

where x - feature vector.

To determine the losses, we use the binary cross-entropy function. This is defined according to the expression:

$$H(y,y') = -\frac{1}{m}\sum_{l=1}^{m}[y\cdot\log(y') + (1-y)\cdot\log(1-y')] \tag{6}$$

where y - prediction Conv.Net, y'_j - true value.

Additional we use the method of optimization Adam, proposed in work [7], with learning rate equal 0.0005. Training it took approximately 45 epochs. We concatenated to one and used the following descriptors as input for neural network: CLBP [8], HOG [9], and LCP [10].

Texture CLBP operator is an extension of work [11]. The main difference lies in the preservation of not only the sign, but also the magnitude of the component:

$$d_p = s_p * m_p, \begin{cases} s_p = sign(d_p) \\ m_p = |d_p| \end{cases}, \quad s_p = \begin{cases} 1 & d_p \geq 0 \\ -1 & d_p < 0 \end{cases} \tag{7}$$

where d_p is the vector of difference between the Central pixel and its neighbors, s_p is the vector of the sign component, m_p is the vector of the magnetic component.

The use of two components allows achieving the accuracy of the texture description higher than the original LBP method, which uses only the sign component. Texture operator LCP is a combination of feature vectors of the LBP method and weighted coefficients (MiC). Coefficients are calculated for adjacent pixels to the center, in the LBP pattern. The formula for calculating the coefficients is written as follows:

$$E(a_0,\dots,a_{P-1}) = \left| g_c - \sum_{r=0}^{P-1} a_r g_r \right| \tag{8}$$

where g_c is the center pixel, g_r is the adjacent pixel, a_r is the weight coefficient calculated for each g_r.

To calculate the weights a_r, the least-squares method is used:

$$C_L = V_L A_L \tag{9}$$

$$A_L = (V_L^T V_L)^{-1} L_L^T C_L \tag{10}$$

where C_L is the vector of the types of patterns of interest, V_L is the intensity (brightness) of adjacent pixels, A_L is the vector with unknown values of the weight coefficients a_r, L is the type of pattern.

Image rotation stability is achieved by applying the Fourier transform to a vector A_L:

$$H_L(k) = \sum_{i=0}^{P-1} A_L(i)\cdot e^{-j2\pi ki/P} \tag{11}$$

where $A_L(i)$ is i the vector element A_L.

Leaving the amplitude component of the vector, we get an additional vector (MiC) for the original LBP method. Texture operator HOG based on counting the number of gradient directions in local areas of the image, on a dense grid of uniformly distributed cells. To increase the accuracy of the description of the texture features of the image, normalization of local contrast is used. The algorithm for constructing the feature vector consists of several stages. At the first stage, the values of the gradients are calculated. To calculate the gradients, a one-dimensional differentiating mask is used in the horizontal and vertical directions. At the next stage, the calculated gradients are used to form a histogram for each of the cells, passing the weighted voting procedure.

In order to take into account, the brightness characteristics of the image, the gradients are locally normalized. For this, cells are grouped into larger connected blocks. Normalization of blocks occurs in accordance with the expression:

$$f = \frac{v}{\sqrt{\|v\|_2^2 + e^2}} \tag{12}$$

where f is the normalization factor, v is the non-normalized vector, $\|v\|_2$ is its norm, e is a certain small constant.

As a result, the resulting descriptor has a length of 469 bins, of which 108 values correspond to the values of the CLBP operator, which is calculated for each color component of the color image, 36 and 289 values, correspond to the texture operators HOG and LCP, respectively, calculated for the original image in grayscale, and also 36 values correspond to the values of the CLBP operator calculated for the image after morphological filtering. This composite descriptor is calculated for a local window with 20×20 the pixel size taken around each pixel marked by a unit after threshold processing.

3. Results and discussion

To compare the effectiveness of detecting damage in the image, the proposed method is compared with methods based on machine learning: convolutional neural networks (CNN) and support vector machine method (SVM). Three color components of the image are used as input to the convolutional neural network. The first convolutional layer has 10 feature maps, the second layer has 20 feature maps, the third layer has 30 feature maps. The size of the convolution filters is 5×5 pixels. ReLU is used as an activation function in hidden layers. The following parameters were also used: the learning rate is 0.001, the size of the training data batch is 20 samples, for the training was using the ADAM method. The support vector machine used a linear separating hyperplane, with an acceptable error of 5%. The previously described texture descriptors are used as a descriptor.

Figure 3 illustrates the result of the proposed method for detecting defects, a method based on support vector machine and the method based on the use of convolutional neural networks.

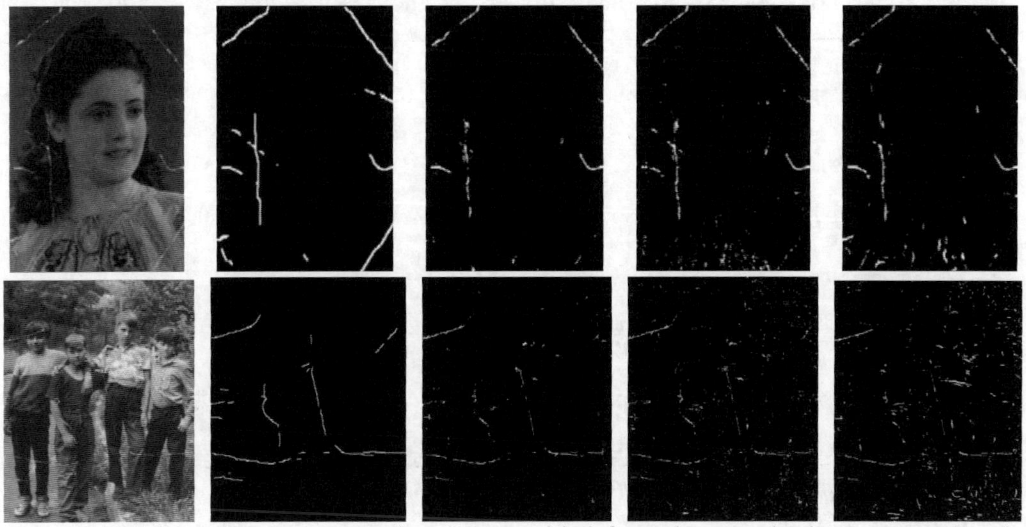

Figure 3. The first column corresponds to images with defects; the second column – ground truth; the third column – proposed method; the fourth column – SVM; the fifth column – CNN.

Approximately 2,500 samples (20×20 pixels) containing cracks as well as 2,500 undamaged samples (20×20 pixels) were used for training all methods. Table 1 shows the results for 5 test images. Test images obtained from free access.

Table 1. The results of determining the iron content in the wear particles.

	Probability false alarm			Probability correct detection
	Proposed method	CNN	SVM	
Test image 1	0.035	0.056	0.053	0.92
Test image 2	0.027	0.073	0.063	0.83
Test image 3	0.051	0.053	0.115	0.79
Test image 4	0.00215	0.00247	0.00237	0.6
Test image 5	0.0074	0.0110	0.0138	0.5
Averaged	0.0251	0.0398	0.0432	0.65

From the estimates given in table 1, it can be concluded that the proposed method reduces the number of false positives by an average of 1.6 times, while maintaining the probability of correct detections on a par with known techniques. The disadvantages of the classifier based on the convolutional neural network include the fact that it has a low generalizing ability with not enough training data. Also, an excessive increase in the training parameters can further reduce the generalization ability. The main drawback of the support vector machine classifier is that it is not efficient enough to classify using a multi-part descriptor that includes multiple texture operators.

4. Conclusion

In this work it was proposed method for the automatic detection of defects on the image. For a preliminary localization of the crack on the image in this work, we used morphological filtering. To reduce the number of false alarms is used a neural network with concatenating vector descriptors. of the proposed method with the various known methods, showed high efficiency in the use of neural networks for detection of defects in the image. The result probably can be improved depending on particular tasks, for example by increasing the size of the training data, modifying the architecture neural networks, changing set descriptors or set parameter more accurate, the resolution of the data used can also be increased, increasing contrast, etc.

5. Acknowledgments

This work was supported by Russian Ministry of Education and Science in accordance to the Government Decree № 218 from April 9, 2010 (project number № 074-11-2018-013 from May 31, 2018 (03.G25.31.0284)).

References

[1] I Giakoumis, N Nikolaidis, I Pitas. Digital image processing techniques for the detection and removal of cracks in digitized paintings. *Department of Informatics Aristotle University of Thessaloniki*

[2] S G Schirripa, F Somma. 2010 Virtual restoration of cracks in digitized image of paintings. *International Conference on Defects in Insulating Materials, Journal of Physics: Conference Series* **249** 012059

[3] A Gupta, V Khandelwal, A Gupta, M C S Thammasat 2008 Image Processing Methods for the Restoration of Digitized Paintings. *Int. J. Sc. Tech.* **Vol.** 13, No.3

[4] N Otsu 1979 A Threshold Selection Method from Gray-Level Histogram. *IEEE Transaction on Systems, Man, and Cybernatics,* **vol.** SMC-9, no.1, pp. 62-66
[5] B Cornelis, T Ruzic, E Gezels, A Dooms, A Pizurica, L Platisa, J Cornelis, M Martens, M De Mey, I Daubechies. 2013 Crack detection and inpainting for virtual restoration of paintings: The case of the Ghent Altarpiece. *Signal process.,* **vol.** 93, no3, pp. 605-619
[6] V Voronin, V Marchuk, R Sizyakin, N Gapon, M Pismenskova, S Tokareva. 2016 Automatic image cracks detection and removal on mobile devices. *Mobile Multimedia/Image Processing, Security, and Applications*
[7] Kingma D and Ba J Adam 2014 A method for stochastic optimization. *CoRR*
[8] Z Guo, L Zhang and D Zhang 2010 A completed modeling of local binary pattern operator for texture classification. *IEEE Transactions on Image Processing*, **vol.** 19, no. 6, pp. 1657–1663
[9] N Dalal and B Triggs. 2005 Histograms of oriented gradients for human detection. **INRIA**
[10] Yimo Guo, Guoying Zhao, and Matti Pietikaeinen. 2011 Texture Classification using a Linear Configuration Model based Descriptor. *Proceedings of the British Machine Vision Conference*, pp. 119.1-119.10
[11] Ojala T, Pietikaeinen M, Maenpaa T 2002 Multiresolution Grayscale and Rotation Invariant Texture Classification with Local Binary Patterns. *IEEE Transactions on pattern analysis and machine intelligence*, **vol.** 24, no. 7

IOP Conf. Series: Materials Science and Engineering **680** (2019) 012042 doi:10.1088/1757-899X/680/1/012042

Modelling of Electronically–Controlled Filters of Microwave (SHF), SubTHz and THz–Bands Based on Graphene Meta–Surfaces

L V Cherckesova[1], P N Bashly[2] and D A Bezuglov[2]

[1] Don State Technical University, Gagarin Square 1, Rostov-on-Don 344000, Russia
[2] Russian customs academy, Budennovsky 20, Rostov-on-Don 344002, Russia

E-mail: bezuglovda@mail.ru

Abstract. Development of generating and receiving radiation methods and means by microwave, sub–Hz and THz ranges is connected with the creation of new high–speed elements of radioelectronics based on new nanomaterials. These includes graphene and nanotubes, in order to develop highly efficient tunable nanophotonics devices for signal processing using new principles, materials and structures. The creation of transmission lines, filters, parametric generators and amplifiers based on multilayer graphene structures is very perspective. However, for today, the physical phenomena and effects in graphene nanostructures are investigated poorly; there are no engineering methods of devices calculation, which is the main difficulty in the way of radioelectronic devices creating of new generation with new element base on the ground of graphene. For investigation of new nanoscale properties and functional possibilities of devices based on nanostructured materials and components – microresonators, transmission lines, antennas, etc. – in the microwave, sub–Hz and THz ranges, a unified approach to their mathematical modelling is required. To overcome the limitations associated with the calculation methods and design, the authors are grounding on the ideas of computational electrodynamics, successfully implemented in computer – aided modelling and design systems in the SHF technique.

1. Introduction

High mobility of charge carriers in graphene makes it the promising and perspective material for using in various applications, as one among the future bases for nanoelectronics and possible replacement of silicon in integrated circuits; it contributes to the creation of radioelectronic devices and arrangements of microwave, sub–Hz and THz–bands. Progress in the production and formation of figured graphene presents great opportunities for the creation of tunable microwave, sub–THz, THz metamaterials and integrated plasmonic devices with potential applications in the filters and polarizers [1].

Using of figure graphene represents enough interesting for overcoming the limitations of existing technologies, in terms of operating frequencies, complexity of electric displacement due to the known effect of graphene controllability of by electric field, as well as integration and nanominiaturization [2].

New directions, perspectives and applications appear in the field of graphene–based metamaterials and their 2D versions, metasurfaces and ultra–thin structures. Graphene–based meta-materials with artificial structure of the resonance elementary cell make it possible to combine functions, such as in microwave devices with control of absorption frequency, modulation, polarization and reflection / absorption at switching (Figure 1) [3].

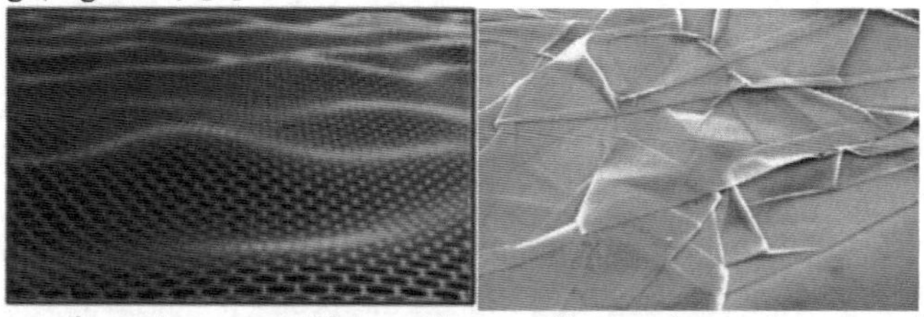

Figure 1. Layers of graphene on silicon substrate (image from electron microscope with magnification of 5000 times).

Research promotion of in the field of microwave, sub–Hz and THz frequencies was carried out both from the side of range millimeter waves and from the side infrared radiation spectrum. In the course of mastering, methodological approaches and technical solutions peculiar to both microwave technology and optics were introduced into research process of the THz range. As a result, both the lower and upper bounds of sub–THz range turned out to be "blurred", and the methods of research and practical development were different for low – frequencies and high – frequencies areas of the THz band. These ranges are the area of electronics and photonics convergence, significantly different as the theoretical base, and the technique of generation, implementation of reception and processing of electromagnetic waves. Development and implementation of microwave, sub–Hz and THz systems for various purposes are directly depending on success in the field of modern electronics and photonics [4].

The obvious advantages of THz range radioelectronics: the absence of ionizing influence effects, the large information capacity, the ability to penetrate through non–transparent objects, the possibility of high directional radiation and some others – have determined to the fast development of THz technology in the whole world. These advantages determine the attractiveness of practical application of THz waves for the creation and construction of high–speed communication lines; high–precision radiolocation stations (radars), capable of operating in the complex electromagnetic environment; systems of image reception with very high resolution; arrangements of chemical substances distance identification and other military and civilian equipment [5].

Modern telecommunication systems are intended for the organization of digital information networks for data, voice, and video transmission. The most prospective are the fiber –optical systems of information transmission, which means that the increase in the efficiency of fiber–optical systems of information transfer is largely determined by the general trends in the development of optical technologies. This is due to the constant increasing of the amount of information transmission via telecommunication systems. Therefore, the increase in the transmission capacity and the speed of information transmission along the fiber–optical communication lines are the main tasks and problems to improving the efficiency of information transmission systems [6].

The speed and transmission capacity of fiber–optical information transmission systems are determined by the capabilities of optical waveguides, the operation speed of the element base of radioelectronics and optoelectronics. The improvement of the element base depends on the level of development of the production technology and on the successes in the creation of new materials and substances.

The operation speed of the radioelectronics element base is determined by the time of electron flight, consequently, its speed and the length of the gate of the transistor. Therefore, the maximal operating frequency of information processing and transmission devices depends on the electronic characteristics of materials and substances and on the geometric dimensions of the microelements. With decreasing in the geometric dimensions, the precision of the element base fabrication should be increased. However, there are technological limits related to parameter spread, power consumption

IOP Conf. Series: Materials Science and Engineering **680** (2019) 012042 doi:10.1088/1757-899X/680/1/012042

and heat generation, as well as physical limits of size (dimension) decreasing. To further improving of the transmission velocity and operation speed, the transition to low–dimensional structures is needed.

Future technologies will require from photonics the devices creating with high degree of integration and energy efficiency, much more advanced the volumetric optical components in the silicon photonics. Such integration can be achieved by combining signal processing functions and waveguide properties at the material level, creating a new concept of meta–structures and meta–devices. It can be assumed that strength (durable) and reliable meta–devices will allow to photonics competing with electronics not only in telecommunication systems, but also at the level of consumer products, such as mobile phones, TVs and cars. The main problems in achieving this prospect are the development of economically efficient production and technologies for meta–devices integration [7].

The aim of the research is to create high–speed elements grounded on graphene in order to develop highly efficient tunable nanophotonic devices for signal processing using new principles, new materials, substances and structures.

The research tasks consist in mathematical modelling of energetic–controlled filters of microwave, sub–THz and THz ranges based on multilayer structures of "graphene – dielectric".

Scientific novelty of research consists in establishing the fact that the studied periodic multilayer microstructures of the “graphene–dielectric” type can be used to create wideband filters of microwave, sub–THz and THz ranges of the planar design, controlled by electric field and fast tunable with small changes in the Fermi’s energy level of graphene.

2. Graphene–based filters and their characteristics

The structure low–frequencies filter of nth order based on graphene consists of graphene band and several polysilicon gate elemental area located below it, as shown in the figure 2.

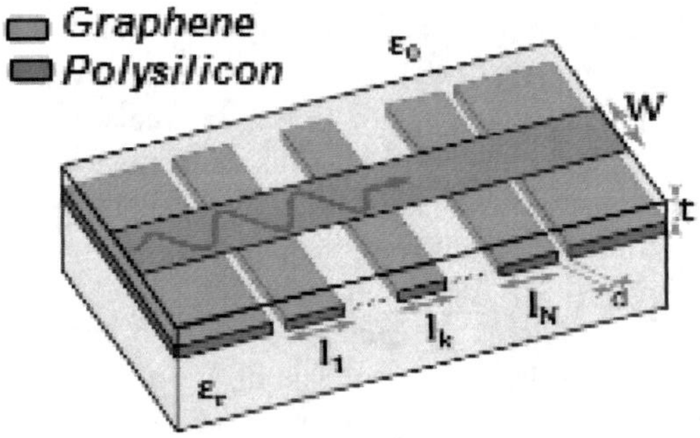

Figure 2. Low – frequencies filter of nth order of microwave range based on graphene [8].

The band promotes the propagation of localized transverse magnetic plasmons; the resistance can dynamically change along the entire length of the structure by applying of different bias voltages (displacement) to the gates. If all the gates are displaced equally, the structure behaves as a simple plasmonic transmission line. When different bias voltages (displacement) are applied to the gate elemental areas, the resistance increases due to the field effect in graphene. This conception is used to the implementation of lower frequencies filters with step change in resistance, which consist of cascade of transmission lines with high and low wave resistance. The filters cut frequency can be dynamically tuned (adjusted) by simultaneously changing of the constant displacement field applied to the gates.

S–parameters of the filter scattering matrix are shown in the figure 3. The filter is intended to obtaining of cut frequency in 2.3 THz. The results were obtained using the procedure of ideal synthesis (Figure 3, a). The restructuring of the filter is occured by changing of the constant (bias) field of displacement on the various gate elemental areas (Figure 3, b).

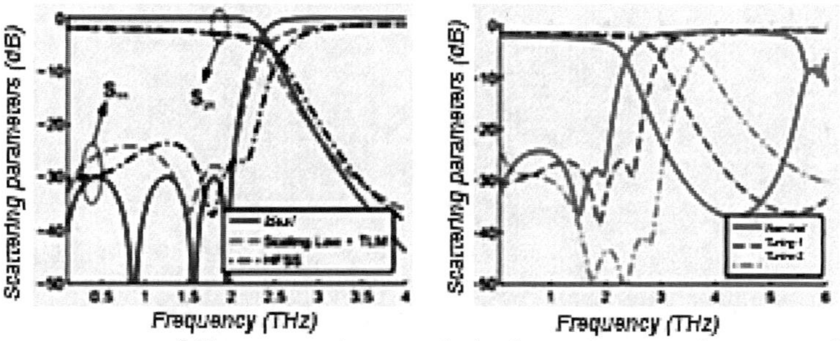

Figure 3. S – parameters of filter scattering matrix in dependence on frequency: W = 150 nm, dielectric constant of substrate $\varepsilon_R = 1.8$ [9].

Rejective (band–stop) plasmonic filters of Sub–THz and THz ranges made using steaks from five layers of coated micro – disks of graphene with high filling coefficient [10].

On the figure 4, a, there is shown the spectrum of THz wave transmission through two grating (lattices) of disks with the same diameter *d* (4.4 mm), but with different periods of grating (lattices 4.8 and 9 mm), made of graphene monolayer.

The single–layer lattice of graphene disks has very high maximal absorption. It also features enough low plasmonic resonant frequency. This is consequence of Coulomb's interactions between disks within the same layer.

The resonance weakens when a smaller grating (lattice) period is used, which is due to interaction of graphene disks in the same layer. In the figure 4, b, is shown the spectra of THz waves passing through filters with different diameters *d*of disks. The grating (lattice) periods a are greater than the diameters d of the disks by 400 nm. At using of such gratings (lattices) of micro–disks, the maximum of relative absorption of 85% is achieved, which makes it possible to create the rejective filter with attenuation (dissipation) coefficient of 8.2 dB.

The maximum attenuation can be achieved by changing the diameters *d* of the micro – disks, as it is shown the in figure 4.

The resonance frequency is modified by changing the diameters of these micro–disks [11].

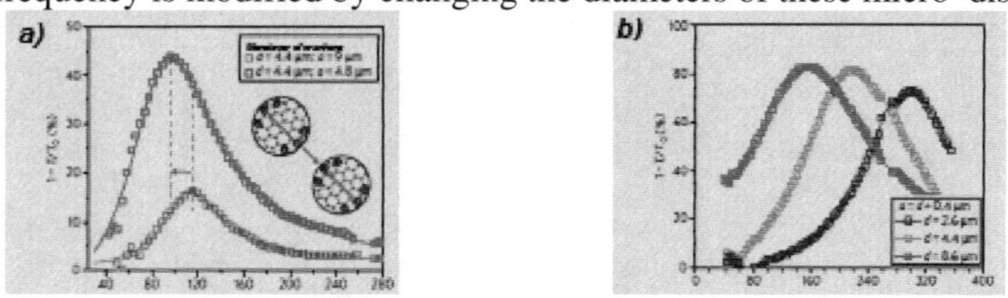

Wave number (cm–1) Wave number (cm–1)

Figure 4. Transparent filters of microwave range [11].

In ultra–narrow graphene tapes, only the surface wave propagates, which can simplify the electromagnetic coupling (connection) between objects. This feature is used for the production of filter construction (design) of two graphene nano– tapes, related with graphene ring cavity (annular) resonator [12].

In the figure 5, the structure of planar band–pass filter based on graphene monolayer with obstructing (blocking) device, in which the different values of the bias voltage (displacement) in the different zones are used, for creation of inhomogeneous conductivity is demonstrated.

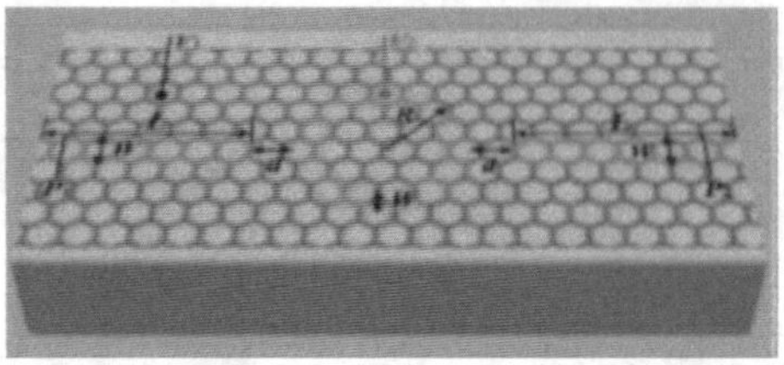

Figure 5. Structure of planar band–pass filter consisting of two graphene waveguides connected by graphene ring resonator of graphene monolayer, where L – length of graphene waveguide; W – width of graphene ring; R – external radius [13].

Plasmon filters based on single– and double– layer doped graphene metasurfaces in microwave, sub–THz and THz ranges were investigated in the work [13] using the finite differences in the temporal domain method (FDTDM).

The cell structure of meta–surface element consists of two parallel symmetrical and asymmetrical structures, fulfilled from the graphene nano–tape.

At the first step the meta–surface is considered, which cell consists of two parallel symmetric graphene nano–tapes (figure 6, a). The geometry of the meta–surface cell of element, which is located in the center of the quartz substrate ($n = 1.5$), is schematically depicted in figure 6, b. Normal incident THz radiation, polarized perpendicularly to the nano–tapes is used for exciting of surface plasmons regime.

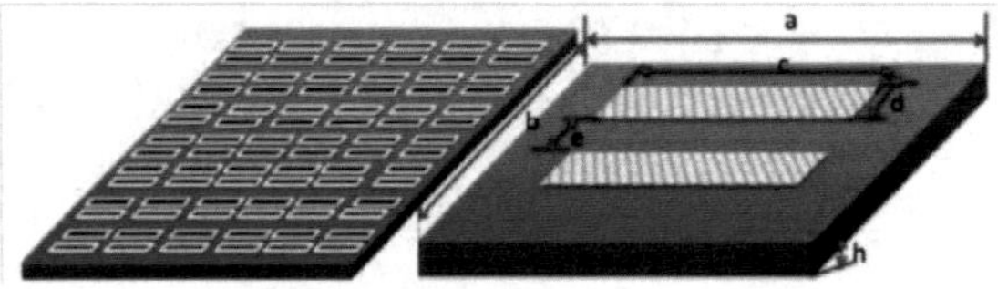

a)

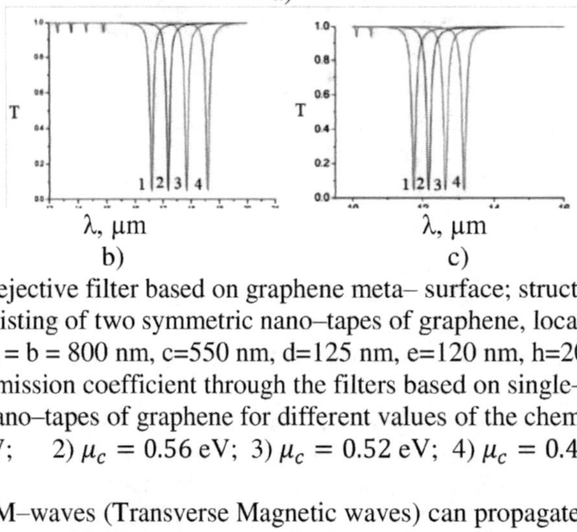

b) c)

Figure 6. Plasmonic rejective filter based on graphene meta– surface; structure of elementary cell of meta–surface, consisting of two symmetric nano–tapes of graphene, located parallelly with geometric dimensions: a = b = 800 nm, c=550 nm, d=125 nm, e=120 nm, h=200 nm; the frequency dependencies of transmission coefficient through the filters based on single– and double–layer structures of symmetric nano–tapes of graphene for different values of the chemical potential. Curves: 1) $\mu_c = 0.6$ eV; 2) $\mu_c = 0.56$ eV; 3) $\mu_c = 0.52$ eV; 4) $\mu_c = 0.48$ eV [14].

Surface plasmons of TM–waves (Transverse Magnetic waves) can propagate perpendicularly to the direction of the nano–tapes with low losses, and plasmonic modes are forming the standing (stationary) wave in the nano–tapes. Periodic boundary conditions in the x and y directions are used for modelling of the experimental investigated sample.

On the figure 6, c, below, the results of modelling – the calculated frequency dependences of the transmission coefficient of the plane wave through single – layer and two – layer structures of symmetric graphene nano – tapes, respectively, for different values of the chemical potential μ_c (Fermi energy) are demonstrated. The resonance frequency of the filter can be controlled. Almost ideal perfect filtering (T_{min}) can be achieved in the tuning range.

In comparison with a single – layer meta – surface, in the structures with double graphene layer it is possible to increase additionally the resonance frequency in the direction of higher frequency area of the THz spectrum.

On the next step the meta – surface is considered, which cell consists of two parallel asymmetrical nano–tapes of graphene (figure 7). The geometry of the meta – surface element cell is schematically shown in figure 7, a; geometric dimensions: g=450 nm, other parameters are the same as in figure 6, b.

On the Figure 7, b, c, there are demonstrated the results of modelling – calculated frequency dependences of transmission coefficient of plane wave through single–layer and two – layer structures from the asymmetric nano – tapes of graphene, respectively, for different values of the chemical potential (for the case of incident of SubTHz / THz radiation polarized along the graphene sheets). The results show that the resonance frequency clearly changes in the direction of higher frequencies with an increase in the chemical potential μ_c (Fermi energy).

Comparing two frequency dependences of transmission coefficient in the figure 7, c, it can be discovered that there are two deep minima in the filter characteristics for two–layer asymmetric structure. Polarization of electrical field of incident THz radiation has influence on its transmission through the long and short graphene nano – tapes.

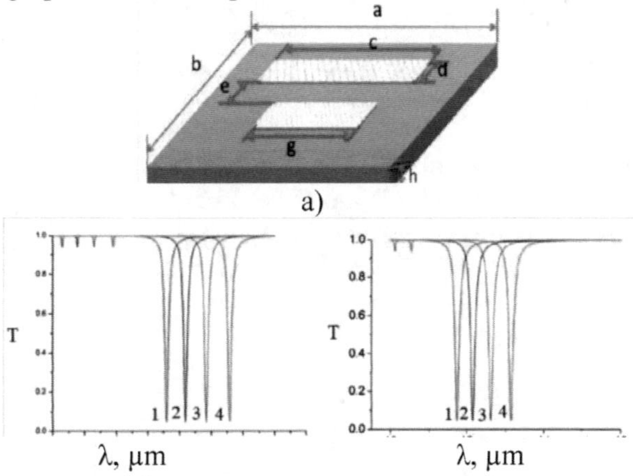

b) c)

Figure 7. The cell structure as element of meta–surface composed of asymmetric graphene nano–tapes of parallel orientation. Frequency dependence of coefficient of transmission (passing) through the filters based on single– and double–layer structures of asymmetric graphene nano–tapes for different values of chemical potential. Curves: 1) $\mu_c = 0.6$ eV; 2) $\mu_c = 0.56$ eV; 3) $\mu_c = 0.52$ eV; 4) $\mu_c = 0.48$ eV [14].

Results of modelling show that significant displacements of resonance frequency can be achieved with small changes in the concentration of graphene nano – tapes doping.

Meta–surfaces with symmetrical single – layer or asymmetrical double–layer graphene can be used as highly sensitive refractive sensors with sensitivity up to 5100 nm and double–circuit switches. These prospects pave the way for the creation of superfast plasmonic devices based on graphene in the microwave, sub–THz and THz ranges.

3. Electronically controlled filters of thz range based on graphene metasurfaces

In [15], graphene metasurfaces for transmission of radiation induced by plasmons in THz range, whose cell consists of the structure based on graphene ring and graphene nano–tape, were investigated. Both graphene rings and graphene nano – tapes create radiation transmission regimes (modes) – transparency spectral windows induced by electric dipole resonances.

Weak hybridization between the two elements leads to appearance of new window of spectral transparency induced by plasmons, which can be controlled by changing the geometric dimensions of

IOP Conf. Series: Materials Science and Engineering **680** (2019) 012042 doi:10.1088/1757-899X/680/1/012042

the graphene – based meta –surface cell structure. The resonance frequency of the transparency spectral window can be dynamically tuned in a wide range of THz frequencies, changing the chemical potential (Fermi energy) of graphene by applying an external electric field (electrostatic sampling, gating) instead of re – producing of the structures [16].

Figure 8 shows the structures of graphene–based meta–surface cells to demonstrate the phenomena of radiation induced by plasmons. The basic elemental cell of the meta–surface consists of graphene ring and a graphene nano–tape. The inner and outer radiuses of graphene ring are $r_1 = 2$ μm and $r_1 = 3.2$ μm. Length on the nano–tape is $L = 9$ μm and width $W = 0.7$ μm. The dielectric substrate is photopolymer with relative dielectric permeability $\varepsilon = 2.4$ and thickness $h = 0.5$μm. The structure is located along the x and y directions with period $p = 10$μm. The incident wave is perpendicular to the $x - y$ plane with polarization E_x.

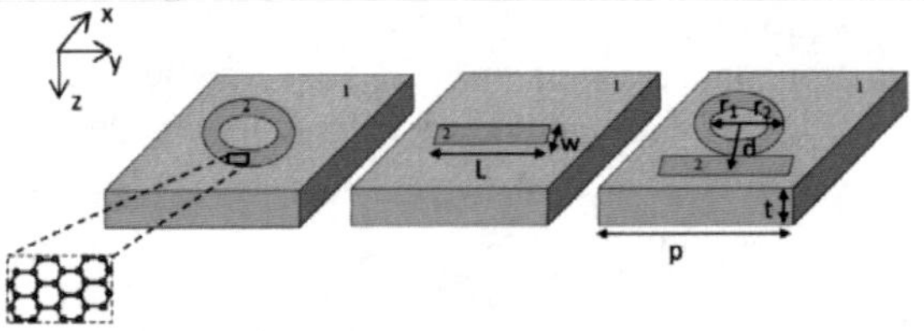

Figure 8. Meta–surfaces cells of different configurations: graphene ring operating in the regime of window spectral transparency for the radiation, induced by plasmons [16].

To demonstrate the effect of radiation induced by plasmons, the frequency dependences of the transmission coefficient of passing through the filters based on graphene meta–surface with the elementary cell were calculated. This is composite structure based on graphene ring and graphene nano–tape (curve 3 in figure 9); single graphene nano–tape (curve 2); and single graphene ring (curve 1); for the value of the chemical potential (Fermi energy) $\mu_c = 0.5$ eV.

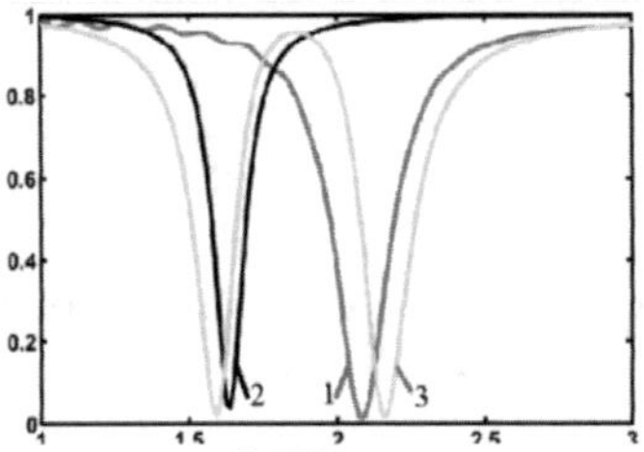

Figure 9. Frequency dependence of transmission coefficient through the filters based on graphene meta–surfaces for various configurations of meta– surface elements. Curves: 1) graphene ring; 2) grapheme nano–tape, 3) structure based on graphene ring and graphene nano–tape [16].

From figure 9 it is obviously, that when at association of graphene rings and graphene nano–tape in the composite structure, oriented along the direction of the electrical field, the window of spectral transparency for THz radiation, induced by plasmons, with transmission coefficient more than 95% is observed at frequencies $f_1 = 1.68$ THz and $f_2 = 2.14$ THz.

To demonstrate the physical mechanism of plasmon– induced radiation effect, on the figure 10 is shown the calculated distribution of electric field for three resonant frequencies of the filter $f_1 = 1.68$THz, $f_2 = 1.86$ THz, $f_3 = 2.14$ THz, which correspond to minimum of transmission coefficient (figure 9).

In figure 10, a, at the first resonance at the frequency $f_1 = 1.68$THz, it is seen that the nano–tape is excited strongly by the incident electric field, and the ring is excited weakly.

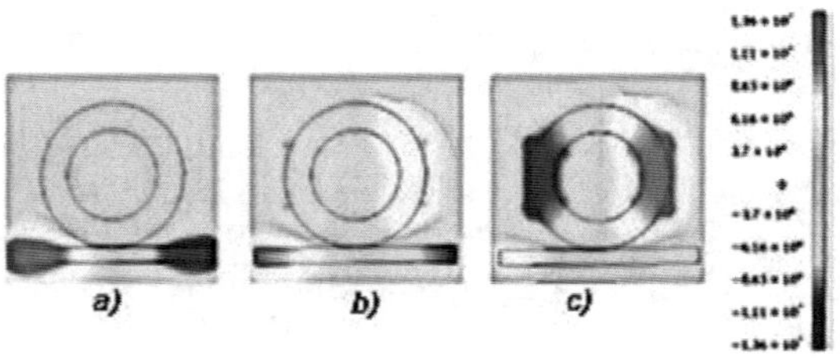

Figure 10. Electric field distribution for frequencies [17]: a) $f_1 = 1.68$THz – first minimum of transmission coefficient; b) $f_3 = 1.86$ THz – second minimum of transmission coefficient; c) $f_2 = 2.14$THz –window of spectral transparency of radiation.

At the second resonance at the frequency $f_2 = 2.14$THz (figure 10, b), the optical response of the ring and nano–tape is mainly concentrated on the ring, which behaves as optical antenna and oscillates with external electric field. Both resonance creates window of spectral transparency for THz radiation induced by electric dipole oscillations. Figure 10, b is shown the electric field at the third resonance frequency $f_3 = 1.86$THz. Quadrupole oscillation appears on the ring and nano–tape, and the phase resonance emerge in two component composite parts: graphene ring and nano–tape ones.

Influence of geometric dimensions on the response of THz radiation, induced by plasmons, was numerically investigated. Result is the fact, that with increasing length L of nano–tape, the first resonance $f_1 = 1.68$THz is rapidly changing towards lower THz frequencies [18].

The second resonance $f_2 = 2.14$THz does not change; thus, the interval between two resonances and maxima transmission coefficient increases. The dependence of resonance frequency on graphene nano–tape length L is plotted. As the length L increasing, the frequency displacement of second resonance keeps invariance and the frequency f_1 of the first resonance changes very quickly. Similarly, at increasing the width W of the graphene nano–tape the first resonance is displaced towards higher THz frequencies, while the second resonance keeps invariance, and windows of spectral transparency for THz radiation, induced by plasmons, is narrowed sharply.

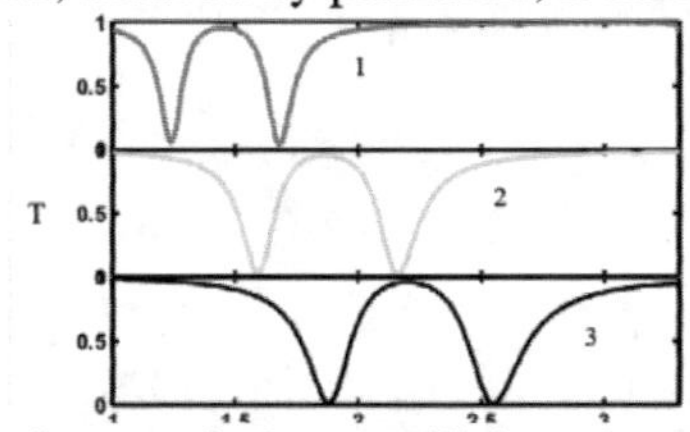

Figure 11. Frequency dependencies of transmission coefficient through the filter at different value of the chemical potential of graphene. Curves: 1) $\mu_c = 0.3$eV; 2) $\mu_c = 0.5$eV; 3) $\mu_c = 0.7$eV .

The surface conductivity of graphene can be changed by controlling its chemical potential (Fermi energy) μ_c. Control can be realized by changing the chemical doping (alloying) or electrostatic gating (sampling). Dynamic control of the windows spectral transparency of the meta–surface is realized without geometry reconstruction or introduction of other controlled materials by changing the chemical potential.

On figure 11, the frequency dependences of transmission coefficient at different values of graphene μ_c are shown. The window of spectral transparency can be tuned effortlessly in a wide range of frequencies by small changing of the chemical potential μ_c (Fermi energy). When μ_c is increasing from 0.3 to 0.7 eV, the resonance frequencies rise too, from 1.47 to 2.25 THz, thanks to what the window of spectral transparency rebuilds completely [19].

4. Electrodynamic calculation of THz range filters characteristics based on multilayer structure of type "graphene – dielectric"

Principles of construction of THz–band filters, controlled by the electric field, are investigated, and the electrodynamic calculation of their characteristics is carried out by the FAB (Autonomous Floquet Blocks) method [20].

The calculation model of sub–THz / THz range planar filter based on the periodic multi–layer structure of type "graphene–dielectric" is demonstrated in the figure 12.

Using the computational algorithm, based on FAB method, the electrodynamic calculation of elements scattering matrix of the filters of sub–THz / THz range, based on periodic multi–layer structures "graphene–dielectric" type in THz range.

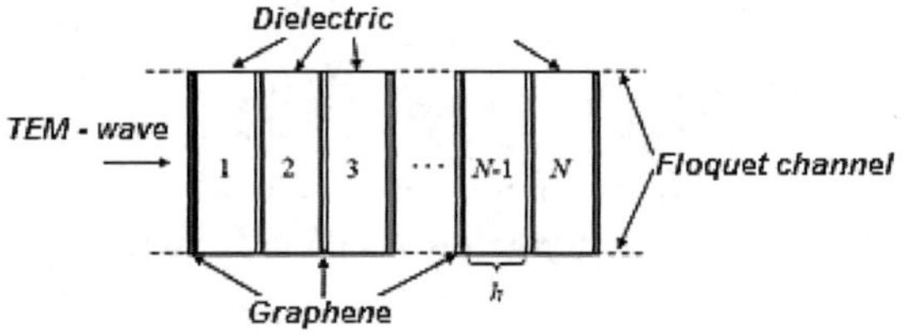

Figure 12. Calculation model of the planar filter THz range based on periodic multi–layer structure of type "graphene–dielectric".

On the figure 13 are shown the calculated frequency characteristics – dependences of the element |S21| scattering matrix – transmission coefficient of TEM–wave (Transverse Electromagnetic wave) through the THz range filter based on multilayer structure "graphene–dielectric" with different number N of graphene sheets (N = 33, 47, 60) for the values of chemical potential $\mu_c = 0$eV, $\mu_c = 1$ eV [21].

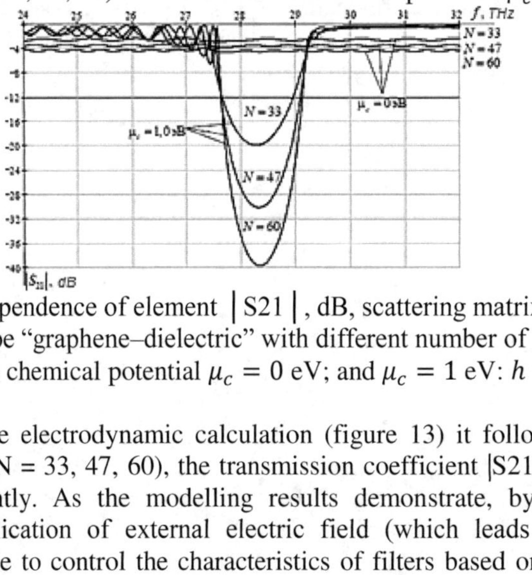

Figure 13. Frequency dependence of element | S21 |, dB, scattering matrix of the filters based on multilayer structures of type "graphene–dielectric" with different number of graphene sheets (N = 33, 47, 60) for values of the chemical potential $\mu_c = 0$ eV; and $\mu_c = 1$ eV: $h = 3.65$ μm; $\varepsilon_d = 2.2$.

From the results of the electrodynamic calculation (figure 13) it follows that with increase of graphene sheets number (N = 33, 47, 60), the transmission coefficient |S21| in the non–transmission band decreases significantly. As the modelling results demonstrate, by changing the chemical potential μc by the application of external electric field (which leads to change of graphene conductance), it is possible to control the characteristics of filters based on multilayer structures of type "graphene–dielectric".

So, at chemical potential value $\mu_c = 0$ eV (if external electric field absents), THz radiation attenuation in the non-transmission band is 2 – 4.5 dB (depending on the number N = 33 – 60, respectively). At application of external electric field, which corresponds to chemical potential $\mu_c = 1$ eV, the attenuation of radiation increases to 20 – 40 dB (with change in the number N of graphene sheets N = 33 – 60).

Thus, from the results of modelling it follows that the investigated periodic layered microstructure of type "graphene–dielectric" can be used to create the broadband filters of sub–THz and THz band of planar construction, controlled by electric field and quickly tunable with small changes of the level of graphene Fermi energy [21].

5. Conclusions

Filters based on multilayer structures of type “graphene– dielectric” are controlled by changing the chemical potential by electric field displacement. In the absence of electric field, the filter transmits the corresponding radiation, in the presence of electric field– the attenuation of microwave, sub – THz or THz radiation can reach –40 dB and above, depending on the number of graphene layers.

Model of the filter based on the multilayer structure of type "graphene – dielectric" is constructed. It is shown that such filter has stability and resistance in relation to the spread of the values of the surface conductance of graphene sheets included in the investigated filter.

The researched structures can be apply for creating of controlled wideband filters of microwave, sub–THz and THz ranges of frequency of planar design and construction. Rejection filters built on the base of multilayer structures of the "graphene–dielectric" type are particularly promising.

As modelling results show, changing the chemical potential by means of external electric field, which leads to change in the graphene conductivity, it is possible to control the transmission coefficient of radiation of the corresponding range through filters based on multilayer structures of the "graphene–dielectric" type.

References

[1] Chen P Y Soric J Padooru Y R Bernety H M Yakovlev A Al`u A 2013 Nanostructured graphene meta–surface for tunable terahertz cloaking *New Journal of Physics* **15** 123029

[2] Fallahi A et al 2012 Design of tunable biperiodic graphene metasurfaces *J Physical Review* **86** 195408

[3] Zhang Y Feng Y Zhu B Zhao J Jiang T 2014 Graphene based tunable meta–material absorber and polarization modulation in terahertz frequency *J Optics Express* **22** 19 pp 22743–22752

[4] Chernozatonskiy L A Sorokin P B Artyukh A A 2014 New Graphene–Based Nanostructures: Physico–Chemical Properties and Applications *Advances in chemistry* pp 251–279 (In Russian)

[5] Isaev V M Kabanov V V Komarov V P 2014 Modern Radioelectronic Systems of THz Range *Reports of TUSUR* **4 (34)** pp 5 – 21 (In Russian)

[6] Kivshar Yu S 2015 From meta–materials to meta–surfaces and meta–devices *Nanosystems: Physics, Chemistry, Mathematics* **6** 3 pp 346–352

[7] Shi S F Zeng B Han H L Hong X Tsai H Z Jung H S Zettl A Crommie M F Wang F 2014 Optimizing broadband terahertz modulation with hybrid graphene /meta–surface structures *Nano Letters* **15** pp 372–377

[8] Innocenti R D Jessop D S Shah Y D Sibik J Zeitler J A Kidambi P R Hofmann S Beere H E Ritchie D A 2014 Terahertz optical modulator based on meta–material split ring resonators and graphene *J Optical Engineering* **57** 108 pp 1–5

[9] Golovanov O A 1991 Diffraction on the Nonlinear Dielectric in the System of Connected Strip *J Lines Radio Engineering* **7** pp 65–70. (In Russian)

[10] Horng J Chen C F Geng B Girit C Zhang Y Hao Z Bechtel H A Martin M Zettl A Crommie M F Shen Y R Wang F 2010 *J Physical Review* **83** 165113

[11] Yan H Li X. Chandra B Tulevski G Wu Y, Freitag M Zhu W Avouris P Xia F 2012 Tunable infrared plasmonic devices using graphene / insulator stacks *J Nature Nanotechnology* **7** 5 pp 330–334

[12] Khromova I Andryieuski A Lavrinenko A 2014 Multilayer graphene for waveguide terahertz modulator *Meta Conference* p 3

[13] Li H J WangL L Zhang H Huang Z R Sun B Zhai X Wen S Ch 2014 Graphene–based mid–infrared, tunable, electrically controlled plasmonic filter *J Applied Physics Express* **7 (2)**

024301–1–024301–4
[14] Zhongchao W Xianping L Jianjun Y Rong H Yuebo L Wei W Hongzhan L Hongyun M 2016 Active plasmonic band–stop filters based on graphene meta-material at THz wavelengths *J Optics Express* **24** **(13)** pp 14344 –143451
[15] Zhang H Cao Y Liu Y Li Y Zhang Y etc 2017 A novel graphene meta–material design for tunable terahertz plasmon induced transparency by two bright mode coupling *J Optics Communications* **391** pp 9–15
[16] Wang J etc 2016 Recent advances in graphene–assisted nonlinear optical signal processing *Journal of Nanotechnology* **2016** pp 275–292
[17] Wu Y Yao B C Feng Q Y Cao X L etc 2015 Generation of cascaded four–wave–mixing with graphene– coated microfiber *J Photonics Research* **3** 2 pp 64–68
[18] Ji M Cai H Deng L Huang Y Huang Q Xia J Li Z Yu J Wang Y 2015 Enhanced parametric frequency conversion in compact silicon– graphene micro–ring resonator *J Optics Express* **123** 14 pp 18679–18685
[19] Sun Y Qiao G Sun G 2014 Direct generation of graphene plasmonic polaritons at THz frequencies via four wave mixing in the hybrid graphene sheets waveguides *J Optics Express* **22** 23 pp 22880–27891
[20] Golovanov O A 2006 Autonomous Blocks with Virtual Floquet Channels and Their Application for Solving Applied Problems of Electrodynamics *J Radio Engineering and Electronics* **51** 12 pp 1423 – 1430 (In Russian)
[21] Golovanov O A Makeeva G S Varenitsa V V Electrodynamic calculation of transmission coefficients of the TEM–wave through the multilayer periodic structures monolayer of type "graphene–dielectric" in THz range *J Izvestiya vuzov 4 (32)* pp 123 – 137 (In Russian)

IOP Conf. Series: Materials Science and Engineering **680** (2019) 012043 doi:10.1088/1757-899X/680/1/012043

Multi-View Images Fusion Model

M M Zhdanova[1,2], V V Voronin[1,2], R A Sizyakin[1], M S Minkin[1] and A A Zelensky[2]

[1]Lab. "Mathematical methods of image processing and intelligent computer vision systems" Don State Technical University, Gagarin sq. 1, Rostov-on-Don 344000, Russia
[2]Moscow State University of Technology "STANKIN", Vadkovsky line 1, Moscow 127055, Russia

E-mail: mpismenskova@mail.ru

Abstract. The tasks of recognition actions and classification objects are fundamental in computer vision systems. Even subtasks, such as recognition of atomic motion and single objects form the basis for understanding the situation in the work area and the scene in general. This is especially important in video surveillance systems designed to ensure security. Thus, the effectiveness of recognition and classification methods is one of the primary tasks of computer vision. But the visual methods implemented in similar video surveillance systems, encounter some difficulties, such as inhomogeneous background, uncontrolled operating environments, irregular illumination, etc. To address these drawbacks, the paper presents a model for combining visible range images and depth images. This model allows to improve the quality of recognized images, provides the construction of a more informative descriptor, which also positively affects the recognition efficiency. Our results show that it has good performance in fusion visible image and depth map.

1. Introduction

Through the high-tech development of modern society, computer vision is implemented in such systems as security systems, monitoring and control systems, authentication systems, automation of various production processes, quality control, and inspection systems, etc. Recognition of objects and classification actions are fundamental in solving almost all of the above problems. Automated systems for recognizing objects and actions encounter a number of difficulties when implemented in real conditions, such as a different background, uncontrolled operating environments, irregular illumination, etc. In this regard, the topical task is to combine information from cameras in the visible range and depth sensors to obtain informative signs of objects located in the frame.

In a number of previously published papers [1–4], it has been shown that, the fusion of vision and depth cameras or infrared sensors data improves the accuracy of recognition. The focus is on depth cameras and inertial sensors as these two types of sensors are cost-effective, commercially available, and more significantly they both provide 3D human action data and scene structure.

In paper [1] presented thermal infrared and visible data fused for face recognition. Thermal face recognition technology has achieved high resistance to changes in lighting, and therefore, scientific and business circles are studying its application [4]. An infrared thermal image of a face was first introduced as a reliable face signature in [5], because the infrared "appearance" consists of a temperature picture of the face and branches of a blood vein; therefore, it is considered a reliable and distinguishable physiological biometric feature [1].

In [2, 3, 6] presented techniques for human action recognition based on depth sequences. It was shown that fusing depth and vision information leads to more robust recognition.

In paper [7] presented a new method for object recognition from RGB-D data, In particular, authors focused on making recognition robust to imperfect sensor data. The proposed architecture consists of two convolutional network streams operating on color and depth information. The network automatically learns to fusion these two streams information.

As literature shows, Infrared images provide a picture of the heat of the human body or scene affected by ambient temperature, airflow conditions, exercise, illness, and more [2]. In its turn, depth sensor technology an exhibit less sensitivity to lighting changes and background clutter.

In the paper presents a model for combining visible range and depth images

2. Fusion model

The most notable benefit of the joint use of depth and visible sensors is the complementary nature of different modalities that provide the depth and visible-light information of the scene. And the complementarity of information captured by the different modalities can increase the reliability and robustness of recognition or surveillance [8]. As a prototype of our model in this paper, a new image aware HVS based multi-view images fusion technique that makes use of the Parameterized Logarithmic Image Processing (PLIP) model is proposed [1]. The PLIP model [9], a generalized version of LIP model, views images in terms of their gray-tone functions and processes them using new arithmetic operators replacing the standard arithmetical operators. The gray-tone function Ψ, for a given image, is generated using:

$$\Psi = \alpha - 1, \tag{1}$$

where α is a model parameter. Value α is the maximum intensity value in the image I.

The fused image can be obtained using the following,

$$fusedImage = \widehat{\Psi}_1 \tilde{\oplus}\, \widehat{\Psi}_2 \,\tilde{\oplus}\, \widehat{\Psi}_3, \tag{2}$$

where Ψ_1 is the visible image; Ψ_2 is the depth image; Ψ_3 is the maximum-combined image obtained by selecting the maximum intensity value of depth and visible image at each pixel location; $\tilde{\oplus}$ is the PLIP addition operator and $\widehat{\Psi}_i$ is the result of PLIP scalar multiplication as defined below.

$$\Psi_1 \tilde{\oplus} \Psi_2 = \Psi_1 + \Psi_2 - \left(\frac{\Psi_1 \Psi_2}{\max(\Psi_1, \Psi_2)}\right), \tag{3}$$

$$\widehat{\Psi}_i = \left(\Omega_i \tilde{\otimes} \Psi_i\right) = \max(\Psi_i) - \max(\Psi_i)\left(1 - \frac{\Psi_i}{\max(\Psi_i)}\right)^{\Omega_i}. \tag{4}$$

The constants Ω_i were chosen as represented in [1] that $\sum \Omega_i = 1$, $\Omega_1 = 0.2989$, $\Omega_2 = 0.5870$ and $\Omega_3 = 0.1141$.

3. Results and discussion

NYU-Depth V2 dataset [10] has been used for experimental protocol design and testing. This dataset consists of 1449 pairs of RGB and depth images.

The results of the depth and visible image fusion using the prosed fusion method are shown in figure 1-5.

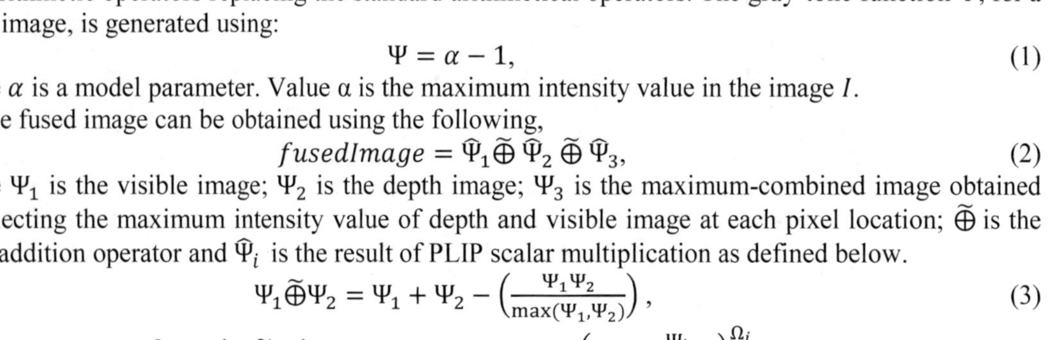

(a) (b) (c)

Figure 1. Depth-visible image fusion. (a) visible image; (b) depth image; and (c) fused image

IOP Conf. Series: Materials Science and Engineering **680** (2019) 012043 doi:10.1088/1757-899X/680/1/012043

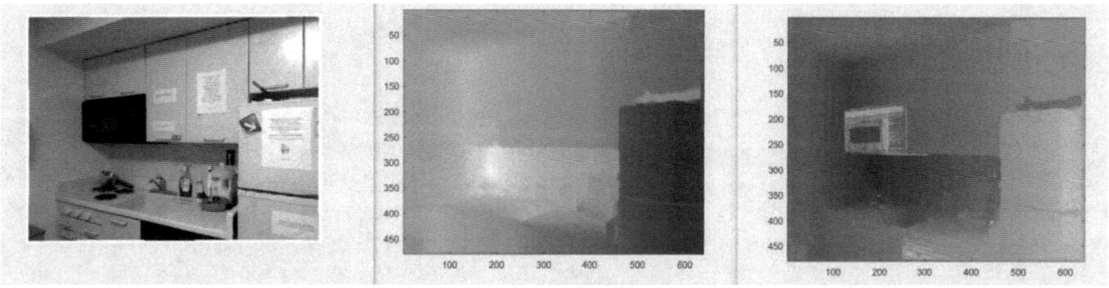

(a) (b) (c)

Figure 2. Depth-visible image fusion. (a) visible image; (b) depth image; and (c) fused image

(a) (b) (c)

Figure 3.. Depth-visible image fusion. (a) visible image; (b) depth image; and (c) fused image

(a) (b) (c)

Figure 4. Depth-visible image fusion. (a) visible image; (b) depth image; and (c) fused image

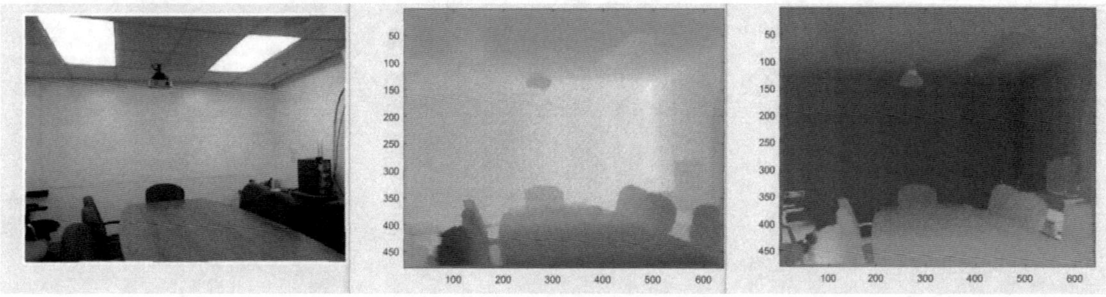

(a) (b) (c)

Figure 5. Depth-visible image fusion. (a) visible image; (b) depth image; and (c) fused image

Successful image fusion is an important step for the increase in the accuracy of the recognition system. The fused image contains the characteristics of both visible and the depth images combining the advantages of both the modalities. This model is planned to use in further studies for object recognition and classification of human actions.

4. Conclusions

It has been proposed the model for combining visible range images and depth images. This model allows to improve the quality of recognized images, provides the construction of a more informative descriptor, which also positively affects the recognition efficiency. The fusion approach tested on several databases and showed good results using data in visible range and depth maps.

5. Acknowledgments

This work was supported by Russian Ministry of Education and Science in accordance to the Government Decree № 218 from April 9, 2010 (project number № 074-11-2018-013 from May 31, 2018 (03.G25.31.0284)).

References

[1] Wan Q Rao S P Kaszowska A Voronin V Panetta K Taylor H A Agaian S 2018 Face description using anisotropic gradient: thermal infrared to visible face recognition *Mobile Multimedia/Image Processing, Security, and Applications* 10668

[2] Chen C Jafari R Kehtarnavaz N 2014 Improving human action recognition using fusion of depth camera and inertial sensors *IEEE Transactions on Human-Machine Systems* **45 (1)** pp 51-61

[3] Chen C Jafari R Kehtarnavaz N 2015 A real-time human action recognition system using depth and inertial sensor fusion *IEEE Sensors Journal* **16 (3)** pp773-781

[4] Mostafa E Hammoud R Ali A et al 2013 Face recognition in low resolution thermal images *Computer Vision and Image Understanding* **117(12)** pp 1689-1694

[5] Prokoski F J Riedel R B and Coffin J S 1992 Identification of individuals by means of facial thermography pp 120-125

[6] Yang X Zhang C Tian Y L 2012 Recognizing actions using depth motion maps-based histograms of oriented gradients *Proceedings of the 20th ACM international conference on Multimedia* pp 1057-1060

[7] Eitel A et al 2015 Multimodal deep learning for robust RGB-D object recognition *IEEE/RSJ International Conference on Intelligent Robots and Systems (IROS)* pp 681-687

[8] Serrano-Cuerda J Fernández-Caballero A and López M T 2014 Selection of a visible-light vs. thermal infrared sensor in dynamic environments based on confidence measures *J Applied Sciences* **4(3)** pp 331-350

[9] Panetta K Wharton E and Agaian S 2007 Parameterization of logarithmic image processing models *IEEE Tran. Systems, Man, and Cybernetics, Part A: Systems and Humans*

[10] Silberman N et al 2012 Indoor segmentation and support inference from rgbd images. *European Conference on Computer Vision* pp 746-760

IOP Conf. Series: Materials Science and Engineering **680** (2019) 012044 doi:10.1088/1757-899X/680/1/012044

Mathematical Modeling of the Image Processing in Television Systems

D A Bezuglov[1] and V V Voronin[2]

[1] Russian customs academy, Budennovsky 20, Rostov-on-Don 344002, Russia
[2] Don State Technical University, Gagarin Square 1, Rostov-on-Don 344000, Russia

E-mail: bezuglovda@mail.ru

Abstract. Modern vision systems include a television image processing system. Improving the technical characteristics and expanding the capabilities of modern devices for recording television images makes it necessary to develop and create highly efficient algorithms for processing television images. Such systems process large amounts of information, therefore, the requirements for the reliability of processing large amounts of data at limited time intervals are constantly increasing. The rapid development of high-performance computing systems has recently created the conditions for expanding the range of tasks that can be solved with the help of vision systems. The research for improving the efficiency of methods and algorithms for digital image processing significant. Currently, the processing of television images is divided into several steps. The first step is the formation of a digital representation of the image (discretization, quantization and input of the image into the computer memory). The second step is the pre-processing of images (restoration and filtering). The third step is the formation of a graphic preparation of the image (segmentation and contour detection). One of the most complex and relevant is the scientific and technical task of the contour detections of objects in television images on the background with additive noise. The purpose of this study is the development of the contour detection method on the image with impulse noise using mathematical modeling.

1. Introduction

Modern complexes of technical vision of autonomous robots include television image processing systems. Improving the technical characteristics and expanding the capabilities of modern devices for recording television images makes it necessary to develop and create highly efficient algorithms for processing television images. Such systems process large flows of information, and the requirements for the reliability of processing large amounts of data at limited time intervals are constantly increasing. The fast development of high-performance computing systems in recent years has created the conditions for expanding the range of tasks solved with the help of vision systems. This indicates the relevance of research in the field of improving the efficiency of methods and algorithms for digital image processing [1-4].

Currently, television image processing is divided into stages. The first stage is the formation of a digital representation of the image: discretization, quantization, and save of the image into the computer memory. The second stage is the pre-processing of images: recovery and filtering. The third stage is the formation of a graphic preparation of the image: segmentation and contouring. In the future, analysis, classification, image recognition and the using of the information obtained in modern systems of technical vision are carried out [5-10].

One of the most complex and relevant is the scientific and technical task of contour extraction of objects in television images on the background of additive interference.

Objective: to study the efficiency of contours extraction of the vision system under the influence of impulse noise by reducing the standard deviation (RMS) of the error using mathematical modeling methods.

2. Methods

Mathematical modeling - is the simulate physical process with a system of mathematical relationships, the solution of which allows getting an answer to the question about the behavior of an object without creating a physical model. But often it's expensive and inefficient.

Mathematical modeling is a way of studying a real system by replacing its a mathematical model that is more convenient for experimental research using a computer. The mathematical model is an approximate representation of real objects, processes or systems, expressed in mathematical terms and preserving the essential features of the original. In our case, we will consider a digital representation of a grayscale image as a mathematical model.

The expression for the parabolic B-spline is written as follows [16-19]:

$$F_c(x) = \sum_{i=0}^{N+1} b_i B_3^i(\chi_{i3}), \quad \chi_{i3} = \frac{1}{h}\left(x - x_{i+\frac{n+1}{2}}\right) \tag{1}$$

where b_i - spline coefficients; $x_{i+\frac{n+1}{2}}$- coordinates of the middle of the B-spline carrier; n - degree of spline; h – grid pitch.

We get the expression for the spline approximation of the derivative of the one-dimensional signal:

$$F_{ca}(x) = \sum_{i=1}^{N-1}\left(\frac{\chi^2}{2h}(b_{i+2} - 3b_{i+1} + 3b_i - b_{i-1}) + \frac{\chi}{h}(b_{i+1} - 2b_i + b_{i-1}) + \frac{1}{2h}(b_{i+1} - b_{i-1})\right) \tag{2}$$

Let a matrix *S (i, j)* of a black-and-white image of size *N*×*N* be given. To solve the problem of contours extraction, we divide the image matrix into rows and columns, differentiate by the proposed method, and then calculate the operator for selecting contours. In contrast to the known approaches [1-4], the differentiation will take into account information about the intensity in the entire line of the image.

The expression for the intensity gradient *G(S(i,j))* can be written as follows:

$$G(S(i,j)) = \left[\begin{array}{l}\sum_{i=1}^{N-1}\left(\frac{\chi^2}{2h}(b_{i+2}^n - 3b_{i+1}^n + 3b_i^n - b_{i-1}^n) + \frac{\chi}{h}(b_{i+1}^n - 2b_i^n + b_{i-1}^n) + \frac{1}{2h}(b_{i+1}^n - b_{i-1}^n)\right)^2 + \\ + \sum_{j=1}^{N-1}\left(\frac{\Upsilon^2}{2h}(b_{j+2}^m - 3b_{j+1}^m + 3b_j^m - b_{j-1}^m) + \frac{\Upsilon}{h}(b_{j+1}^m - 2b_j^m + b_{j-1}^m) + \frac{1}{2h}(b_{j+1}^m - b_{j-1}^m)\right)^2\end{array}\right]^{\frac{1}{2}} \tag{3}$$

Thus, the proposed new noise-resistant method of gradient selection of the contours of objects in digital halftone television images, involves the following operations [16-19]:

1. Calculation of coefficients of spline approximation by normalized smoothing one-dimensional cubic B-splines for all rows and columns of digital grayscale image S.
2. Calculation of two matrices of components of the brightness gradient at each point in the image by the inverse transform. Where analytical functions describing the derivatives of the inverse transform spline functions used in the corresponding coordinates are used as the transformation kernel.
3. The calculation of the module of the brightness gradient at each point in the image by calculating the square root of the sum of squares and the components of the brightness gradient of the image is an expression (3).
4. Performing threshold conversion of the brightness gradient module, during which the contours of objects are formed by highlighting the elements in black on a new white matrix. The gradient modulus of which in the corresponding coordinates exceeds the transformation threshold.

To obtain quantitative estimates of the proposed algorithm, we will use mathematical modeling methods. Afterward, 2 types of noise were superimposed on the original image S: impulse noise “broken pixels”; impulse noise "salt and pepper". Moreover, in the course of mathematical modeling, the influence of the spline smoothing coefficient on the accuracy characteristics was studied.

As quantitative criteria for the accuracy characteristics of the contouring algorithm, we use the following:

1. The standard deviation e_{RMS}:

$$e_{RMS} = \sqrt{\frac{1}{MN}\sum_{i=0}^{N-1}\sum_{j=0}^{M-1}\left(MK_{ij} - \widehat{MK}_{ij}\right)^2} \quad (4)$$

where: *MK* - test image, *MK* - image obtained by the proposed method.

2. Peak SNR:

$$SNR = \frac{255-\mu}{\sqrt{\frac{1}{MN}\sum_{i=0}^{N-1}\sum_{j=0}^{M-1}\left(\hat{MK}_{ij}-\mu\right)^2}},\ \mu = \frac{1}{MN}\sum_{i=0}^{N-1}\sum_{j=0}^{M-1}\left(\hat{MK}_{ij}\right)^2 \quad (5)$$

where: μ is the average value $\hat{MK}_{ij}$.

3. Peak SNRF (using background RMS calculations):

$$SNRF = \frac{255-\mu}{\sigma_{\text{background}}},$$

$$\sigma_{background} = \sqrt{\frac{1}{\left(N_{background}\right)^2}\sum_{i=n_1}^{n_1+N_{background}}\sum_{j=m_1}^{m_1+N_{background}}\left(\hat{MK}_{ij} - \mu_{background}\right)^2}$$

$$\mu_{\text{background}} = \frac{1}{\left(N_{\text{background}}\right)^2}\sum_{i=n_1}^{n_1+N_{\text{background}}}\sum_{j=m_1}^{m_1+N_{\text{background}}}\left(\hat{MK}_{ij}\right) \quad (6)$$

where $\sigma_{background}$ - standard deviation of the background; $\mu_{background}$ - is the average value of the background; n_1, m_1 - coordinates of the selected background area with the size $N_{background} \times N_{background}$ in the image under study $\hat{MK}_{ij}$.

3. Experiments and Results

The essence of the computational experiment in mathematical modeling was as follows. The original *Lena* image shown in Figure 1a was processed by the Canny border detector from the Mathcad 14 package (Figure 1b). In the future, all received images will be compared with this image *MK*.

IOP Conf. Series: Materials Science and Engineering **680** (2019) 012044 doi:10.1088/1757-899X/680/1/012044

(a) (b)

Figure 1. Test image: a) the original image; b) the result of processing the test image with a Canny border detector.

Then the test image was exposed to 2 types of noise: impulse noise “broken pixels” (Figure 2a); pulse noise “salt-pepper” (Figure 2b).

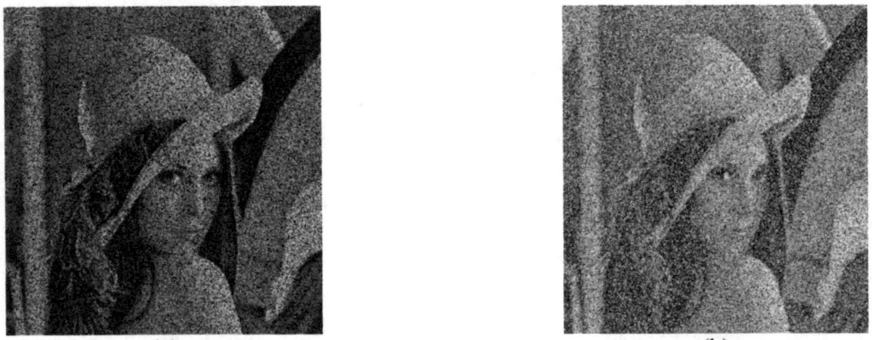

(a) (b)

Figure 2. Test image exposed to pulsed noise: a) “broken pixels”; b) "salt and pepper".

Figure 3 shows the results of processing by the Sobel operator of the images presented in Figure 2, respectively.

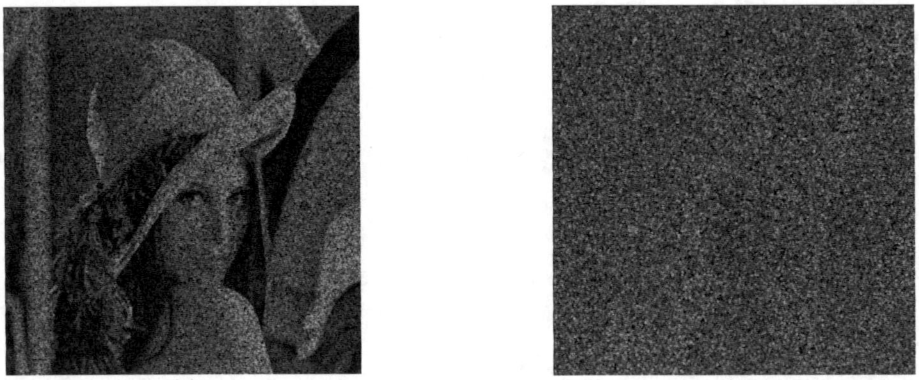

(a) (b)

Figure 3. Results of processing by a Sobel operator of a test image exposed to noise: a) “broken pixels”; b) "salt and pepper".

Figure 4 shows the results of processing by the spline differentiation method for the images shown in Figure 2, respectively.

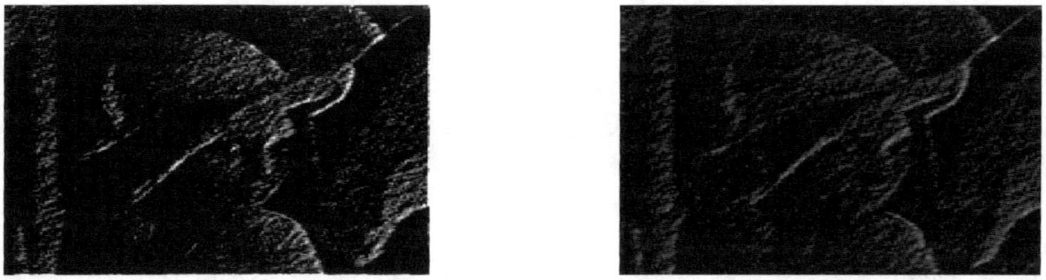

(a) (b)

Figure 4. Results of processing by spline differentiation of a test image exposed to impulse noise: a) "broken pixels"; b) "salt and pepper".

The results of mathematical modeling of the proposed method for selecting contours are given in tables 1-2.

Table 1. e_{RMS}, SNR and SNRF, broken pixel noise.

Criterion 1, RMS e_{RMS} gain, in dB.								
Noise probability	Spline function smoothing factor							
	1	5	10	50	100	200	300	500
0,3	0,82	1,52	1,63	1,73	1,73	1,72	1,71	1,71
0,5	0,95	1,75	1,90	2,03	2,04	2,05	2,04	2,04
Criterion 2, SNR gain, in dB.								
Noise probability	Spline function smoothing factor							
	1	5	10	50	100	200	300	500
0,3	1,96	4,27	4,97	6,16	6,46	6,68	6,81	6,92
0,5	2,19	4,66	5,48	6,75	7,11	7,43	7,57	7,69
Criterion 3, SNRF gain, in dB.								
Noise	Spline function smoothing factor							
probability	1	5	10	50	100	200	300	500
0,1	3,13	5,97	6,80	8,75	9,62	10,48	10,96	11,52
0,3	3,08	5,72	6,68	8,72	9,53	10,33	10,83	11,43
0,5	3,43	6,07	7,05	8,87	9,60	10,38	10,81	11,35

Table 2. e_{RMS}, SNR and SNRF, noise "salt and pepper".

Criterion 1, RMS e_{RMS} gain, in dB.								
Noise	Spline function smoothing factor							
probability	1	5	10	50	100	200	300	500
0,3	0,99	1,48	1,54	1,55	1,55	1,54	1,52	1,51
0,5	1,44	2,26	2,41	2,54	2,56	2,56	2,56	2,55
Criterion 2, SNR gain, in dB.								
Noise	Spline function smoothing factor							
probability	1	5	10	50	100	200	300	500
0,3	2,73	5,09	5,65	6,45	6,61	6,76	6,82	6,86
0,5	3,02	5,47	6,19	7,09	7,37	7,53	7,61	7,67
Criterion 3, SNRF gain, in dB.								
Noise	Spline function smoothing factor							
probability	1	5	10	50	100	200	300	500
0,3	3,40	6,16	7,02	8,95	9,76	10,64	11,12	11,70
0,5	3,36	6,09	7,06	8,88	9,72	10,49	10,95	11,49

IOP Conf. Series: Materials Science and Engineering **680** (2019) 012044 doi:10.1088/1757-899X/680/1/012044

4. Conclusions

The conducted experimental studies using special test images against the background of impulse noise showed the efficiency and high efficiency of the developed method for isolating the vision system under the influence of impulse noise by reducing the standard deviation (RMS) of the error using mathematical modeling methods.

5. Acknowledgments

This work was supported by Russian Ministry of Education and Science in accordance to the decree of the Government of April 9, 2010, №218 (project number 03.G25.31.0284).

References

[1] Soifer V A Gashnikov M V Glumov N I 2003 Methods of computer image processing (Moscow: Fizmatlit) p 784

[2] Pratt W K 1991 Digital Image Processing (New-York NY: John Wiley and Sons)

[3] Gonzalez R Woods R 2005 Digital image processing (Moscow: Tekhnosfera) (in Russian)

[4] Zavyalov Yu S Kvasov B I Miroshnichenko V L 1980 Methods of spline functions (Moscow: Nauka 1980) p 350

[5] Bezuglov D Kuzin A Voronin V 2016 Contour detection based on wavelet differentiation. In Mobile Multimedia *International Conf Image Processing, Security, and Applications* Vol 9869 98690

IOP Conf. Series: Materials Science and Engineering **680** (2019) 012045 doi:10.1088/1757-899X/680/1/012045

Solving the problem transform of size images on series base points found on frames in the problem of stitching images

E Semenishchev[1], I Shraifel[1], V Voronin[1], A Nadykto[2], V Marchuk[1]

[1]Don State Technical University, sq. Gagarin 1 Rostov-on-Don, 344000, Russia
[2]Moscow State University of Technology “STANKIN”,ln. Vadkovsky 1 Moscow, 127055 Russia

E-mail: sea.sea@mail.ru

Abstract. Stitching several images into a single content is an important task for many areas, including video analytics, machine vision, security, etc. The operation of combining is based on the solution of a number of operations related to the allocation of local features, the selection of transformation matrices of objects, changing the scale and saturation of the image. The process fixing frames is he's with a number of difficulties, includes the appearance of a noise component in the image, blurring, various non-linear effects of the optical system, as well as problems associated with incorrect operation of data preprocessing software. To select the key points used as alignment points is a complex problem, the solution of which is the subject of many works. However, the transformation of coordinate systems of images with combination frames is a difficult problem, since the system is nonlinear.

The article discusses the solution to the problem of changing the position of points in clusters, under conditions of their nonlinearity. The method and algorithm that implements the procedure of nonlinear transformations size over the entire image is presented. Operations include vector analysis, as well as the results of changing the scale of a group of pixels. The set of test data shows the possibility of applying these approaches to change one of the images according to the specified combination points by the operators.

1. Introduction

Digital image processing is an important element of many control and monitoring systems. Solving the problem of stitching images is an important particular task of many operations of subsequent analysis and decision making [1,2]. Combining data into a single space allows us to reduce computational procedures, perform complex analysis, search and manage processes. Possible fields of application for cross-linked images are: medicine, when combining x-ray data; microbiology, when combining images of microorganisms; genetics, when constructing DNA structures from a series of images by electron microscopes; access control, when analyzing the intersection of the territory; security systems, when generating access keys by fingerprints; geology and cartography, in the analysis of images obtained from UAC, etc. More often than not, information about the camera’s displacement in space is unknown, and camera into settings are also being rebuilt between images [3]. All these processes, including the addition of a noise component to the image, complicate the process of finding correspondences between frames and make the task of stitching images complex and multi-stage [4].

The paper considers the problem of finding correspondences between pairs of images. The method is based on multiple changes in the dimension of the image, averaging the elements inside this block and finding matches between a series of enlarged blocks. As a result of the sequential analysis of enlarged blocks, it is possible to determine the correspondence between the elements of the input images.

2. Model images

As input data, a series of images obtained by a digital camera with fixed parameters is used. Figure 1 shows the image model and the process of large-scale change in their dimension.

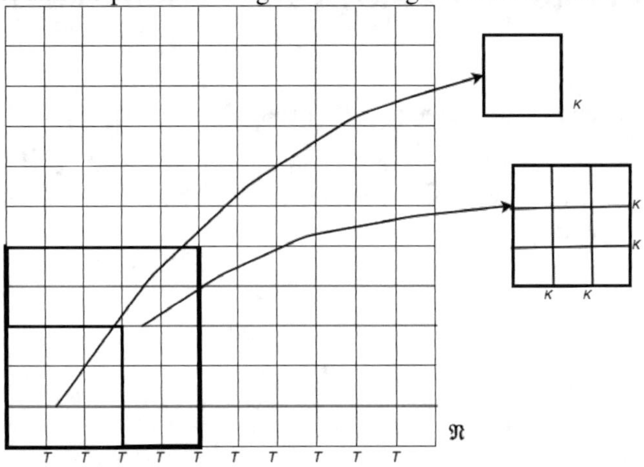

Figure 1. Model of digital images and transform of size images.

A simplified mathematical image model can be represented as [5]:

$$Y_{i,j} = S_{i,j} + \eta_{i,j}, \quad i = \overline{1,n}, \quad j = \overline{1,m}$$

when: $Y_{i,j} = \mathfrak{N}$ – input image, $S_{i,j}$ - pixel brightness intensity for ideal image, $\eta_{i,j}$ - noise component.

A digital image is a set of pixels, the value of each of the pixels reflects the intensity of the object. Since all the pixels of the image are equal, it is advisable to take the arithmetic mean s_i of the brightness of the ordinary pixels constituting it as $i = 1,2,\dots,n$ as the brightness of a large pixel K_i.

3. The algorithm transform of size images on series base points

We fix an arbitrary non-negative number ε. Let $m(\varepsilon)$ denote the smallest possible number of clusters into which the set $\mathfrak{N}$ can still be partitioned, provided that the fineness of the resulting partition does not exceed ε.

The use of the algorithm begins by setting the value $\varepsilon \geq 0$ - the upper limit of fineness of the desired partition of the set $\mathfrak{N}$. Then we find m - the smallest possible number of clusters for which there is still a partition $\mathfrak{N}$ of fineness $\leq \varepsilon$. For this we know, it suffices to construct a ε -partition $\mathfrak{N}$ and take as m the number of its clusters $m(\varepsilon)$. The initial partition $S_{i,j} \in \xi = \{T_1, T_2, \dots, T_m\}$ of the set $\mathfrak{N}$ is chosen so that the numbers of large pixels in any two of its clusters differ by no more than one. Under this condition, the smallest number of cluster elements is $s = [n/m]$. Suppose that the quantity $r = m(s+1) - n$ clusters of the initial partition will contain s elements each, and (for $n \vdots m \Leftrightarrow r = m$) the remaining $m - r$ clusters will contain (s+1) elements. We assume that is s -element clusters $T_1, T_2, \dots, T_r$. Then $\xi = \langle n_1, n_2, \dots, n_{m-1} \rangle$, when $n_j = js$ and $j = 0, 1, \dots, r$ and (for $r \neq m$) $n_j = j(s+1) - r$ for $j = r+1, r+2, \dots, m$. The algorithm proposed in Figure 2 is implemented as:

1) Appropriation $\gamma := 0$;
2) Appropriation $j := 1$;
3) Value calculation $i := i_{n_{j-1}+1, n_{j+1}}$ for

$$i_{pq} = p + \sum_{k=p+1}^{q-1} \mathrm{sign}(1 + \mathrm{sign}(\tilde{\alpha}_k - \frac{\tilde{\alpha}_p + \tilde{\alpha}_q}{2})).$$

Calculation $\tilde{\alpha}_i = \frac{\tilde{\alpha}_{n_{j-1}+1} + \tilde{\alpha}_{n_{j+1}}}{2} < \frac{\tilde{\alpha}_{i-1} + \tilde{\alpha}_{i+1}}{2}$, and $i := i - 1$;

4) For $i = n_j$, appropriation $\gamma := \gamma + 1$, if $i \neq n_j$ – appropriation $\gamma := 0$ и $n_j := i$;

5) For $\gamma = m - 1$ the calculation is completed, and its final result is the current $\langle n_1, n_2, \dots, n_{m-1} \rangle$. If $\gamma < m - 1$, then $j < m - 1$ and $j := j + 1$ и, go to step (3), if $j = m - 1$ – go to step (2).

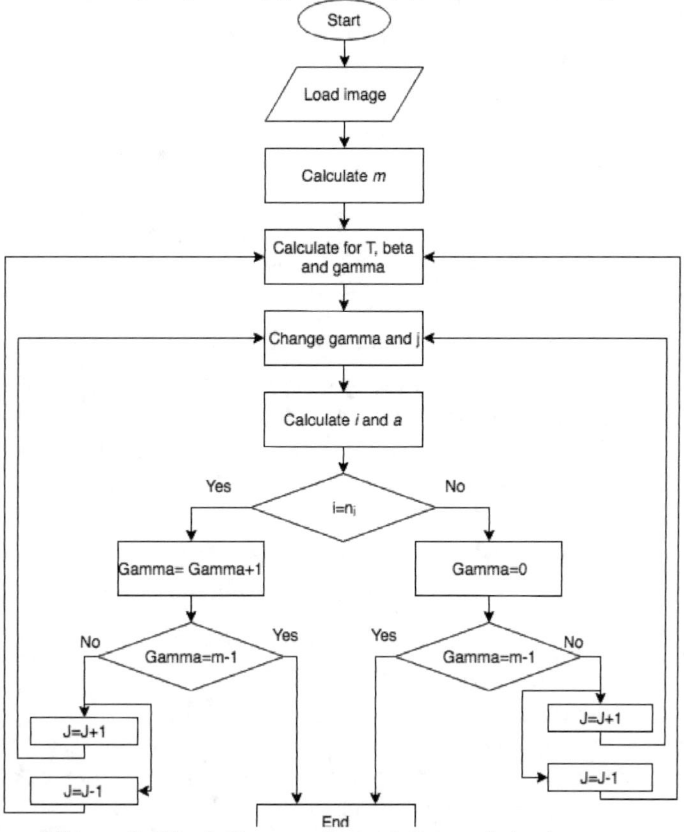

Figure 2. Block diagram for transform of size images.

Comments for algorithm.

a) At each step of the computational procedure, one of the numbers is recounted or remains the same from n_j, $1 \leq j \leq m - 1$. In case, the relation $\tilde{\theta} \prec \tilde{\eta}$, где $\eta = (\Delta_j, \Delta_{j+1}), \theta = (\delta_j, \delta_{j+1})$ (hereinafter, the following notation is used: $\Delta = (\Delta_1, \Delta_2, \dots, \Delta_m), \delta = (\delta_1, \delta_2, \dots, \delta_m)$ – cluster sweep vectors of the current partition of the set $\mathfrak{N}$, respectively, before and after the next passage of step (4)).

b) Because $\tilde{\theta} \prec \tilde{\eta} \Rightarrow \tilde{\delta} \prec \tilde{\Delta}$, that is, each step of the procedure leads to an improvement in the partition $\mathfrak{N}$ in the sense accepted for the algorithm. This ensures alignment of the ranges of the next pair of neighboring clusters. In this case, the value γ of the counter of consecutively ineffective steps is reset. If the next step is ineffective, $\delta = \Delta \Rightarrow \tilde{\delta} = \tilde{\Delta}$, which gives only the relation $\tilde{\delta} \preccurlyeq \tilde{\Delta}$. The counter γ after such a step increases by one.

c) The achievement by γ of the value of $m - 1$ indicates the onset of stabilization in the calculation, that is, about the exhaustion of the stock of "effective" steps. The calculation ends here, since it is impossible to further improve the partition of the set $\mathfrak{N}$ in the framework of this procedure. The described algorithm does not always lead to an ideal decomposition of the set $\mathfrak{N}$ into m clusters (in the sense that the corresponding vector $\tilde{\Delta}$ is the smallest element of the $(V_{m1n}, \preccurlyeq)$). At the same time, the result of using the algorithm has the necessary feature of such an ideal partition - the balance of any two of its "neighboring" clusters. This guarantees a high degree of alignment of the of all m clusters, i.e., if not optimal, then a sufficiently high-quality result.

In case of failure to fulfill condition (5). We define an equivalence relation on the set $\mathfrak{N}$: $K_i \sim K_j \Leftrightarrow \alpha_i = \alpha_j$. Let be $\{D_1, D_2, \dots, D_N\} -$ factor set of the set $\mathfrak{N}$ with respect to $\sim$. In the framework of the case under study, at least one of the sets D_j contains more than one element. We assume that the equivalence classes D_j are numbered in descending order of the brightnesses of their representatives (this time we are talking about a decrease in the strict sense). In each set D_j there is a unique element $K_i = K_{\theta_{j'}}$ for $i = \max\{p: K_p \in D_j\}$. From that $1 \le 1' < 2' < \cdots < N' = n$. Denote the set $\{K_{\theta_{1'}}, K_{\theta_{2'}}, \dots, K_{\theta_{N'}}\}$ that $\mathfrak{N}'$. Instructions for finding the numbers j' can be, for example, such. From that $0' = 0, \varphi_0 = 0$ and sequentially giving the variable j values $1, 2, \dots, N,$ we calculate the auxiliary quantity

$$\varphi_j = n - \varphi_{j-1} - \sum_{i=\varphi_{j-1}+2}^{n} \operatorname{sign}(\tilde{\alpha}_{\varphi_{j-1}+1} - \tilde{\alpha}_i),$$

and numberj$' = (j-1)' + \varphi_j$, and change value j. When using this procedure, the number N is determined automatically: sooner or later, the next value j' coincides with n, after which it remains only to perform the assignment $N := j$. Note that the sequence $(\tilde{\alpha}_{1'}, \tilde{\alpha}_{2'}, \dots, \tilde{\alpha}_{N'})$ of brightness of elements of the set $\mathfrak{N}'$ strictly decreases. Clustering it using the above algorithm, we obtain a partition,

$$\{T_j' = \{K_{\theta_{(N_{j-1}+1)'}}, K_{\theta_{(N_{j-1}+2)'}}, \dots, K_{\theta_{(N_j)'}}\}, 1 \le j \le M\},$$

from $\mathfrak{N}'$ (here $N_0 = 0, N_M = N$). The corresponding partition of the set $\mathfrak{N}$ has the form $\langle (N_1)', (N_2)', \dots, (N_{M-1})' \rangle$ and is the ultimate goal of applying the algorithm.

4. Conclusion

As a result of the research, the following result was obtained:

1. An approach has been developed that allows performing the image simplification operation. The simplification operation is performed by incrementally zooming the enlarged cluster. The replacement operation is carried out using the calculation of the average value between the minimum and maximum intensity values inside the simplified block.

2. An approach has been developed that allows performing the operation for the search of the relation between sets of clusters of image pairs. Ranked values indicate whether there are similar blocks highlighted at different scales.

5.Acknowledgments

This work was supported by Russian Ministry of Education and Science in accordance to the Government Decree No 218 from April 9, 2010 (project number No 074-11-2018-013 from May 31, 2018 (03.G25.31.0284)).

References

[1] Semenishchev E, Voronin V, Shraifel I 2018 The solution of the problem of simplifying the images for the subsequent minimization of the image bit depth. *26th International Conference in Central Europe on Computer Graphics, Visualization and Computer Visionin co-operation with EUROGRAPHICS Association*, p. 67-72

[2] Brown M, Lowe D G 2007 Automatic panoramic image stitching using invariant features. *International journal of computer vision*, **74(1),** 59-73

[3] Raviv D 1991 A quantitative approach to camera fixation. In Proceedings. *IEEE Computer*

Society Conference on Computer Vision and Pattern Recognition. pp. 386-392
[4] Xiong Y, Pulli K 2010 Fast panorama stitching for high-quality panoramic images on mobile phones. *IEEE Transactions on Consumer Electronics*, **56(2)**, 298-306
[5] Semenishchev E, Marchuk V, Shrafel I, Dubovskov V, Onoyko T, Maslennikov S. 2016 Image denoising using a combined criterion. *In Mobile Multimedia/Image Processing, Security, and Applications,* **Vol.** 9869, p. 98690E
[6] Semenishchev E, Shraifel I, Marchuk V, Svirin I, Makov S 2016 A multi-criteria method for noise reduction. *In 2016 IEEE East-West Design & Test Symposium (EWDTS).* pp. 1-6

IOP Conf. Series: Materials Science and Engineering **680** (2019) 012046 doi:10.1088/1757-899X/680/1/012046

Ways to optimize the prediction of vascular events in patients with acute coronary syndrome and atherosclerotic lesions of the renal arteries

N Kobzeva[1], V Terentev[1], I Zolotuhina[2]

[1] Rostov State Medical University, lane 29 Nakhichevanskyi, Rostov-on-Don 344000, Russia
[2] Don State Technical University, Gagarin's square 1, Rostov-on-Don 344000, Russia

E-mail: kobzeva.nataliya1@gmail.com

Abstract. Using the logistic regression equations obtained during the work, forecast tables were developed to assess the risk of developing vascular events in the long term. The proposed forecasting model is an independent method for assessing the risk of fatal and non-fatal vascular events. The consistency of the model has a scientific and practical justification. The proposed original risk model can be used in a comprehensive examination of patients with acute coronary syndrome and atherosclerotic lesions of the renal arteries.

1. Introduction

Predicting the course of diseases in patients is one of the important and difficult tasks that practicing cardiologists solve. An assessment of the risk of life-threatening cardiovascular events makes it possible to identify patients with a low and high risk of complications and death[1].

At low risk, routine outpatient drug therapy is usually performed to control symptoms and slow organ damage. In cases of a high risk of complications and death, active drug therapy is indicated, often in the intensive care unit, as well as invasive treatment [2,3].

The assessment of the prognosis of diseases and the corresponding treatment tactics are included in almost all modern recommendations for the treatment of cardiovascular diseases. [4]. Convenient tools for assessing risks in clinical practice are special scales that quantify the risk of adverse events. There is a tendency to use scales as mandatory diagnostic tools, along with traditional laboratory tests. [5, 6]. Scales are developed on the basis of analysis of data from large groups of patients with an assessment of the frequency of development of cardiovascular events and death during long-term follow-up. The scales usually include well-studied, easily identifiable indicators of the patient's health status. Using mathematical methods, for example, regression analysis, the most significant factors for forecasting are selected and a scale adapted for practice is constructed [7].

In order to evaluate the effectiveness of a predictive system, C-statistics is usually used, which includes an estimate of the area under the ROC (receiver operating characteristic) curve. The ROC curve is a function of the frequency of true positive results (sensitivity) versus the frequency of false positive results (100 specificity). Each point of the ROC curve represents a sensitivity / specificity pair corresponding to the accepted threshold [8,9]. The area under the curve (AUC) allows you to evaluate how well the scale distinguishes between diagnostic groups. C-statistics <0.7 characterizes an inadequate distinction between data, a value from 0.7 to 0.8 is acceptable, and from 0.8 to 0.9 an excellent distinction [10,11].

2. Methods

The study included 323 patients with acute coronary syndrome who were treated in the cardiology department of the regional vascular center, of which 217 (67.1%) were men, 106 (32.9%) were women. The average age of the examined patients was 59.6 ± 9.2 years.

All patients underwent standard biochemical studies, including the assessment of lipid metabolism: total cholesterol; high and low density lipoprotein cholesterol; atherogenic index; triglycerides, creatinine and urea levels.

The complex of instrumental studies included recording ECG at rest, Holter ECG monitoring, ultrasound of the heart, ultrasound examination of the kidney, coronary angiography to assess the degree of damage to the coronary vessels, selective angiography of the renal vessels.

Statistical analysis of the results was carried out using a set of applied statistical programs Microsoft Office Excel 2010 (Microsoft Corp., USA) and STATISTICA 10.0 (StatSoft Inc., USA). In order to assess the type of data distribution, Kolmogorov-Smirnov analysis was used; for values of $p > 0.05$, the distribution was considered not different from the normal one.

Descriptive statistics were carried out to determine the following features: the data were presented as M ± SD (M is the arithmetic mean, SD is the standard deviation) with a normal distribution and in the form Me [Q1; Q3] (Me is the median, Q1 and Q3 are the first and third quartiles) with abnormal distribution. In the normal distribution of the sample, when comparing two independent samples, the Student criterion was used, and when different from the normal one, the Mann-Whitney test and $\chi 2$ or the Leuven test with the determination of F. We also used logistic regression analysis with the calculation of relative risks (OR) and determination $\chi 2$, the relationship was considered statistically significant at a value of $p < 0.05$.

ROC analysis with the construction of operational characteristic curves and the calculation of the curve area coefficient (AUC) was used to assess the predictive information content of the methods used.

3. Results

In a two-factor logistic regression analysis, stenosis of the renal arteries or bilateral stenosis of the renal arteries was chosen as one factor, as the other signs that previously demonstrated their influence on the risk of developing vascular events. Atherosclerotic lesions of the renal vessels together with the size of the ejection fraction or size, left atrium, right ventricle, and also the thickness of the interventricular septum influenced the risk of non-fatal vascular events ($\chi 2 = 7.46$, $p = 0.024$; $\chi 2 = 7.37$, $p = 0.025$; $\chi 2 = 11.1$, $p = 0.004$; $\chi 2 = 9.52$ $p = 0.009$, respectively). Also, the discussed risk increased in the case of an increase in Δ GFR after coronary angiography, $\chi 2 = 9.83$, $p = 0.007$ (table 1).

Table 1- Logistic regression analysis of the effect of renal artery stenosis and other clinical signs on the risk of non-fatal vascular events

Another clinical symptom	Assessment stenosis/ another sign	Odds ratio, stenosis / another sign	Odds ratio, stenosis / other trait	χ^2	Degree of freedom	P
Throw Faction	0,99 / -0,05	2,69 / 0,95	2,69 / 0,10	7,46	2	0,024
Left pres	0,92 / 0,11	2,51 / 1,12	2,51 / 12,44	7,37	2	0,025
Right ventricle	1,36 / 0,17	3,91 / 1,19	3,91 / 2609	11,1	2	0,004
Interventricular	0,89 /	2,43 /	2,43 /	9,52	2	0,009

septum	0,19	1,22	7119			
Δ GFR	1,14 / 0,03	3,11 / 1,03	3,11 / 168	9,83	2	0,007

The obtained logistic regression equations made it possible to develop forecast tables, one of which is presented as an example in table 2.

Table 2- Forecast table for assessing the risk of non-fatal vascular events depending on the presence of renal artery stenosis and the value of the left atrium

Left atrium, mm and renal artery stenosis	38	40	42	44	46	48	50	52	54
The amount of risk,%	11	14	17	22	24	28	35	38	44
Left atrium, mm and no renal artery stenosis	38	40	42	44	46	48	50	52	54
The amount of risk,%	3	5	7	9	11	13	15	17	19

According to the findings of echocardiography, it was found that in the presence of renal vascular stenosis and left atrium size = 38 mm, the risk of non-fatal vascular events was 11%, with left atrium size = 46 mm, this risk was 24%, while in patients without renal vascular lesions - 3% and 9%, respectively. When the ejection fraction = 60% in the examined patients with renal vascular stenosis, the risk discussed was 17%, while the ejection fraction = 45% was 37%, while in patients without atherosclerotic lesion of the renal vessels, these indicators were - 6% and 21%, respectively.

When factors such as renal artery stenosis and the size of the right ventricle are included in the risk table, the probability of developing non-fatal vascular events can reach 84%, while in patients without renal vascular damage this indicator was significantly lower and amounted to 63%. When analyzing cardiovascular risk in patients with atherosclerotic lesions of the renal arteries and interventricular septum thickness = 12 mm, the specified risk is 9%, in patients without lesion - 5%, with interventricular septum thickness = 20 mm - 28% and 15% .

In the course of the work, it was found that among patients with renal vascular stenosis and Δ GFR of 60 ml / min / 1.73 m2, the risk of developing non-fatal vascular events was 57%, while in patients without renal artery disease, it was 14% (table 3).

Table 3- Prediction table for assessing the risk of developing non-fatal vascular events depending on the presence of renal artery stenosis and Δ GFR.

Δ GFR and renal artery stenosis	-80	-60	-40	-20	0	20	40	60	80
The amount of risk,%	3	5	7	9	15	27	38	57	75

IOP Conf. Series: Materials Science and Engineering **680** (2019) 012046 doi:10.1088/1757-899X/680/1/012046

Δ GFR and renal artery stenosis no	-80	-60	-40	-20	0	20	40	60	80
The amount of risk,%	1	3	4	5	7	9	10	14	22

The presence of bilateral renal vascular stenosis, as well as a number of signs, such as the size of the left atrium, thickness of the interventricular septum, and total blood cholesterol increased the risk of fatal vascular events (table 4).

Table 4- Logistic regression analysis of the effect of bilateral renal artery stenosis and other clinical signs on the risk of fatal vascular events

Another clinical symptom	Assessment stenosis/ another sign	Odds ratio, stenosis / another sign	Odds ratio, stenosis / other symptom	χ^2	Degree of freedom	P
Left atrium	1,22 / 0,09	3,44 / 1,1	3,44 / 8,7	6,43	2	0,04
Interventricular septum	1,24 / 0,13	3,25 / 1,14	3,25 / 256	7,96	2	0,018
Total cholesterol	1,35 / 0,12	0,86 / 1,13	0,86 / 6,17	6,59	2	0,037

So, the risk of developing fatal vascular events in patients with bilateral renal artery stenosis and a total cholesterol value of 6 mmol / L was 42%, in patients without renal vascular stenosis - 11% (table 5). If the total cholesterol value is increased to 10 mmol / l, the discussed risk is 54 and 16%, respectively.

Table 5- Forecast table for assessing the risk of fatal vascular events depending on the presence of bilateral renal artery stenosis and the value of total cholesterol

Total cholesterol, mmol / l and bilateral renal artery stenosis	6	8	10	12	14	16
The amount of risk,%	42	47	54	59	65	72
Total cholesterol, mmol / L and no renal artery stenosis	6	8	10	12	14	16
The amount of risk,%	11	13	16	19	22	27

t was also found that the risk of developing fatal vascular events in patients with bilateral renal vascular lesion and left atrium size = 38 mm was 36%, while in patients without stenosis the risk discussed was significantly lower - 10%.The probability of developing fatal vascular events in patients

IOP Conf. Series: Materials Science and Engineering **680** (2019) 012046 doi:10.1088/1757-899X/680/1/012046

with bilateral renal artery disease and interventricular septum thickness = 12 mm was 28%, while in patients without renal vascular stenosis this risk was 6%.

During a two-factor logistic analysis, it was found that bilateral damage to the renal arteries, as well as factors such as stages of chronic kidney disease ($p = 0.034$, $\chi 2 = 6.78$), glomerular filtration rate before angiography ($p = 0.0001$, $\chi 2 = 18.3$), as well as the severity of stenosis of the proximal and distal sections of the anterior interventricular branch ($p = 0.005$, $\chi 2 = 10.6$) ($p = 0.027$, $\chi 2 = 7.2$) influenced the risk of fatal and nonfatal vascular events.

Analysis of the obtained evidence showed that patients with bilateral renal vascular disease and stage 2 chronic kidney disease had a 32% fatal and non-fatal vascular event risk; in patients without renal vascular stenosis, the risk discussed was 12%.

It was also found that in patients with bilateral renal artery damage and the presence of hemodynamically significant stenosis in the proximal section of the anterior interventricular branch, this risk was 63%, while in patients without renal artery damage, the risk was 8%. In patients with hemodynamically significant damage to the middle third of the anterior interventricular branch and bilateral renal artery stenosis, the risk of developing fatal and non-fatal vascular events was 49%, in patients without renal vascular stenosis - 7%.

In the course of the work, a ROC analysis was carried out, during which the informativeness of the forecasting tools for the endpoints discussed above was separately analyzed: non-fatal vascular events, fatal vascular events and fatal / non-fatal vascular events. In relation to the first endpoint, a combination of signs : the presence of renal artery stenosis, ejection fraction, sizes of the left atrium, right ventricle, thickness of the interventricular septum, Δ GFR allowed to increase AUC to 0.78, increasing the sensitivity of the technique to 75%, while maintaining specificity at the level of 70% (Figure 1).

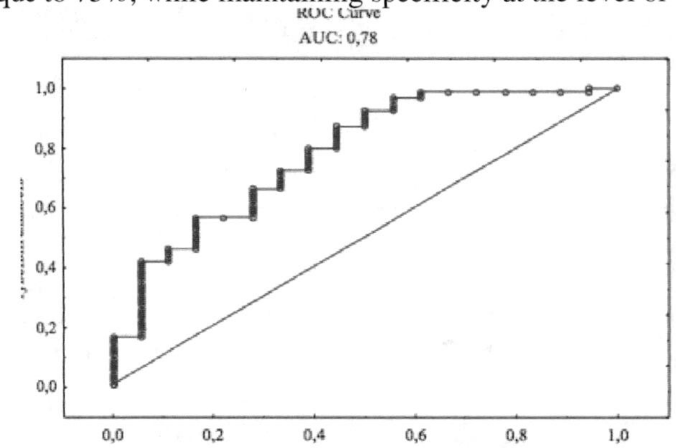

Figure 1.-ROC-curve sensitivity / specificity for renal artery stenosis, parameters of echocardiography and Δ GFR (CKD-EPI) in predicting the risk of non-fatal vascular events

ROC analysis of the influence of factors on fatal vascular events showed a higher informativity than during the analysis discussed above. The combination of such factors as the presence of bilateral renal vascular stenosis, the size of the left atrium, the thickness of the interventricular septum and the value of total blood cholesterol made it possible to increase the AUC to 0.93, the sensitivity of the method to 82%, and the specificity to 95%. (Figure 2).

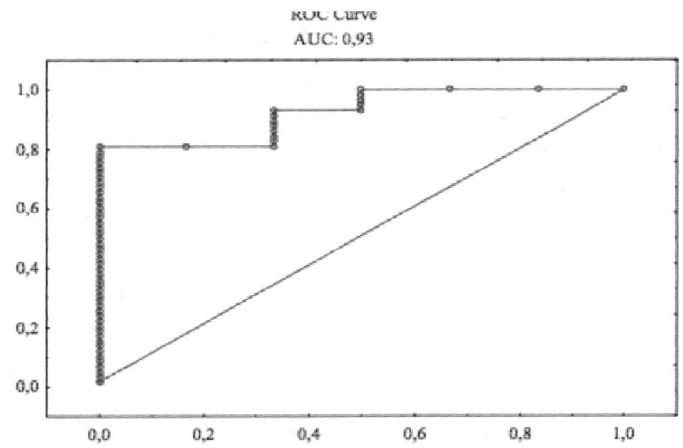

Figure 2. ROC-curve sensitivity / specificity for bilateral renal artery stenosis, sizes of LP and MJP and level of cholesterol in predicting the risk of developing non-fatal vascular events.

The combined point of the presence of bilateral renal vascular stenosis, the stage of chronic kidney disease, the severity of proximal and middle UCW stenosis showed an AUC of 0.92, with a sensitivity of 90% and a specificity of 90% (Figure 3).

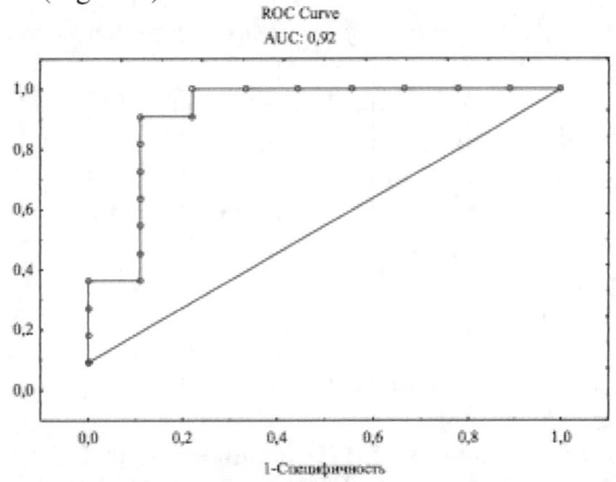

Figure 3. ROC curve sensitivity / specificity for bilateral renal artery stenosis, stages of CKD and the severity of stenosis of the proximal and middle sections of the UML in predicting the risk of fatal and non-fatal vascular events.

4. Discussion

With the introduction of interventional diagnostic methods into practical cardiology, the definition of tactics for treating a patient with acute coronary syndrome and prediction of cardiovascular risk has become the most successful and effective 8-11. Among people with renal artery stenosis, there is a very high risk of fatal and non-fatal cardiovascular events 12.

According to the literature we have analyzed to date, no methods have been proposed to predict outcomes in the long term in the group of patients under discussion.

The proposed forecasting method is based on the fact that, using an adequate mathematical apparatus, a complex of factors was determined that affect the risk of developing fatal and non-fatal vascular events in patients with acute coronary syndrome and atherosclerotic lesion of the renal arteries.

Using the logistic regression equations obtained during the work, forecast tables were developed to assess the risk of developing vascular events in the long-term period.

5. Conclusion

The proposed forecasting model is an independent method for assessing the risk of fatal and non-fatal vascular events. The consistency of the model has a scientific and practical justification. The proposed original risk model can be used in a comprehensive examination of patients with acute coronary syndrome and atherosclerotic lesions of the renal arteries.

Acknowledgements
The study had no sponsorship.

References

[1] Piepoli MF, Hoes AW, Agewall S ea.2016 European Guidelines on cardiovascular disease prevention in clinical practice. Eur Heart J. 2016 Aug 1; 37 (29):2315-81.
[2] Kaasenbrood L, Boekholdt SM, van der Graaf Y ea. Distribution of Estimated 10-Year Risk of Recurrent Vascular Events and Residual Risk in a Secondary Prevention Population. Circulation. 2016 Nov 8; 134 (19):1419-1429
[3] Task Force on Myocardial Revascularization of the European Society of Cardiology (ESC) and the European Association for Cardio-Thoracic Surgery (EACTS)l; European Association for Percutaneous Cardiovascular Interventions (EAPCI), Wijns W., Kolh P., Danchin N., Di Mario C., Falk V., Folliguet T., et al. Guidelines on myocardial revascularization. Guidelines on myocardial revascularization Eur. Heart J 2010; 31:2501-2555. DOI: 10.1093/eurheartj/ehq277
[4] Levey AS, Stevens LA, Schmid CH, et. al. for the CKD-EPI (Chronic Kidney Disease Epidemiology Collaboration). A New Equation to Estimate Glomerular Filtration Rate. Ann Intern Med. 2009 May 5; 150:604-612. PMID: 19414839 PubMed Logo
[5] Pencina MJ, D'Agostino RB Sr., Larson MG, Massaro JM, Vasan RS. Predicting the 30-year risk of cardiovascular disease: the Framingham Heart Study. Circulation 2009;1 19:3078-84
[6] Conroy RM, Pyorala K, Fitzgerald AP, et al. Estimation of ten-year risk of fatal cardiovascular disease in Europe: the SCORE project. Eur Heart J 2003;24:987-1 003
[7] Ridker PM, Paynter NP, Rifai N, Gaziano JM, Cook NR. C-reactive protein and parental history improve global cardiovascular risk prediction: the Reynolds Risk Score for men. Circulation 2008;1 1 8:2243-51
[8] Ridker PM, Paynter NP, Rifai N, Gaziano JM, Cook NR. C-reactive protein and parental history improve global cardiovascular risk prediction: the Reynolds Risk Score for men. Circulation 2008;1 1 8:2243-51
[9] Palaniappan L.P., Araneta M.R.G., Assimes T.L. et al. Call to Action: Cardiovascular Disease in Asian Americans: A Science Advisory From the American Heart Association. Circulation. 2010; 122: 1242-1252
[10] Palaniappan L.P., Araneta M.R.G., Assimes T.L. et al. Call to Action: Cardiovascular Disease in Asian Americans: A Science Advisory From the American Heart Association. Circulation. 2010; 122: 1242-1252
[11] Stroke Risk in Atrial Fibrillation Working Group. Comparison of 12 Risk Stratification Schemes to Predict Stroke in Patients With Nonvalvular Atrial Fibrillation. Stroke. 2008; 39: 1901-1910
[12] Gofman E.A., Martsevich S.Y., Deev A.D., Malysheva A.M., Polyanskaya Y.N., Tolpygina S.N., Mazaev V.P. FIRST RESULTS OF CHD PROGNOSIS STUDY. *The Clinician*. 2012;6(1):56-62. (In Russ.)

AUTHOR INDEX